AF430880

Distant Hybridization in Horticultural Crops

THE EDITORS

Dr. M.R. Dinesh was born on 05th May 1959 in Kodagu district of Karnataka. He studied at University of Agricultural Sciences and obtained B.Sc. (Hort.) M.Sc. (Hort.) and PhD (Hort.). He started his career in the 'Division of Fruit Crops' at the Indian Institute of Horticultural Research as Scientist S1 and presently he is heading the ICAR-IIHR as Director. He has worked on the breeding of papaya, guava and mango for more than two decades. He started his career with papaya breeding programme and has released two gynodioecious varieties 'Arka Surya' and 'Arka Prabhath'. He was one of the team members in mango and guava breeding programmes, which resulted in the development of one guava variety 'Arka Mridula' and one mango variety 'Arka Neelkiran'. He has released one pink pulp dual-purpose guava variety 'Arka Kiran'. He has collected, evaluated and characterized more than 500 varieties of mango. He has handled several foreign aided projects during his career so far. He has to his credit more than 50 research papers on fruit breeding in national and international journals of repute. Dr. Dinesh has guided seven M.Sc. and two Ph.D students. He has published five catalogues, two books and he has contributed chapters for six books especially on fruit breeding. He is the fellow of Horticultural Society of India and has been conferred with several National and International Awards.

Dr. M. Sankaran, was born on 10th May, 1974, in Dharmapuri District of Tamil Nadu. He did B.Sc. (Hort.) and M.Sc. (Hort.) from TNAU, Coimbatore and PhD (Pomology) from IARI, New Delhi. He is a BOYSCAST fellow and did his Post doctoral research at International Centre for Tropical Agriculture, Cali, Colombia.He has associated in developing and releasing 12 high yielding varieties in various horticultural crops. Surveyed and collected 20 accessions of mango, 16 accessions of tuber crops and 35 coconut accessions from A&N Islands which includes three dwarf type and also a makapuno type, multiple spicata type and a persistant petiole type. Characterized the 24 Pacific Ocean accessions and 6 Nicobar Island accessions and a descriptor has been prepared according to COGENT Descriptor (64 traits). He has published more than 35 research papers on improvement of horticultural crops in national and international journals of repute. Dr. Sankaran, has guided seven M.Sc. and two Ph.D students. He has also published 6 books, 12 book chapters, 8 technical bulletins and 3 database. He was awarded with Fakhruddin Ali Ahmed Award for outstanding contribution for tribal farming improvement by ICAR.

Distant Hybridization in Horticultural Crops

M.R. Dinesh

M. Sankaran

2017

Daya Publishing House®

A Division of

Astral International Pvt. Ltd.

New Delhi – 110 002

© 2017 EDITORS

ISBN 978-93-86071-66-8 (International Edition)

Publisher's Note:

Every possible effort has been made to ensure that the information contained in this book is accurate at the time of going to press, and the publisher and author cannot accept responsibility for any errors or omissions, however caused. No responsibility for loss or damage occasioned to any person acting, or refraining from action, as a result of the material in this publication can be accepted by the editor, the publisher or the author. The Publisher is not associated with any product or vendor mentioned in the book. The contents of this work are intended to further general scientific research, understanding and discussion only. Readers should consult with a specialist where appropriate.

Every effort has been made to trace the owners of copyright material used in this book, if any. The author and the publisher will be grateful for any omission brought to their notice for acknowledgement in the future editions of the book.

All Rights reserved under International Copyright Conventions. No part of this publication may be reproduced, stored in a retrieval system, or transmitted in any form or by any means, electronic, mechanical, photocopying, recording or otherwise without the prior written consent of the publisher and the copyright owner.

Published by : **Daya Publishing House®**
A Division of
Astral International Pvt. Ltd.
– ISO 9001:2015 Certified Company –
4736/23, Ansari Road, Darya Ganj
New Delhi-110 002
Ph. 011-43549197, 23278134
E-mail: info@astralint.com
Website: www.astralint.com

Dr. K.L CHADHA
Ph.D (Hort.), D.Sc. (Honoris causa), FNAAS, FHSI, FISHS
(Padma Shri Awardee)
President
The Horticultural Society of India
F-1, Societies Block, NASC Complex,
New Delhi-110 012
Tel.: (011-25842127
E-mail.:-hsi42@rediffmail.com

Formerly
National Professor (Hort.) ICAR, New Delhi
Deputy Director General (Hort.) ICAR, New Delhi
Horticultural Commissioner, Min. of Agri., GoI, New Delhi
Executive Director, NHB, Min. of Agri., GoI, New Delhi
Director, IIHR (ICAR), Bangalore
Project Coordinator (Fruits) ICAR, Lucknow

FOREWORD

India is blessed with diverse agro-climatic conditions, which favour the cultivation of a wide range of horticultural crops such as fruits, vegetables, flowers, plantation crops, tuber crops, spices, condiments, medicinal plants, aromatic crops and mushrooms. At present the area under horticulture crops is 23.69 million hectares with a production of 283.47 million tonnes surpassing food grain production being 257.07 million tonnes. Fruit crops 7.1 million ha and vegetable crops 8.95 million ha account for 92% of total horticulture production. In recent years, there has been growing awareness about climate change and its impact on agriculture. One of the strategies to counter this effect is the development and adoption of varieties with wide genetic base to withstand various biotic/ abiotic stress tolerance. This cannot be achieved by conventional breeding. Distant hybridisation involving interspecific and intergeneric crosses is one of the important approaches for introgressing the economic traits in breeding populations or hybrids in a crop. It is pertinent to use the wild relatives, which possess unique genes conferring the biotic and abiotic stresses that may be transferred to the cultivated plants so that these can adapt to adverse environmental conditions. Interspecific and intergeneric hybridisation has been tried in various fruits, vegetables, ornamental crops; spices, plantation crops, medicinal and aromatic crops. However, the resulting hybrids proved to be sterile and not of much use. However, success has been obtained by the use of techniques viz., mentor pollination, bud pollination, embryo rescue and protoplast fusion which played a significant role in recovering the hybrids/varieties through distant hybridisation in various horticultural crops. Notable interspecific and intergeneric hybrids have been obtained in banana (A & B genome), citrus (limequat, citromelos, tangerines & tangelos), papaya, guava, fig, annona, several temperate fruits, vegetables (solanaceous, leguminous, cucurbitaceous & crucifers) and flowers (lilies, petunias & orchids), etc. Considering the climate change, population explosion and shrinkage of natural resources, there is an urgent need to develop climate resilient crops, which is possible only through gene introgression or gene pyramiding from the related species or genera.

I compliment Dr. M.R. Dinesh and Dr. M. Sankaran for bringing out this useful compilation on *"Distant Hybridisation in Horticultural Crops"* from the lead lectures given by eminent scientists covering distant hybridisation working on various horticultural crops during the seminar on "Distant Hybridization for Improvement of Horticultural Crops" held at Bengaluru. I hope that this book will be of great interest and value to the students, research scholars and scientific fraternity.

(K.L.Chada)

PREFACE

Distant hybridisation involving Interspecific and intergeneric cross is one of the important research approaches for introgressing the economic traits in breeding populations or hybrids in a crop. Wild species of cultivated crops are the sources of potential genes which confers both biotic and abiotic stresses tolerance.These species have emerged naturally by evolution in the centres of diversity. The desirable traits like resistance against pathogens or abiotic stress within varying environments, composition of metabolites or morphological traits and their responsible genes are often found only in the wild species, related species and genera of the cultivars. The concept of gene pools has proved to be useful to both plant breeders and those working with plant genetic resources. With the advent of new technologies like DNA markers like SSR, SNPs; next generation sequencing, development of mapping or training population and precise phenotyping have increased our understanding of genome structure of the crops which will certainly accelerate the traditional breeding. The genomic information combined with precise phenotyping methods, provide a powerful and rapid tool for identifying the genetic basis of agriculturally important traits and for predicting the breeding value of individuals in a breeding population. The development of improved breeding lines for horticultural crops is a time consuming and expensive. With the deployment of genomics-assisted breeding, the generation of such lines is easier and faster. In order to meet the demands of the ever increasing human population and climate change, breeders have to accelerate the pace of our current breeding practices and apply genomics-based selection approaches.

In this book, there are 18 chapters delaing with distant hybridization in various horticultural crops. We hope that this book will be useful to the researchers, students and policy makers in understanding the problems and prospects of wide hybridization in horticultural crops.

-Editors

CONTENTS

LIST OF CONTRIBUTORS

Anandhan, S
ICAR-Directorate of Onion & Garlic Research, Rajgurunagar, Pune-510505, Maharastra, India

Balamohan, T.N
Horticultural College & Research Institute, Periyakulam-625604, Theni District Tamil Nadu, India

Belavadi, V.V
Department of Entomology University of Agricultural Sciences Bangaluru-560024 Karnataka, India

Benke, A
ICAR-Directorate of Onion & Garlic Research, Rajgurunagar, Pune-510505, Maharastra, India

Bhat, K.V
ICAR - National Bureau of Plant Genetic Resources, Pusa Campus, New Delhi - 110012, India

Dutta, O.P
Division of Vegetable Crops ICAR-Indian Institute of Horticultural Research, Hesaraghatta, Bangaluru-560 089 Karnataka, India

Devasia, J
Central Coffee Research Institute, Coffee Board, CRS (P.O) - 577 117, Chikmagalur Dist., Karnataka, India

Dinesh, M.R.
Division of Fruit Crops ICAR-Indian Institute of Horticultural Research, Hesaraghatta, Bangaluru-560 089, India

Ganeshan, S
Division of Plant Genetic Resources, ICAR-Indian Institute of Horticultural Research, Hesaraghatta, Bangaluru-560 089, India

Gopal, J
ICAR-Directorate of Onion & Garlic Research, Rajgurunagar, Pune-510505, Maharastra, India.

Jollikop, S.H
Division of Fruit Crops ICAR-Indian Institute of Horticultural Research, Hesaraghatta, Bangaluru-560 089, India

Joseph John, K
ICAR - National Bureau of Plant Genetic Resources, Regional Station, Thrissur - 680656, India

Latha, M
ICAR - National Bureau of Plant Genetic Resources, Regional Station, Thrissur - 680656, India

Mahajan, V
ICAR-Directorate of Onion & Garlic Research, Rajgurunagar, Pune-510505, Maharastra, India.

Mathur, R.K
ICAR-Indian Institute of Oil Palm Research, Pedavegi-534 450, West Godavari District, Andhra Pradesh, IIOPR-Research Centre, Palode, Kerala, India

Mukherjee, A
ICAR- Central Tuber Crops Research Institute, Regional Centre Bhubaneswar-751019, Odisha, India

Murugesan, P
ICAR-Indian Institute of Oil Palm Research, Pedavegi-534 450, West Godavari District, Andhra Pradesh, IIOPR-Research Centre, Palode, Kerala, India

Pitchaimuthu, M.
Division of Vegetable Crops ICAR-Indian Institute of Horticultural Research, Hesaraghatta, Bangaluru-560 089, India

Prakash, N.S
Central Coffee Research Institute, Coffee Board, CRS (P.O) - 577 117, Chikmagalur Dist., Karnataka, India

Raghuramulu, Y
Central Coffee Research Institute, Coffee Board, CRS (P.O) - 577 117, Chikmagalur Dist., Karnataka, India

K. Soorianathasundaram
Professor (Horticulture)
Horticultural College and Research Institute, TNAU, Coimbatore-641 003, India

Nazeer Ahmed
Director
ICAR-Central Institute for Temperate Horticulture, Srinaga-190007 Jammu and Kashmir, India

Rajasekharan, P.E
Division of Plant Genetic Resources, ICAR-Indian Institute of Horticultural Research, Hesaraghatta Bangaluru- 560 089, India

Roy, Y.C.
ICAR - National Bureau of Plant Genetic Resources, Regional Station, Thrissur - 680656, India

Sankaran, M
Division of Fruit Crops ICAR-Indian Institute of Horticultural Research,Hesaraghatta, Bangaluru- 560 089 Karnataka, India

Sudha, M
Section of Economics and Statistics ICAR-Indian Institute of Horticultural Research,Hesaraghatta, Bangaluru- 560 089 Karnataka, India

Sukanya, D.H
Section of Medicinal Plants ICAR-Indian Institute of Horticultural Research Hesaraghatta, Bangaluru-560 089 Karnataka, India

Sunilkumar, K.
ICAR-Indian Institute of Oil Palm Research, Pedavegi-534 450, West Godavari District, Andhra Pradesh IIOPR-Research Centre, Palode, Kerala, India

Sureshkumar, V.B
Central Coffee Research Institute, Coffee Board, CRS (P.O) - 577 117, Chikmagalur Dist., Karnataka, India

Padmini, K,
Principal Scientist
Division of vegetable crops, ICAR-Indian Institute of Horticultural Research,Bangalore-560089, India

Singh, T.H
Principal Scientist
Division of vegetable crops,
ICAR-Indian Institute of Horticultural
Research,Bangalore-560 089, India

Dhananjay. V. Naik
Research Scholar
Division of vegetable crops,
ICAR-Indian Institute of Horticultural
Research, Bangalore-89.

1

An Overview of Distant Hybridization in Horticultural Crops

M.Sankaran and M.R. Dinesh

*Division of Fruit Crops, ICAR-Indian Institute of Horticultural Research,
Hesaraghatta, Bengaluru-560089*

Introduction

The genetic improvement of a crop for yield and quality is potentially constricted by the narrow genetic base which has been a great challenge for the plant breeders. In order to overcome this problem, the wide hybridization is resorted to broaden the genetic base in breeding populations targeted for selection for both qualitatively and quantitatively inherited traits which are of economic importance. The main purpose of distant hybridization is not just to produce true hybrids but to introgress the potential genes and it includes a progeny containing significant traits of both parental genomes, but rather to obtain a plant that is virtually identical to the original crop except for a few genes contributed by the distant relative species. In some cases, it may even be possible to use wide crossing to obtain a plant that is almost identical to an elite variety of a crop except for the presence of a new trait or gene transferred from a wild relatives. The wild species are good sources of potential genes which confers both biotic and abiotic stress tolerance. These species emerged naturally by evolution in the centres of diversity and the evolution is a continuous and long term process which resulted lot of new species and genera in the living being which includes plants and animals. The utility of wild germplasm as a trait donor will obviously depends on the degree of genetic relationship with their cultivated relatives (Iyer, 1991). Except for special manipulation involving the addition or substitution of chromosomal segments, the successful introgression requires some degree of similarity between the genome of wild (donor) and cultivated (recipient) without it, there is no basis for genetic recombination to effect the transfer. The greater the homology, the more likely it is that chromosomal conjugation will occur in the critical F1 generation. The crossability of two species is often related to phylogenetic relationship and chromosome number. Crosses between closely related species with the same chromosome number are more likely to hybridize and produce fertile progeny, whereas crosses between distantly related species with different chromosome numbers are more likely to either not hybridize or to produce sterile offspring resulting from genetic incongruity or meiotic errors (Sybenga, 1992). Harlan and De wet (1971) has introduced a concept of gene pools which is highly useful to both plant breeders and those working with plant genetic resources. i. A species or accessions in the primary gene pool of the crop which cross easily and resultant hybrids are fertile, ii.Species in the secondary gene pool cross with crops with difficulty and their hybrids are partially sterile and iii.species in the tertiary gene pool can be crossed with the crop only by use of special technique such as embryo rescue and their hybrids are partially or completely sterile. Breeders face further difficulties once an interspecific hybrids has been obtained because of genetic recombination in the F1 and subsequent backcross generations can seldom be controlled precisely segments of chromosome or even whole chromosome, not just a single gene are usually introduced in to the genotype of the crop. Breaking linkages with unwanted wild type genes and resorting the genotype associated with commercial quality may take very long time. Moreover, after the successful transfer of a gene or genes, further time consuming crossing and selection are needed to produce, evaluate and multiply a new commercial cultivar. Therefore, improvement through interspecific hybridization often takes longer than inter varietal crossing.

Variations may also be created through induced mutation with the help of mutagens but most of the mutants are recessive and not stable. Thus, conventional interspecific hybridization via sexual process is always useful to the plant breeders. Based on the chromosome behaviors of wide hybrids and the resulting chromosome constitutions in their progenies, chromosome manipulation of wide hybrids for crop improvement is classified into three main categories viz., a).Incorporation of singe-chromosome or chromosome fragment from a wild species into an existing crop in order to enhance crop genetic diversitys (Qi *et al.*, 2007), b). Incorporation of all the alien chromosomes by chromosome doubling in order to produce amphidiploid and partial amphidiploid and c). Elimination of all alien chromosomes in order to induce crop haploid which can be doubled to enhance breeding efficiency or facilitate genetic analysis (Pratap *et al.*, 2010).

The utilization of wild relatives depends on the degree of relationship between the wild relatives and cultivated. For the successful introgression requires some degree of homology between the cultivated genome and the wild donor genome. The greater degree of homology, the more likely its is that chromosomal conjugation will occur in the F_1 generation. Distant hybridization has been attempted in fruits, vegetables, flowers, coffee, medicinal and aromatic plants. In this chapter, an attempt has been made to give an overview of intergeneric and interspecific hybridization in various horticultural crops.

INTERSPECIFIC HYBRIDIZATION IN FRUIT CROPS

The introgression of genes/ gene pyramiding from wild relatives is one such approach for developing the diseases & pests resistant high yielding varieties which can substantially reduce the production cost. Despite the development of chemical and cultural control measures for the pests and diseases, yet there is significant losses to from production to harvest are still a reality, especially in years where extreme weather is experienced due to climate change. In mango, one of the most significant diseases is anthracnose caused by the fungal pathogen *Colletotrichum gloeosporioides*. The loss due to anthracnose in mango has been studied by various workers and reported to be 25-30%. Bompard (1993) suggested the use of *M. laurina* for incorporating resistance to anthracnose, a fungal disease, and utilizing the genes available in *M. orophila* from Malaysis and *M. dongnaiensis* from Vietnam for developing varieties to make it distant possibility to grow mango in Mediterranean region, since these species grow well in the mountain forests at 1000-1700 metres above mean sea level.*M. altissima* is reported to be free from mango leaf hoppers and tip and seed borers (Angeles, 1991). Besides, there are several Mangifera species possess resistant to fruit fly (*M. camptosperma* & *M. andamanica*) and tolerant salinity which can be used for breeding purposes.. The PRSV (Papaya Ring Spot Virus) is a major problem in papaya production, in order to address this issue; the intergeneric hybridisation has been attempted. The intergeneric hybrids of *C. papaya V. quercifolia* validated for hybridity and resistant to 'papaya ringspot virus' (Mendoza-Garces *et al.*, 2010) and similarly the *C.papaya x V.cauliflora* has been validated for hybridity and resistant to PRSV in India (Dinesh *et al.*, 2013). There are several well known interspecific and intergeneraic hybrids in citrus (Tristeza, phytophthora, salinity, alkalinity & drought), grapes (mildew, nematode, salinity & alkanity), banana (viruses & fusarium wilt) and in many temperate fruits. More emphasis should be given on these aspects in the coming years to breed the varieties which are having resistant to major pests and diseasesIn cultivated banana, *Musa* spp., there are four known genomes, A, B, S, and T. These correspond to the genetic constitutions of wild *Eumusa* species *M. acuminata, M. balbisiana, M. schizocarpa* and the *Australimusa* species, respectively. Genomic *in situ* hybridization (GISH) could be used to differentiate the chromosomes of these four genomes (Hont *et al.*, 2000). Cultivated varieties are mainly triploids (2n = 3x = 33), derived from intraspecific hybridisations within *M. acuminata* (A genome) and interspecific hybridisation between *M. acuminata* and *M. balbisiana* (B genome). Jenny *et al.*, (2013) observed that the progenies involving an in vitro synthesized tetraploid 'Kunnan' (AABB) and several AA and BB clones have higher productivity, disease resistance and better fruit quality. A population of AA x BB (wild acuminata x *M.balbisiana*) showed segregation for pseudostem colour, bunch orientation and flower colour (Rekha, *et al.*, 2009). In temperate fruits, several attempts have been made to develop intergeneric and interspecific hybids but the resultant hybrids were proved to be sterile namely Mule (Troth early peach x Wild goose plum), Kamdesa (Peach x Sand cherry), Pyronia (Pear x Quince), Royal walnut (Persian walnut x California walnut) and Paradox walnut (Persian walnut x Eastern black). However, in apple, the *M. baccata* and *M. prunifolia* for frost resistance; *M. floribunda, M. atrosanguinea, M. baccata, M. micromalus, M. prunifolia etc.* for apple scab resistance, *M. zumi, M. robusta, M.sargentii, M. baccata jackii etc.* for powdery mildew resistance *etc.* which are being used breeding programme. In the interspecific crosses, apple crab species transmitted in addition to desirable traits of the wild parent, the undesirable ones, especially very small size and low fruit taste. Interspecific hybridization was previously used on different *Malus* species in order to induce resistance to apple scab (e.g. *M. floribunda, M. atrosanguinea, M. baccata, M. micromalus, M. prunifolia, etc.*) as well as to induce resistance to powdery mildew (e.g. *M. zumi, M. robusta, M. sargentii, M. baccata jackii etc.*) (Sestras, 2010). Catalina *et al.*, (2010) observed that the desired fruit size are greater when in interspecific hybridization are involved as genitors *M. zumi* and *M. floribunda*, compared to *M. niedzwetzkyana* and *M. prunifolia*. Sestras *et al.*, (2010) found that the M. *coronaria* and *M. prunifolia* proved to have descendents with valuable response to scab attack (based on genetic gain and expected selection response), compared to *M. zumi* off springs. The intergeneric/ interspecific hybrids such as Aprium, Apriplum, Plumcot, Peacotum, Pluot, and Nectaplum are becoming popular due to their yield and quality.

INTERSPECIFIC HYBRIDIZATION IN VEGETABLE CROPS

Wide hybridization has been attempted in Crucifers, Cucurbits, Okra, bulbs and tuber crops due to presence of potential genes in wild relatives (Table 1 , 2 & 3). First classical work on intergeneric hybridization was done Crucifers. The Raphanobrassica is an intergeneric hybrid between *R.sativus* and *B.oleraceae* was reported by Karpechenko (1928). Subsequently, a number of intergeneric and interspecific hybridization between Brassica species of the 'U' triangle and *R.staivus* have been produced; the U triangle refers to the ancestral A,B and C genomes that have been rise to modern Brassica crop species. Matsuzawa *et al.*, (1996) suggested a system for the use of interspecific and intergeneric hybridizations to develop five types of hybrid lines; synthetic diploid lines, alien gene introgression lines, alloplasmic lines, monosomic alien chromosome addition lines and monosomic alien chromosome substitution lines. These hybrids lines would be valuable genetic resources both for breeding more productive cultivars with novel agronomic traits and for research to better understand each chromosome and gene in this hybrid. Sharma *et al.*, (2008) reported that the *Solanum pimpinellifolium* is the only red-fruited wild species of tomato, and the only species from which natural introgression into the cultivated tomato has been self-compatible and bi-directionally cross-compatible with the cultivated tomato. Because of the close phylogenetic relationship between the two species, there is little or no difficulty in initial crosses in subsequent generations of pre-breeding. and breeding activities. Furthermore, *S. pimpinellifolium* harbors numerous desirable genes for disease resistance, abiotic stress tolerance and good fruit quality. *S. pimpinellifolium* and *S. peruvianum* contain genes which confer resistance to *Fusarium* wilt and early blight. Crossability barriers between *S. lycopersicum* and *S. peruvianum* have hindered the efficient introgression of important characteristics into the cultivated tomato gene pool. Both pre-zygotic and post-zygotic barriers prevent interspecific hybridization between these two distantly related species . The use of wild forms in breeding crop plants, particularly to obtain vigour and resistance has been well studied. Identification of resistance is possible through quantifying the biochemical components present in the genotype. The biochemical constituents like glycoalkaloid (solasodine), phenols, phenolic oxidase, enzymes namely poly phenol oxidase and peroxidase are available in brinjal and these biochemical constituents possess insect resistance properties

(Kalloo, 1988). As the cultivated brinjal is found to be highly susceptible to shoot and fruit borer, some other wild species possessing resistance were crossed with the brinjal to incorporate the resistance to shoot and fruit borer in the hybrids. In brinjal, interspecific crosses were done with wild forms of *Solanum* species namely *S. incanum* L. and forwarded the generation till F_4 (Rao 1981), *S.macrocarpon* L. by Gowda *et al.* (1990), *S.indicum* L. by Preneetha (2002) to obtain high yield with shoot and fruit borer resistance. Among these wild relatives, the *S. viarum* Dunal is a potential species, which yields solasodine and it is cultivated in India. Nassar (1989) hybridized two *Manihot* species, *M. neusana* and *M. anomala*, with cassava, *M. esculenta*, through controlled crosses using insects.

Table 1. Wild relatives in different vegetables

Crop	Wild Species	Resistance to
Tomato	*S. pimpinellifolium'* (Pearl Harbor)	Spotted Wilt
	S. peruvianum	Tomato Leaf Curl Virus
	S. pimpinellifolium	Tomato Leaf Curl Virus
	S. hirsutum f. glabratum B 6013	Tomato Leaf Curl Virus
	S. pimpinellifolium (LA 121)	Yellow Leaf Curl Virus
	S. cheesmanii	Yellow Leaf Curl Virus
	S. hirsutum f. glabratum	Yellow Leaf Curl Virus
	S. peruvianum	Yellow Leaf Curl Virus
	S. peruvianum	Yellow Leaf Curl Virus
	S. peruvianum PI 128655	Tomato Yellow Top Virus
	S. peruvianum PI 128655	Big Bud disease
Okra	*Abelmoschus manihot* ssp. manihot	Yellow Vein Mosaic Virus
	A. tetraphyllus	Yellow Vein Mosaic Virus
	A. angulosus ssp. grandiflorus	YVMV
	A. ficulneus	Cercospora leaf spot

The importance of wild Cucumis species for cucumber (*Cucumis sativus* L.) and melon (*C.melo* L.) improvement has long been recognized because they possess resistance to pests and diseases (Leppick,1996). Whitaker (1930) reported the occurrence of interspecific cross compatabilities in *Cucumis*. However, successful cross between the cucumber and melon has been reported (De Ruiter,1973). Chen *et al.*, (1995) reported that based on the isozyme analysis of *C.hysteria, C.melo* and *C.sativus* led to the hypothesis that a triangular poly genetic relationship occurs among the species.

Table 2. Resistance Source for insect pests:

Crop	Wild species	Resistance to
Tomato	*L. hirsutum, L. hirsutum f. glabratum*	Leaf miner & Fruit borer
Brinjal	*S. khasianum, S. sisymbriifalium, S. melongena, S. incanum, S. gilo, & S indicum.*	Resistant to shoot and fruit borer
	S. sisymbriifolium, S. mammosum	Aphis gossypii
Okra	*A.caillei* and *A.tetraphyllus*	Fruit borer and Jassids
	A. moschatus A. crinitus	Jassids
Muskmelon	*C.melo* var.*callosus*	Fruit fly and leaf catterpillar
Onion	*A.fistulosum*	Thrips

Table 3. Abiotic stress tolerant species in vegetable crops

Crop	Wild species	Resistance to
Tomato	*Solanum cheesmanii, S. pimpinellifolium, S. chilense, S. pennelli*	Drought
	Superior seedling growth at 0°C was introgressed from high-altitude strains of *Solanum hirsutum,* Other species: *S.peruvianum, S.chilense, S.hirsutum.f.glabratum*	Low temperature
Brinjal	*S. macrocarpum, S. gilo*	Drought
Tapioca	*Manihot glaziovii, M. dichotoma, M. saxicola*	Drought

DISTANT HYBRIDIZATION IN FLOWER CROPS

In breeding ornamental crops, interspecific hybridization is the most important source of genetic variation. The flower crops wild species and their resistance to biotic stresses has been given in table 4. Many of the cultivars have originated from complex species crosses which have given rise to a broad range of shapes and colours to plants and flowers. Hybridisation between species of different sections of Lilium is difficult due to pre and post fertilization barriers. However, successful interspecific hybrids have been developed with the use *in vitro* embryo, ovule or other rescue methods. Interspecific crosses among Petunia species generally produce fertile progeny (Watanabe *et al.* 1996, 2001; Ando *et al.* 2001), though the percentage of successful crosses varies widely across parental combinations. The cultivated petunia (Petunia x hybrida) is derived from a hybrid between *P. axillaris* and *P. integrifolia* (Stehmann *et al.* 2009). Petunia species have been useful for introgressing genes for novel traits such as growth habit (*P. altiplana*) and flower color (P. exserta) into the cultivated petunia (Griesbach 2007). The interspecific hybrids are useful for assessing species relationships through the analysis of chromosome pairing during meiosis using genomic in situ hybridization (GISH) will be of great help (Lim *et al.*, 2001; Van Tuyl *et al.*, 2005). Fedorova *et al.*, (2010) studied the differentiation and hybridization in wild rose populations (*Rosa spinosissima, R.gallica* & *R.canina*) in Western Ukraine. It was found that *Salvia transsylvanica* may be a useful donor for improving floral size of the popular herbaceous perennial *Salvia nemerosa* (Joseph and

Warner, 2011). Successfully developed and characterized the intergeneric hybrids between *Dichroa febrifuga* x *Hydrangea macrophylla*. (Sandra and Jones, 2008). Orchid species (Orchidaceae) are particularly prone to hybridization most likely because of the high number of sympatric species, in combination with a general lack of complete reproductive barriers (Mallet, 2005). Genetic incompatibility between closely related species have been discovered in Orchidaceae (Sanford, 1964), even then, several intergeneric and interspecific hybrids have been developed in orchids.

Table 4. Some of the flower crops wild species and their resistance to biotic stresses

S.No.	Crop	Wild species	Resistant to
1.	Jasmine	*Jasminum flexile* and *J. calophyllum*	insect pests
		J. auriculatum cv. Parimulli	Mite resistance
2.	Gladiolus	*G. psittacinus* and cv. Dhiraj released from IIHR (Beauty spot X *G. psittacinus*)	Fusarium resistace
3.	Anthurium	*A. antioquiense*	Bacterial blight
4.	Carnation	*Dianthus capitatus spp. andrzejowskianus* and *D. nenteri*	Bacterial wilt caused by *Pseudomonas caryophylli*
5.	Zinnia	Zinnia marylandica	Alternaria zinnia
6.	Rose	*R. rugosa* and *R. bracteata*	Immune to black spot
		R. wichuraina	Blackspot
		R. macrophylla and R. spinosissima	Rust
		Rosa multiflora and *R. indica*	Nematodes

One concern for the future of wide crossing is that many potentially beneficial donor species or local populations of wild plants are being destroyed every year by habitat degradation, industrialization and agricultural expansion. This illustrates the need for an inventory and/or the improved conservation of wild plants that could possibly contribute useful genes to major crops such as that influencing disease resistance.

BARRIERS IN DISTANT HYBRIDIZATION

The techniques needed to produce intergeneric / interspecific hybrids will depend very much on the nature of barriers involved in each cross. Various barriers in wide hybridization are as follows;

Barriers in stigma and style

An interspecific or intergeneric fertilization frequently fails due to incompatible pollen and pistil. The pollen tube growth stops growing before it reaches the ovule. There are two different views on the relatiship between intra and interspecific compatibilities put forward by Pandey (1981) believes that both phenomena are pleiotrophic effects of the S gene (Super gene) which controls the incompatibility. On the contrary, Hogenboom (1984) considers that alien pollen is rejected when pollen and pistil of different species do not complement each other properly.

Structure and function of stigma

Before the germination of pollen grains on stigma, the pollen must adhere and take water from the surface of the wet stigma (Heslop-Herrison ,1987). The Solanaceae plants produces wet stigma which secretes fluid so that the pollen fallen on the stigma can readily hydrate itself. Similarly, the Cruciferous plants have dry stigma which do not have fluid secretion at pollination but the compatible pollen grains apparently hydrate by withdrawing water from the stigma through discontinuities in the cuticle. In legumes, whose flowers requires tripping before they will set fruit, such as *P.coccineus* and most cultivars of *Vicia faba*, a stigmatic secretion is accumulated under cuticle and is released only when the cuticle is ruptured.

The failure in the interspecific crosses may be due to the failure of fertilized ovules to develop into mature seeds. The causative factors include the presence of lethal genes and failure or early breakdown of the endosperm (Khush and Brar, 1992). Arrest of normal development of embryo and endosperm may also result from unfavorable interaction between the embryo and surrounding ovular tissues. Often, embryo breakdown is associated with proliferation of the nucellar or integumentary cells (Cooper and Brink, 1940). Failure of endosperm leading to death of potentially viable hybrid embryos is the common result of causes between closely related species (Taylor and Smith, 1979).

This may also be due to post-fertilization barriers, in which hybrid embryos obtained after fertilization degenerate during their development, can sometimes be overcome by embryo rescue techniques such as embryo or ovule culture. In some ornamental plants, such as *Lilium* spp. (Van Tuyl *et al.*, 1991), Primula spp. (Kato *et al.*, 2008), *Cyclamen* spp. (Ishizaka, 2008), interspecific hybrid cultivars have so far been produced *via* embryo rescue. In Colchicaceous ornamentals, intergeneric hybridization has recently been performed for widening the variability in horticultural traits: Santonia 'Golden Lights', an intergeneric hybrid cultivar of *Sandersonia aurantiaca* and *Littonia modesta*, was developed *via* ovule culture (Clark *et al.*, 2005 and Burge *et al.*, 2008); intergeneric hybrids between *S. aurantiaca* and *Gloriosa* spp. were produced *via* ovule culture (Nakamura *et al.*, 2005; Kuwayama *et al.*, 2005; Burge *et al.*, 2008).

METHODS TO OVERCOME BARRIERS IN WIDE HYBRIDIZATION

The nature of cross incompatibility between cultivated species and its commercial varieties with their wild relatives is not well understood. Incompatibility is characterized by failure of pollen germination, slow growth of the pollen tube, inability of the pollen tube to reach the ovary. These barriers are due to genic or ploidy differences between the species. The post pollination barriers includes; hybrid invariability and weakness leading to chromosome elimination, lethality and embryo abortion, hybrid sterility as well as hybrid break down with weak or sterile individuals in second hybrid population (F_2) owing to recombination of the gene complements of the parental species.

Traditional approaches in interspecific hybridization have been applied to overcome the hybridization barriers in are; mechanical removal of style followed by pollination of the exposed stylar end, bud pollination, bridge cross, back crossing, use of mixed or mentor pollen, chemical treatment as well as use of growth hormones such as GA3, IAA, NAA *etc.*

Biotechnological techniques such as *In vitro* fertilization, protoplast fusion, chromosome doubling before hybridization, Embryo rescue have also been suggested as possible tools for overcoming interspecific hybridization barriers.

SUMMARY

Global climate change and its impact on agriculture is going to a major threat to the biodiversity including the human populations. The development varieties with broad genetic base will be highly useful to mitigate the various biotic and abiotic stresses. Considering this point view, the interspecific and intergeneric hybridisation is being tried in various horticultural crops such as fruits, vegetables, ornamental crops; spices, plantation crops medicinal and aromatic crops and most of the resulting hybrids were proved to be sterile and not much useful. However, success has been obtained with the use of techniques viz., mentor pollination, bud pollination, embryo rescue and protoplast fusion have played significant role in recovering the hybrids/varieties through the distant hybridisation in various horticultural crops. The notable interspecific and intergeneric hybrids have been obtained in banana (A &B genome), citrus (limequat, citromelos, tangerines & tangellos), papaya, guava, fig, annona, several temperate fruits, vegetables (Solanaceous, leguminous, cucurbitaceous & crucifers) and flowers (lillies, petunias & orchids), *etc.* The utility of wild germplasm as a trait donor will obviously depends on the degree of genetic relationships with the cultivated relatives. Nevertheless, the conservation and utilization of potential wild relatives and land races will be of great important for the genetic improvement of horticultural crops.

REFERENCES

Angeles, D.E., 1991. *Mangifera altissima. In:* Verheij, E.W.M. and R.E. Coronel (eds.), *Edible Fruits and Nuts,* pp: 206–7. Plant Resources of South East Asia 2. PUDOC. Wegeningen .

Ando, T., Nomura, M., Tsukhra, J., Watanabe, H., Kokubun, H., Tsukamoto, T., Hashimoto,G., Marchesi, E and L.J.Kitching.2001. Reproductive isolation in native population of Petunia. *Ann.Bot.,* 88:403-413.

Bompard, J.M., 1993. The genus Mangifera rediscovered: the potential contribution of wild species to mango cultivation. *Acta Horticultuae,* 341: 69–77.

Burge, G.K., E.R. Morgan, J.R. Eason, G.E. Clark., J.L. Catley and J.F. Seelye, 2008. *Sandersonia aurantiaca*: Domestication of a new ornamental crop. *Sci. Hort.,* **118:** 87–99.

Chen, J.F., S. Isshiki, Y. Tashiro, and S. Miyazaki. 1995. Studies on a wild cucumber from China (*Cucumis hystrix* Chakr.). I. Genetic distances between *C. hystrix* and two cultivated *Cucumis* species (*C. sativus* L. and *C. melo* L.) based on isozyme analysis. J. Jpn. Soc. Hort. Sci. 64(suppl. 2):264–265.

Clark, G.E., G.K. Burge, E.R. Morgan and C.M. Triggs, 2005. Effects of planting date and environment on the cut flower production of Santonia 'Golden Lights'. *Acta Hort.*, 673: 265–271.

Cooper, D.V. and R.V. Brink, 1940. Somatoplastic sterility as a cause of seed failure after inter-specific hybridization. *Genet.*, **25**: 593-617.

Dinesh, M.R., G.L. Veena, C. Vasugi, M. Krishna Reddy ,K.V. Ravishankar.2013. Intergeneric hybridization in papaya for 'PRSV' tolerance. *Sci. Hort.*, 161(24) 357–360.

Eason, J.R., E.R. Morgan, A.C. Mullan and G.K. Burge, 2001. Postharvest characteristics of Santonia 'Golden Lights' a new hybrid cut flower from *Sandersonia aurantiaca, Littonia modesta. Postharvest Biol. Technol.*, **22**: 93–97.

Farooqi, A.A. and B.S. Sreeramu, 2004. Glory lily. In: Cultivation of Medicinal and Aromatic Crops. Universities Press Private Limited, Hyderabad, pp. 131-138.

Fedrova, A.V., I.A.Schanzer and A.A.Kagalo. 2010. Local differentiation and hybridization in wild rose populations in Western, Ukraine, Wulfenia **17**: 99 –115.

Gowda, P. H. R., K. T. Shivashankar and S.Joshi.1990.Interspecific hybridization between *Solanum melongena* and *Solanum macrocarpon*: study of the F_1 hybrid plants, *Euphytica*, 48(1): 59-61.

Goishizaka, H. 2008. Interspecific hybridization by embryo rescue in the genus *Cyclamen. Plant Biotechnol.*, 25: 511-519.

Griesbach JR (2007). Petunia in: Flower Breeding and Genetics, ed.N.O. Anderson, Springer. 301-336 pp.

Harlan, J.R. and J.M.J. Dewet.1971. Toward a rational classification of cultivated plants, Taxon, 20(4):509-517.

Hant, D.A., Paget-Goy, A., J.Escoute and F.Carred.2000. The interspecific genomic structure of cultivated banana, Musa spp, revealed by genomic DNA insitu hybridization. *Theor.Appl.Genet*, 100:177-183.

Heslop-Harrison J. 1987 . Pollen germination and pollen-tube growth. *International Review of Cytology,* **107** : 1-78.

Iyer, C.P.A., 1991. Recent advances in varietal improvement in mango. *Acta Hort.*, 291: 109–32.

Janick J, Cummins JN, Brown SK, Hemmat M (1996). Apples. In: J. Janick and J. N. Moore (eds.). Fruit breeding. Vol. I: Tree and tropical fruits, p. 1-77. John Wiley and Sons Inc., New York.

Janson J., Reinders M.C., Van Tuyl J.M., and Keijzer C.J., 1993. Pollen tube growth in *Lilium longiflorum* following different pollination techniques and flower manipulations. *Acta Bot Neerl*, 42: 461-472.

Jenny, C., Holtz, Y., J.P.Horry and F.Bakry, 2013. Synthesis of new interspecific hybrids from AB germplasm in banana (*Musa* spp.), *Acta Hort*, 986:209-218.

Junji Amano, Daisuke Nakazawa, Sachiko Kuwayama, Yoko Mizuta, Hajime Okuno, Yusuke Watanabe, Toshinari Godo, Dong-Sheng Han, Masaru Nakano. 2009. Intergeneric hybridization among colchicaceous ornamentals, *Gloriosa* spp., *Littonia modesta* and *Sandersonia aurantiaca via* ovule culture. *Plant Biotechnology*, **26:** 535-541.

Kalloo G (1988). Biochemical basis of insect resistance in vegetables. Vegetable Breeding VolumeII, CRC Press Inc Boca Raton, Florida.

Karpechhenko, G.D.(1928). Polyploid hybrids of *Raphanus sativus* X *Brassica oleraceae* L. Z.Ind.Abst.Verebgsl, 48:1-85.

Kato, J., H. Ohashi, M. Ikeda, N. Fujii and R. Ishikawa, 2008. Unreduced gametes are the major causal factor for the production of polyploid interspecific hybrids in *Primula*. *Plant Biotechnol.*, **25:** 521-528.

Khush, G.S. and D.S. Brar, 1992. Overcoming the barriers in hybridization. In: Distant hybridization of crop plants. Monographs on Ther. Appl. Genet., (Eds. Kallo.G. and J.B. Chowdhury) Springer-Verlag, New York, pp. 47.61.

Kirkbride, J.H., Jr. 1993. Biosystematic monograph of the genus *Cucumis* (Cucurbitaceae). Parkway Publ., Boone, N.C.

Kuwayama, S., T. Nakamura, Y. Mizuta, T. Oomiya and M. Nakano, 2005. Cross-compatibility in interspecific and intergeneric hybridization among the Colchicaceous ornamentals, *Gloriosa* spp., *Littonia modesta*and *Sandersonia aurantiaca*. *Acta Hort.*, 673: 421–427.

Leppick, E.E. 1966. Searching gene centers of the genus *Cucumis*. Euphytica 15:323–328.

Lim, K.B., Chung JD, Van Kronenburg BCE, Ramanna MS, De Jong JH, Van Tuyl JM (2000). Introgression of Lilium rubellum Baker chromosomes into L. longiflorum Thunb. a genome painting study of the F1 hybrid, BC1 and BC2 progenies. *Chromosome Res* 8:119-125.

Mallet J. Hybridization as an invasion of the genome. Trends Ecol Evol. 2005;20:229–237. doi: 10.1016/j.tree.2005.02.010.

Matsuzawa, Y., Y.Kanaeko and S.W.Bang (1996). Prospects of the wild cross for genetics and plant breeding in Brassicaceae. Bull.Coll.Agric. Utsunomiya Univ.16:5-10.

McGuire DC, Rick CM (1954). Self-incompatibility in species of Lycopersicon sect. Eriopersicon and hybrids with Lycopersicon esculentum. Hilgardia 23: 101-112.

Mendoza-Garces, A.C, P.M. Magdalita, M.S. Mendioro, F. Cruz, S. Dela, V.N. Villegas. 2010. Morphological, cytological, biochemical and molecular characterization of *Carica papaya* L., *Vasconcellea quercifolia* (St. Hil.) Hieron and their intergeneric hybrid. Philippine Journal of Crop Science, 35 (2) (2010), pp. 1–11

Nakamura, T., S. Kuwayama, S. Tanaka, T. Oomiya, H. Saito and M. Nakano, 2005. Production of intergeneric hybrid plants between *Sandersonia aurantiaca* and *Gloriosa rothschildiana via* ovule culture (Colchicaceae). *Euphytica*, 142: 283–289.

Pandey, K.K., 1981. Evolution of unilateral incompatibility inflowering plants: further evidence in favour of twin specifities controlling intra-and interspecific incompatibility. *New Phytol* 89: 705–728.

Pratap A, Choudhary AK and Kumar J (2010). *In vitro* techniques towards genetic enhancement of food legumes—a review. *J Food Legumes* 23:169–185.

Preneetha, S (2002). Breeding for shoot and fruit borer (*Leucinodes orbonalis* G.) resistance in brinjal (*Solanum melongena* L.). PhD Thesis, Tamil Nadu Agricultural University, Coimbatore.

Qi, L.L., Frieb, B., Peng, Z and B.S.gill.2007. Homoeologus recombination, chromosome engineering and crop improvement. Chromosome Research, 15(1): 3-19.

Ramanna MS, Jacobsen E (2003). Relevance of sexual polyploidization for crop improvement-a review. *Euphytica* 133:3-18.

Rao GR (1981). Results of interspecific cross pollination between *S. melongena* L. and *S. incanum* L. in eggplant breeding. Proceedings of Indian National Science Academy 47, 893-898.

Rekha, A., K.V.Ravishankar and D.S. Ambika.2009. Generation of mapping population for segregation of B genome. A paper presented in ISHS-Promusa banana symposium, Guangzhou, China, 14-18 Sept,2009, Abstract No. 238.pp

Sanford, W. W. 1964. Sexual compatibility relationship in Oncidium and related genera. Am. Orchid. Soc. Bull., 33,1035-1048.

Sestras R, Dan C, Pamfil D, Sestras A, Jäntschi L, Bolboaca S (2010). The Variability of Juvenile Period, Fruits Size and Response to Diseases Attack on F1 Interspecific Apple Hybrids and the Efficiency of Selection. *Not Bot Hort Agrobot Cluj* 38(1):234-240.

Sharma A, Zhang L, Niño-Liu D, Ashrafi H, Foolad MR (2008). A *Solanum lycopersicumxSolanum pimpinellifolium* linkage map of tomato displaying genomic locations of R-genes, RGAs, and candidate resistance/defense-response ESTs. *Int J Plant Genomics* 2008: 926090.

Singh, B.P. 1991. Interspecific hybridization in between new and old-world species of *Luffa* and its phylogenetic implication. Cytologia 56:359– 365.

Stehmann JR, Lorenz-Lemke AP, Freitas LB, Semir J (2009). Petunia, Evolutionary, Developmental and Physiological Genetics, ed. Springer New York.

Sybenga, J.1992. In Book: Cytogenetics in plant breeding, Springer-Verlag, New York, pp1-407.

Van Tuyl J.M., Van Diën M.P., Van Creij M.G.M., Van Kleinwee T.C.M., Franken J., and Bino R.J., 1991. Application of *in vitro* pollination, ovary culture, ovule culture and embryo rescue for overcoming incongruity barriers in interspecific *Lilium* crosses. *Plant Science*, 74: 115-116.

Van Tuyl JM, Maas IWGM, Lim KB (2002). Introgression in interspecific hybrids of lily. *Acta Hort* 570:213-218.

Warwick, S.I.(1993). Guide to the wild germplasm of Brassica and allied Crops. Part IV. Wild species in the tribe Brassicaceae (Cruciferae) as sources of agronomic traits. Centre for Land and Biological Resources Research, , Canada, Ottawa. Ont.Technical Bulletin, 172: 1-19.

Watanabe, H., Ando, T., Iida, S.I., Suzuki, A., Buto,K.I., Tsukamoto, T., Hashimoto,G., and E. Marchesi.1996. Cross compatibility of Petunia cultivars and *Petunia axillaris* with native taxa of Petunia in relation to their chromosomes. *J.Jap.Soc. Hort*.Sci.66:607-612.

Distant Hybridization in Horticultural Crops   *Pages 15–24*
Editors: **M.R. Dinesh & M. Sankaran**
Published by: **ASTRAL INTERNATIONAL PVT. LTD., NEW DELHI**

2

Allied Species Utilization In Fruit Crop Improvement Way Forward

S. H. Jalikop

Division of Fruit Crops, ICAR-IIHR, Hesaraghatta, Bengaluru-560089

INTRODUCTION

An interspecific hybrid is a cross between plants of two different species. Many times they will be from the same genus, but not always. Interspecific hybridization breaks what is known as the species boundaries and barriers for gene transfer and thus makes it possible to transfer the genome of one species to another. Interspecific hybrids have the potential to capture hybrid vigor as well as combine traits that do not occur within a single species (Volker and Orme, 1988). Because breeder want to add new characteristics to the current cultivars, allied species provide a vast gene pool for crop improvement. It has been used in many crops in an attempt to introgress genes from one species, sometimes a wild relative, into the background of another species, usually cultivated or more desirable. Desirable traits like resistance against pathogens or abiotic stress within varying environments, composition of metabolites or morphological traits and their responsible genes are often found only within wild/allied species and genera of the cultivars.

Most of the fruit crops are characterized by large plant size, high level of heterozygosity, long juvenile phase, long-life, vegetative propagation, somatic mutation, amenable for growing on rootstocks, polyembryony, seedlessness and linkage of desirable and undesirable traits. Beside these drawbacks little or no knowledge of inheritance, difficulty in making small heritable changes and exacting delicate fruit traits make developing of new fruit varieties slow and extremely challenging. Hence interspecific hybridization in fruit crops also differs from that of annual crop breeding.   Further, crossing barriers occur frequently when intra- or interspecific crosses are attempted. The difficulty of creating interspecific hybrids increases along with the phylogenetic distance between the parents (Sharma, 1995). Incompatibility of parents in hybridization may be controlled by either of two major mechanisms, incompatibility or incongruity. Incompatibility operates within a species while incongruity is seen between species. Whereas incompatibility promotes outbreeding, incongruity limits the possibilities for species hybridization (McCubbin and Kao, 1996). Frequently, the primary hybrids can be stabilized and their fertility can be restored by in vitro or in vivo polyploidization and the increase of the ploidy level. In fact, more than 60- 70 % are polyploid origin in angiosperm (Soltis and Soltis, 1993). Problems involved in interspecific hybridization have limited its use in many crops. In some genera of fruit crops, such as *Rubus* and *Vaccinium*, interspecific hybridization has been used extensively, while in others, such as *Malus* and *Pyrus*, it has only recently become a prominent method for crop improvement (Scorza, 1986). However, in breeding ornamental crops, interspecific hybridization is the most important source of genetic variation, especially Lilium, because lily is a model plant for style manipulations and *in vitro* culture methods.

PROGRESS MADE IN FRUIT CROPS

Many of our fruit crops have resulted from interspecific hybridization, polyploidy, or both. This is particularly obvious in *Actinidia*, Citrus, *Fragaria*, *Musa*, *Prunus*, *Rubus*, and *Vaccinium*. Breeders have also extensively used allied species to breed new fruits or improve the existing varieties and rootstocks. Using the allied species and at times allied genera new varieties with improved fruit quality, resistance to biotic and abiotic stresses have been achieved successfully. In certain fruit species like citrus natural interspecific hybrids occur, but in recent times a large number of manmade hybrids have been added. Through interspecific hybridization new fruits in Citrus (tangelos, tangors), Prunus (pluots), and Rubus (tayberry) were also created. In future, unlimited opportunities exist to exploit allied species in fruit breeding.

Heterozygous nature of many fruit crops has allowed in deriving more than one useful hybrid variety from same species (parental) combination. Some of the important hybrids in citrus are: Tangelo (*Citrus paradisi* X *C. reticulata*), see Fig.1, Tangor (*C. reticulata* X *C. sinensis*), Citrumelo (*C. paradisi* X *Poncirus trifoliata*), Citrange (*P. trifoliata* x *C. sinensis*), Citrangor (Citrange X *C. sinensis*), Orangequat (*C. reticulata* x *Fortunella margarita*), Citrangequat (*F.* sp. X

Fig. 1 Tangor *c. reticulata x c. sinensis*

citrange), Limequat (*C. aurantifolia* x *F.* sp.). Citranges and citumelo make good rootstocks for various citrus fruits. In grapes, inter-specific varieties have been used to successfully introgress tolerance to pests and diseases, such as powdery mildew, downy mildew or phylloxera (Alleweldt, *et al.*, 1990). These varieties are the result of efforts to combine the quality of traditional European varieties (*Vitis vinifera*) and pyramid different resistance traits typical of American varieties (*V. riparia*, *V. labrusca*, *V. aestivalis*, *V. berlandieri* and *V. amurensis*). Further, wine industries currently use a high percentage of inter-specific varieties with good results (Fisher, 2000).

Most banana varieties are triploid (2n=3X=33) seedless parthenocarpic clones selected by early farmers and are maintained by vegetative propagation (Simmonds, 1962). Bananas are interspecific derivatives from two diploid species *viz.*, *Musa acuminata* and *M. balbisiana* (Perrier *et al.*, 2009). Likewise the modern strawberry is derived from hybrids between two octoploid (2n=56) species *Fragaria virginiana* and *F. chiloensis* (Staudt, 1999).

Though some related species of mango like *Mangifera paiang, M. foeitida, M. caesia, M. altissima* have certain useful attributes (Bompard, 1993) they are still not exploited in interspecific hybridization. In guava attempts have been made to develop wilt resistant dwarfing rootstocks. *Psidium chiennsis* were crossed to *P. molle* and *P. guajava* Dinesh and Vasugi (2010).In order to incorporate the Papaya ring spot virus (PRSV) from *Vasconcellea cauliflora* was successfully crossed with cultivated papaya by Dinesh *et al.* (2007). In *Annonas*, a hybrid between *Annona atemoya* x *A.squamosa* was

Fig. 2 Arka Sahan Annona Atemoya x A. squamosa

developed as ' Arka Sahan' (Fig. 2) variety (Jalikop, 2007). It produces large (>600g) fruits with few small seeds (8-9/100g fruit weight) and sweet pulp (31°B). In China, intergenic hybridization between litchi (*Litchi chinenis*) and longan (*Dimocarpus longan*) are being evaluated (Zhao Yu-hui *et.al.*, 2008). The hybrids between *Passiflora edulis x P. incarnata* exhibited cold hardiness and tolerance to passion fruit woodiness virus (PWV). Recently, Santos (2013) isolated PWV resistant hybrids from *P. edulis* x *P. setacea*.

Interspecific hybridization in apple was carried out to obtain resistant cultivars to stress, as well as diseases and pest attack (Morales and Traveset, 2008) utilizing different species of *Malus* (many ornamentals), *e.g.*: *M. baccata* and *M. prunifolia* for frost resistance; *M. floribund, M. baccata, M. micromalus, M. prunifolia etc.* for apple scab resistance, *M. zumi, M. robusta, M. sarge etc.* for powdery mildew resistance (Sestras, 2004).

Several interspecific hybrids and new fruits of *Pyrus*, *Prunus* and *Rubus* genera are under cultivation. The enormous number of varieties of European pear (*Pyrus communis*), are without doubt derived from one or two wild subspecies (*P. communis* subsp. pyraster and *P. communis* subsp. caucasica). Pluots, apriums, apriplums, or

Fig. 3 Pluots (Dinosaur egg) Prunus Salicina x P. armeniaca

Fig. 4 Red raspberry *Rubus idaeus* x *R. strigosus*

plumcots, are so some of the hybrids (Fig. 3) between different *Prunus* species that are popularly called "Dinosaur Eggs" or interspecific (or IS) plums (http://en.m.wikipedia. org/). They were developed by crossing plum (*P. salicina* or *P. cerasifera*) and apricot (*P. armeniaca*) and are very sweet with a novel taste. Pluots and Apriums are later-generations hybrids; the former has more plum like while the latter shows apricot traits. Recently a cold hardy plum hybrid 'Black Ice' having complex lineage ('Oka' cherry plum x (*P. salicina*); Oka has in its lineage *P. besseyi* which can withstand severe winters) was released (http://www.warf.org). Sweet cherry (*P. avium*) and sour cherry (*P. cerasus*) were crossed to obtain hybrid called Duke Cherries. They are intermediate between the two parents and are grown on a small scale under different names like *P. acida, Cerasus regalis, P. avium* ssp regalis. Recently it was assigned name as *P. xgondouinii* by Faust and Suranyi (1997). Most modern commercial red raspberry (Fig. 4) cultivars were derived from hybrids between *Rubus idaeus* and *R. strigosus* (Huxley, 1992). Purple raspberries have been produced by hybridization of red and black raspberries.The botanical name *Rubus × neglectus* applies to these hybrid plants.

PROBLEMS

Sexual barriers preventing interspecific hybridization have been distinguished into pre- and post-fertilization barriers (Stebbins, 1958). The nature of the barrier governs the method to be used to overcome the specific barrier. An integrated method of *in vitro* pollination and fertilization followed by embryo rescue has been applied in many crosses.

OVERCOMING PRE-ZYGOTIC BARRIERS: There will be several hindrances prior to fertilization starting from the pollen landing on the stigma. A range of techniques, such as bud pollination, reciprocal cross, use of mentor pollen have been applied successfully to overcome pre-zygotic obstacles. **Bridge cross:** When a direct cross between two species, A and B, is not possible, an intermediate crossing with a third species, C, which is compatible with both species, may bridge the crossing barrier. The use of *Vascocellea parviflora* as a bridging species between *Carica papaya* and PRSV-P resistant *V. pubescens* was attempted by Drew *et al.* (1998). **Reciprocal cross:** Hybridization between *Annona atemoya* and *A. squamosa* is successful only when *A. squamosa* is used as a pollen parent. **Bud pollination:** In order to overcome incompatibility barriers posed by the stigma and/or the style, mature pollen is placed on a stigma that is immature or aged to prevent presence of active factors that inhibit pollen tube growth. Alternatively, pollination is performed at the end of the flowering season. Pollinating immature stigmas at the bud stage was shown to be effective for overcoming incongruity in guava, citrus and in pear. **Time of optimum receptivity:** The presence of the optimal level of receptivity of the stigma can vary from several hours (in mango) to more than one week (in lily). **Use of growth promoters:** Plant growth regulators or sucrose solution are applied to promote successful pollination. In an intergeneric cross of papaya (*Carica papaya*) Dinesh *et. al.* (2007) used 5% sucrose solution to improve pollen germination of *Vascocellea cauliflora* which yielded 35 viable hybrid seeds. **Use of pollen mixtures:** A mixture of compatible and incompatible pollen can be used for pollination. The compatible pollen is able to germinate and penetrate the style, thereby 'clearing' the way for the incompatible pollen. With such pretreatment, the 'assisting' compatible pollen is called **mentor pollen**. Pollination can also be performed in succession, with compatible pollen one or two days ahead of the incompatible pollen. In this case, the 'assisting' compatible pollen is called **pioneer pollen**.The mentor pollen technique was shown to work in such diverse species as apple and pear (Van Tuyl

& De Jeu, 1997). The pioneer pollen method overcame SI in apple and pear (Visser, 1983). Distant hybridization between litchi (*Litchi chinenis*) and longan (*Dimocarpus longan*) was effected by using mixed pollens (Zhao *et.al.,* 2008). **Manipulation of the style:** Shortening of the style (cut-style method) may result in removal of inhibiting factors for incompatible pollen tube growth.**In vitro pollination:** Pollen is applied in the laboratory to the stigma of pistils placed in vitro as demonstrated in Citus by Germana and Chiancone (2001).

OVERCOMING POST- ZYGOTIC BARRIERS: Post-zygotic dysfunctions may be expressed as embryo death, endosperm dysfunction, poor seed germination, lack of recombinants, slow seedling growth or seedling mortality, unusual susceptibility to disease, chlorophyll abnormalities, lack of flowering, sterility (Hogenboom,1984). Subsequently at times hybrid breakdown may occur in F_2 or later generations. **Ovule culture: Before** a certain stage of development, it may not be possible to culture embryos (embryo rescue) to circumvent incompatibility with maternal tissues. In such a case, removing the ovule from the ovary and culturing it in vitro is attempted. Ovule culture was successfully applied in grape and other crops. (Van Tuyl & De Jeu, 1997). **Embryo rescue:** Embryos can abort after fertilization due to incompatibility with the endosperm and/or surrounding maternal tissues of the ovule. The technique is applied in various crops to get the viable hybrids. Inter-specific hybridization was made between cultivars of sweet cherry (*Prunus avium*) and Chinese cherry (*P. pseudocerasus*), and embryo rescue technique was successfully applied to recover hybrid plantlets (Liang *et.al.* 2006).

OVERCOMING PRE- AND POST-FERTILIZATION BARRIERS: In many interspecific and intergeneric crosses, integrated techniques to manipulate both pre- and post-fertilization barriers have been applied. In vitro pollination and fertilization is one such technique where pollination brings pollen grains in direct contact with the ovules, and is, therefore, considered more effective.

SOMATIC HYBRIDIZATION: Somatic hybridization (fusion of protoplasts of two different plant species) can be used to combine specific nuclear and cytoplasmic (plastid and mitochondrial) genomes in an efficient manner, that is, avoiding more lengthy backcrossings with further compatible crosses. Several thousand triploid hybrids have been produced using somatic hybrids as the tetraploid parent (Grosser and Gmitter, 2011) in order to secure seedless citrus hybrids.

OVERCOMING F_1 STERILITY: Interspecific F_1 hybrids may display sterility owing to lack of chromosome pairing during meiosis. **Polyploidization:** By doubling of somatic chromosomes it is possible to induce homologous pairing of chromosomes and restore the fertility in interspecific hybrids. Miyashita *et. al.* (2009) developed blueberry amphiploids from pentaploid hybrids between *Vaccinium corymbosum* (4x) and *V. ashei* (6x) by colchicine treatment. **Haploidization:** Chromosome elimination of one parental genome after fertilization of the egg by the sperm of another species can occur (Dunwell, 2010). This phenomenon results in haploid embryo formation of only one of the parents and by doubling the chromosome of such embryos it is possible to produce of doubled haploids. They have the advantage in finding traits with recessive inheritance. **Production of alien addition or substitution lines:** An alien addition line carries one chromosome pair from a different species in addition to the normal somatic complement of parental species. For this, aneuploid plants are produced by repeated backcrossing. Aneuploids contain an extra single chromosome or an extra chromosome pair from a donor plant, called monosomic or disomic addition lines, respectively. In alien substitution lines, the introduced chromosome or chromosome pair has replaced the pair from the recipient species. However, they can also be used for the localization of genes for particular traits to specific chromosomes. Alternatively monosomic (2n-1) or trisomic (2n+1) also serve this purpose. **Transfer of small chromosome segment:** Chromosome translocation (small chromosome segment carrying specific useful gene) is effected by the interchange of parts between non-homologous chromosomes. The translocation can be induced by irradiation of a trisomic addition line.

PARTIAL STERILITY IN F_1 PLANT: Varying degree of sterility may appear in interspecific progeny in F_1 or in the following generations. The partial sterility in *Annona atemoya* x *A. squamosa* was circumvented by assisted pollination with a specific pollen source which substantially improved the fruit-set, -size and -shape. This hybrid exploits interspecific heterosis as well as substantial interspecific xenic or pollen effect (Jalikop and Kumar, 2003; Jalikop, 2010). Now it is commercially grown following assisted pollination under the name, 'Arka Sahan'.

IMPLICATIONS

FACTORS CAUSING STERILITY: Varying degree of sterility we notice in interspecific hybrids may be due to cytogenetic, genetic or cytoplasmic factors. Fertility can be restored by chromosome doubling in situations where rings and chains at metaphase one (indicating translocations), bridges and fragments at anaphase one (indicating inversions) and loops at pachytene (indicating duplication or deletion) occur. Fine structural differences in chromosome (cryptic structural hybridity) can also lead to sterility, despite normal paring and disjunction (Stebbins, 1958). In case of cytoplasmic sterility reciprocal crossing should help in obtaining fertile hybrids.

SEGREGATION PATTERN IN INTERSPECIFIC HYBRIDS: Interspecific hybrids involve two separate species that would differ for several genes. Most of the fruit species are basically heterozygous so tremendous diversity of types appear in F_1 generation itself. Many segregants will be entirely new resembling neither of the parental species. The possible recombinants in segregating generation of interspecific hybrid is too large, but in practice only a proportion of it is recovered due to abnormality in meiosis, impaired recombination, gametic and zygotic elimination (owing to reduced survival), linkage and pleiotropy. The meiotic process rarely functions with complete normality in interspecific hybrids. Even when chromosomes pair, crossing over (recombination between linked genes) is drastically reduced. Due to such disturbances the segregation for different characteristic does not fit Mendelian ratio. Further, the male gametogenesis is more easily disturbed by genetic disharmonies than female gametogenesis. Often male gametes with genes contributed by both parents tend to be inviable. As a result the F_2 offsprings often to be more like backcross types (resembling the pollen species) than true F_2 segregant.

CONCLUSION

The allied species and genera have an array of novel traits of potential commercial interest. They are an important reservoir of genetic variability for various characteristics such as disease and insect resistance, tolerance for abiotic stresses and improved quality. They provide unique and unlimited opportunities in fruit crop breeding programs. However, in most of the fruit crops characterization of allied / wild types has not been carried out either by morphological or by molecular means, thus their significance in genetic improvement of fruit crops has remained unrealized. In many fruit crops breeding through interspecific hybridization is not explored adequately in any sustained program.

Banana and mango are the most important crops of India. Utilization of allied species of *Musa* and *Mangifera* has been hampered due to difficulties encountered in developing interspecific hybrids by conventional hybridization. In these crops trangene technology to transfer genes of interest from the related species holds promise. In certain fruits like pomegranate related lone species, *Punica protopunica*, is yet to be studied for its potential economic significance. Many novel fruits having exotic taste, colour, flavor have been produced and are under cultivation in the fruits of *Rubus, Prunus, Pyrus* genera. Likewise interesting new interspecific fruits can be created by crossing appropriate edible species of different fruit genera. Most of the fruit crops can be raised on rootstocks, thus it is possible to overcome soil related biotic and abiotic stresses by employing appropriate rootstocks. In such situations fruit breeders should prefer developing suitable rootstocks instead of attempting to incorporate resistance traits into scion as that would lead to addition of undesirable traits from the donor related species along with the desirable ones. Moreover that would result in disturbing the delicate fine fruit qualities of the already long standing well adapted scion varieties.

Understanding inheritance pattern of important traits will help in the incorporation of selected genes from allied species, hence it is essential to initiate genetical studies for characteristics of interest. Aneuploids developed in the interspecific crosses should be used to assign certain genes to specific chromosome and double haploids to know the traits with recessive inheritance. When securing interspecific hybrids through traditional crossing programs is difficult, somatic hybridization is an attractive option. Future research projects should also aim at evaluation of related species of several fruits that have gained commercial importance in the recent times like Indian Bael, jackfruit, Indian Gooseberry, and their subsequent utilization in genetic improvement programs. Finally, some of the limitations of conventional fruit breedingin transferring useful genes from allied species could be overcome by marker assisted selection, and transgene technology and the example of virus resistant papaya has demonstrated that success can be achieved.

REFERENCES

Alleweldt, G., Spiegel-Roy, P. and Reisch, B.I. 1990. Genetic resources of Temperate fruit and nut crops. *Acta Horticulturae*, **290**:291-337.

Bompard, J. M. 1993. The genus *Mangifera* rediscovered the potential contribution of wild species to mango cultivation. *Acta Horticulturae*. **341**:319-326.

Dinesh, M.R., Rekha, A., RavishankarK. V., Praveen, K.S. and Santosh, L.C. 2007. Breaking the intergeneric crossing barrier in papaya using sucrose treatment. *Scientia Horticulturae*, **114**:33–36.

Dinesh, M. R. and Vasugi, C. 2010.Guava improvement in India and future needs. *J. of Hort. Sci.*, **5**:94-108.

Drew, R.A., O'Brien, C.M. and P.M. Magdalita. 1998. Development of interspecific *Carica* hybrids. *Acta Horticulturae*, **461**:285-292.

Dunwell, J.M. 2010. Haploids in flowering plants: origins and exploitation. *Plant Biotechnol J.*, **8**:377–424.

Esen, A. and Soost, R.K. 1973. Seed development in citrus with special reference to 2x x 4x crosses. *Am. J. of Bot.*, **60:** 448–452.

Faust, M. and Suranyi, D. 1997. Origin and dissemination of cherry. *Hortic Rev.*, **19**:263–317.

Fisher, K.H.2000. The development of interspecific grapevine hybrids in Ontario, Canada. *Proceedings of the 6ᵗʰ International Congress on Organic Viticolture, IFOAM*, 25-26 August 2000 pp.205-208.

Germana, M.A. and Chiancone, B. 2001. Gynogenetic haploids of Citrus after in vitro pollination with triploid pollen grains. *Plant Cell Tissu. Org. Cult.*, **66**:59–66.

Grosser, J.W., Moore, G. A., Gmitter, Jr. G. F. 1989. Interspecific somatic hybrid plants from the fusion of 'Key' lime (*Citrus aurantifolia*) with 'Valencia' sweet orange (*Citrus sinensis*) protoplasts. *Scientia Horticulturae*, **39**:23–29.

Grosser, J.W., Gmitter, Jr. G. F. 2011. Protoplast fusion for production of tetraploids and triploids: applications for scion and rootstock breeding in citrus. *Plant Cell Tissue and Organ Culture*,**104**:343-357.

Hogenboom, N.G. 1973. A model for incongruity in intimate partner relationships. *Euphytica*, **22**:219-233.

http://www.warf.org/documents/technology-summary/T03019US.pdf.

http://en.m.wikipedia.org/wiki/Pluot.

Huxley, A.1992. New RHS Dictionary of Gardening. Macmillan ISBN 0-333-47494-5.

Jalikop, S.H. and Kumar, R. 2007. Pseudo-xenic effect of allied *Annona* spp. pollen in hand pollination ofcv. 'Arka Sahan' [(*A. cherimola x A. squamosa*) x*A. squamosa*]. *HortScience*,**42**:1534–1538.

Jalikop, S.H. 2010. Tree and fruit traits of progenies from the cross between (*Annona cherimola* Mill. *A. squamosa* L.) *A. reticulata* L. and approaches for the introgression of valuable genes from *A. reticulata* L.*Euphytica*,**173**:161-171.

Keep, E. 1988. Primocane (autumn)-fruiting raspberries: a review with particular reference to progress in breeding. *J. of Hort. Sci.*, **63**:1-18.

Liang, Q., Chen, X., Zhang, L.,Chuan-jin, W.U. andLiu, C. 2006. Effects of parental cultivars on cross compatibility and embryo rescue in interspecific crosses between *Prunus avium* and *Prunus pseudocerasus. J. of Fruit Sci.,* 2006-03.

Miyashita, C., Ishikawa, S. and Mii, M.2009. In vitro induction of the amphiploid in interspecific hybrid of blueberry (*Vaccinium corymbosum* × *Vaccinium ashei*) with colchicine treatment. *Scientia Horticulturae,***122**: 375-379.

McCubbin, A. and Kao, T. 1996. Self-incompatibility and pollen rejection in angiosperms. **In**: Mohan, S., Sopory, S., Veilleux, R. (Eds) Current Plant Science and Biotechnology in Agriculture: InVitro Haploid Production in Higher Plants, Kluwer Academic Publishers, pp. 225-253.

Morales, C. L. and Travesetb, A. 2008. Interspecific Pollen Transfer: Magnitude, Prevalence and Consequences for Plant Fitness. *Crit. Rev. in Plant Sci.,* **27**:221-238.

Perrier, X., Bakry, F. and Carreel, F. 2009. Combining biological approaches to highlight the evolution towards edible bananas. *Ethnobotany Res.& Applications,***7**:199–216.

Qi, L., Friebe, B., Zhang, P. and Gill, B.S. 2007. Homoeologous recombination, chromosome engineering and crop improvement. *Chromosome Res.,***15**:3–19.

Santos, E. A. 2013. Breeding passionfruit (*Passiflora edilis*) aiming to resistance to Cowpea aphid-borne mosaic virus. Universidade Estadual do Norte Fluminense Darcy Ribeiro. Juno.

Scorza, R.1986. Introduction to the workshop: Overcoming barriers to interspecific hybridization of perennial fruit crops. *HortScience,* **21**:41.

Sestras, R.2004. Horticultural plant breeding (in Romanian). AcademicPres, Cluj-Napoca, p. 334.

Sharma, H. 1995. How wide can a wide cross be? *Euphytica,* **82**:43-64.

Simmonds, N.W. 1962. The evolution of the bananas. London: Longman.

Soltis, D.E. and Soltis, P.S. 1993. Molecular data and the dynamic nature of polyploidy.*Crit. Rev. Plant Sci.,***12**:243–273.

Staudt, G. 1999. Systematics and geographic distribution of the American strawberry species. University of California Press, Berkeley, CA.

Stebbins, G.L. 1958. The inviability, weakness, and sterility of interspecific hybrids. *Adv. Genet.***9**:147-215.

Van Tuyl, J.M. and de Jeu. M.J. 1997. Methods for overcoming interspecific crossing barriers. **In**: Shivanna, K.R. and Sawhney, V.K. (Eds.) Pollen biotechnology for crop production and improvement. Cambridge University Press, Cambridge, pp. 273-292.

Visser, T. 1983. A comparison of the mentor and pioneer pollen techniques in compatible and incompatible pollination of apple and pear. **In**: Mulcahy, D.L., Ottaviano, E. (Eds.) Pollen: biology and implications for plant breeding. Elsevier, Amsterdam, pp. 229-236.

Vizzotto, M, L., Cisneros-Z. and Byrne, D.H. 2007. Large Variation Found in the Phytochemical and Antioxidant Activity of Peach and Plum Germplasm. *J. of the Am. Soci. for Hort. Sci.,***132**:334-340.

Volker, P.W. and Orme, R.K. 1988. Provenance trials of *Eucalyptus globulus* and related species in Tasmania. *Aust. Forestry,***51**: 257-265Zhao, Y.H., Hu, Y.L., Guo, Y.S., Zhou, J., Fu, J.X., Liu, C.M., Zhu, J. and Zhang, M.J. 2008. Inter-generic hybrids obtained from cross between litchi and longan cultivars and their molecular identification. *J. of Fruit Sci.,*2008-06.

3

Distant Hybridization for Improvement of Temperate Fruits and Nuts

Nazeer Ahmed, A.K.Gupta, K.L. Kumawat and J.I. Mir

ICAR-Central Institute of Temperate Horticulture
Old Air Field, Rangreth, Srinagar-190007
Email:dnak59@rediffmail.com

INTRODUCTION

Temperate fruits and nuts such as apple, pear, peach, plum, cherry, apricot, walnut, almond *etc.* are the important crops of temperate world, most of them being cross pollinated have tremendous genetic diversity for many horticultural traits which is now diminishing.However from already existing variability, several, genotypes with elite traits have been selected and exploited through asexual reproduction. But for biotic and abiotic resistance and for certain special attributes like higher biomass, nutritional quality, male sterility, shelf life *etc.* These crops still need improvement as genes for such traits are distributed across wild relatives which are reservoir of genetic variability. The distant hybridization in this regard is an important tool may be inter-specific or inter-generic where one or few characters from diverse genomes are transferred for enriching genetic diversity and producing new genotypes. Certain mechanical, biological and chemical manipulation during wide hybridization breaks species barrier for gene transfer and makes it possible to transfer the genome of one species to another resulting in changes in genotypes and phenotypes of the progenies. Wide hybridization also contributed to species evolution and speciation as chromosome doubling of wide hybrids resulting in origin of many allopolyploid species. While combining desirable traits by inter-specific or inter-generic hybridization, several barriers such as pre-fertilization incompatibility and post-fertilization sterility are limitingsuccess in temperate fruits. Selecting more suitable parents for distant hybridization, pollination at the balloon stage of the flower buds, pollen treatment, mixed pollination with pollen of different species, embryo rescue, somatichybridization use of PGR *etc.* have been successfully tried to overcome the preand post fertilization barriers in *Mallus, Pyrus, Prunus, Juglans, Fragaria* species.

The genes for scab resistance, abiotic tolerance and allotriploids in apple; resistance to fire blight, improved shape, flesh, flavour and low chilling genotypes in pear; winter hardiness and late flowering in peach; dwarfness, winter hardiness, resistance to fungal diseases and viruses, flavour and canning quality in plum; resistance to nematode, vigour and anchorage and male sterility in almond; tolerance to stress, crown rot and root lesion nematode in walnut; winter hardiness, high dry mater and sugar, content, salt and drought tolerance in cherry and improved agronomic traits, fungal and bacterial resistance and pentaploids in strawberry were transferred and new genotypes were developed successfully through distant hybridization. With the increase in genetic erosion and climate change the need for resilient genotypes to biotic and abiotic stresses and other special traits has increased. The pre breeding involving hybridization between diverse species having desirable traits is the new hope for transgressing useful genes in to cultivated species for isolating superior cultivars and rootstocks Pre and post fertilization barriers during wide hybridization methods to overcome followed by selection and propagation of elite transgressive segregates are some of the researchable areas where biotechnological tools have greater role in genetic enrichment.

DISTANT HYBRIDIZATION

Hybridization between two different species (Inter-specific hybridization) or between two different genera (Inter-generic hybridization), is generally termed as distant hybridization or wide hybridization which was successfully attempted first by Thomas Fairchild as early as 1717 who obtained hybrid by crossing carnation with sweet william. Generally the objectives of distant hybridization is to transfer one or few characters, such as disease and insect resistance, tolerance for abiotic stresses, male sterility, increased biomass, yield, and improved quality characteristics (Khush and Brar, 1992). Mating between individuals of different species or genera, provides a way to combine diverged genomes into one nucleus resulting in enrichment of genetic diversity (Chen *et. al.,* 2003). Wide hybridization breaks what is known as the species barrier for gene transfer and thus makes it possible to transfer the genome of one species to another, which results in changes in genotypes and phenotypes of the progenies. It is also very important for species evolution and speciation as chromosome doubling of wide hybrids is responsible for the origin of many allopolyploid species (Liu *et. al.,* 2014).

Barriers in Distant Hybridization

Numerous attempts have been made to combine the desirable traits in temperate fruits by interspecific hybridization; however, barriers such as prefertilization incompatibility and post-fertilization sterility (Boavida *et al.*, 2001; Guillaumin *et al.*, 2003) have limited its success.

Overcoming the Barriers in Hybridization

Selecting more suitable parents for distant hybridization, pollination at the balloon stage of the flower buds, and pollen treatment have been suggested by (Suzuki *et al.*, 1988; Len'kov (1990) and Valk *et al.*, (1991) to overcome the prefertilization barriers. Although pollen tubes germinate and successfully fertilize ovaries, post-fertilization barriers may yet prevent the resulting seed from reaching maturity when fruits continue to develop normally but are seedless (Gordillo *et al.*, 2003; Wen Liu *et al.*, 2007). When early or very early Prunus varieties are used as female parents in wide crosses, the embryos seldom develop fully and the seeds rarely germinate probably they are devoid of normal embryos and endosperm (Kuden *et al.* 1999; Infante and Gonzalez, 2002).

Mixing of pollen of different species to increase efficacy of distant hybridization A number of investigators have shown that if mixed pollen of different varieties, belonging to the same botanical species as the plants pollinated, falls on the stigma of apple, plum, gooseberry, or currant, the results appear to be considerably better than when pollination is carried out with the pollen of any single variety, even if it is a good pollinator. The first method gives better fruit and seed setting, and greater vigour and increased growth of the seedlings. In such intraspecific crosses, a pollen mixture ensures a wide gametic selection and improves the genotypic constitution of the progeny. In interspecific hybridization, breeders use pollen mixtures to facilitate crossing and to study the degree of relationship between different species within certain botanical genera. Different variations of pollination with pollen mixtures are used in experiments, depending on the problems involved and on the biological characteristics of the plant (Yenikeyev, 1965). Experiments with stone fruit which possess one ovule per flower showed highly successful results. The American sand cherry (*Cerasus besseyi* Baily), pollinated with a mixture of pollen of *C. besseyi*, *Prunus ussuriensis* Kov. et Kost., *Cerasus tomenosa* Thunb., *Cerasus vulgaris* Mill., *Armeniaca vulgaris* Lam., and *Prunus persica* Sieb. et Zucc. An overwhelming majority of the plants obtained showed characteristics of Cerasus besseyi, and a few plants were hybrids *Cerasus besseyi* x *C. tomentosa* and *C. besseyi* x *Prunus ussuriensis*. Gorshkova (1947) obtained hybrids by pollinating the Siberian apple (*Malus baccata* var. *sibirica* Maxim.) with a mixture of pollen of several West- European pear varieties (*Pyrus communis* L.). The interspecific hybrids possess a combination of the features of both parental species. Tolmachev (1947) achieved success in interspecific hybridization by pollinating the Siberian black currant (*Ribes nigrum* ssp. *sibiricum*) with a pollen mixture from several gooseberry varieties (*Grossularia reclinata* Mill.).

The cross *Pyrus communis* Tonkovetka' (Thin Branch) x *P. regelii* was effected using *P. regelii* pollen alone or mixed with varying amounts of pollen of *P. communis*, 'Tonkovetka' or 'Bessernanka' (Seedless). In the first year, the F comprised mostly plants with the leaf characters of either the seed or the pollen 1 parent; only few seedlings were intermediate. In the third year, some seedlings acquired mixed inheritance by developing leaves of both parental types on a single plant (Lobanov, 1960).

Embryo Rescue

Embryo rescue refers to invitro culturing of embryos at different stages of development. It is normally used in hybridization of distantly related species to overcome post zygotic incompatibility. Embryo rescue technique helps in making the interspecific crosses successful when there is post fertilization dormancy between the embryo and endosperm. The embryo in interspecific cross is removed before abortion and cultured in nutrient medium. Embryo rescue has been applied widely in many field crops (Kato *et al.*, 2001; Kaushal *et al.*, 2005) while very few cases are reported of interspecific hybridization in temperate fruits because of their long breeding cycle, cross-incompatibility, and hybrid sterility (Chopra *et al.*, 1994; Hormaza 1998). To overcome post-pollination incompatibility, embryo rescue technique was successfully used to produce interspecific hybrid by Yang *et al.* (2004) in plum and apricot and Wen Liu *et al.* (2007) in peach as a female parent with apricot and plum.

Somatic hybridization

Crossing of plants through fusion of somatic cells or protoplasts is called somatic hybridization. Somatic hybridization helps to overcome cross incompatibility and where the sexual process is bypassed and hybrids so obtained are called as somatic hybrids. They may be interspecific or intergenric but have three types of somatic hybridization. 1. Symmetrical hybrids: Somatic hybrids that contain all chromosome of both the sp. involved in the fusion of protoplast. 2. Asymmetrical hybrids: Somatic hybrids that contain complete somatic complement of one sp. only a part of somatic complement of other species. 3.Cybrids:Somatic hybrids with normal protoplast of one species and nucleolus protoplasm of other species.

Use of chemical

Growth regulators like GA ; IAA; 2,4-D *etc.* are also used at the flowering stage to increase fruit set in distant hybridization. A method of obtaining fruit from crosses between species of fruit crops with poor cross compatibility is described by Len'kov (1990). It involves spraying the flower buds and flowers at the stage of peak flowering with a gibberellin solution of 500-1000 mg/ litre, followed in 10-15 min by controlled pollination (with pure or mixed pollen) and then after 6-10 days by a further gibberellin spraying. By this method, 3 fruits were obtained from a sterile hybrid between the peach variety Zheleznyi and Kantsler and the almond Posrednik.

DISTANT HYBRIDIZATION IN TEMPERATE FRUITS CROPS

Apple

Domestic crab-apples are hybrids between *Malus pumila* and one of the primitive species *M. baccata* (L.) Borkh. (Chadha, 1995). Which are used as aseedling rootstock for apple. From cross between quince and apple Rudenko andRudenko (1994) found only allodiploid viable hybrids in F_1 generation while in F_2 more than 50 per cent of the seedlings were allotriploids (3x = 2n= 51) and restwere tetraploids (4x = 2n= 68). Most of the tetraploids resembled quincemorphologically. About 15 per cent of the allotriploids displayed apple like characters and had seedless fruits with intermediate taste. All triploids produce slightly viable pollen. Allotetraploid (4x = 2n= 68) hybrids position is intermediate between the diploids and the triploids.

Bus *et al* (2002) identified at least 10 different resistance genes for apple scab and one Vf gene from the ornamental crabapple M floribunda has been used all over the world to create new scab resistant cultivars (Laurens, 1999). This gene has been cloned and shown to confer scab resistance to a transgenic cultivated variety 'Gala' (Belfanti *et al.*, 2004). Scar markers have been identified for Vbj from *Malus baccata jackii* (Gygax *et al.*, 2004) and Vm from *M. atrosanguninea* (Cheng *et al.*, 1998).

At ICAR-CITH, Srinagar ten wild species of Malus were screened for scab resistance and found that among ten wild species six wild species namely *Malus robusta; M.spectabilis, M. baccata*-CITH-2; *M. baccata*-CITH-3; *M. baccata*-CITH-7 and *M.floribunda* showed resistance to apple scab disease with less than 10% infection under open field conditions. Three wild species *M. baccata* Khrot, *M. baccata*-CITH-1 and *M. baccata*-CITH-5 were moderately resistant showing 10-25% infection and one wild species was found susceptible (>25% infection). To transfer scab resistance, regular bearing, improved quality and shelf life an interspecific hybridization was successfully attempted involving *Malus floribinda* and *M. domestica* cultivars (Ambri, Prima, Well Spur and American Apirogue) and their first generation hybrids are being evaluated/screened for targeted traits on different rootstocks. The preliminary results are very encouraging and are expected to give scab resistant genotypes with superior traits (ICAR-CITH Annual Reports 2011- 12 and 2012-13). Wild species of apple are having different characteristic features (Table 1) which can be utilized for hybridization with domestic apple cultivars for transfer of traits.

Table 1. Apple species with some Horticultural Traits

Species	Chr. no. (2n)	Useful Characteristics
Malus angustifolia (Aiton) Michx.	34, 51, 68	Ornamental tree
Malus × atrosanguinea (hort. ex Spath) C. K. Schneid.	34	Resistance to fire blight
Malus baccata (L.) Borkh.	34, 68	Monogenic (V_b, V_{bj}) resistance to apple scab . source of monogenic resistance to powdery mildew, Semi-dwarfing rootstock
Malus coronaria (L.) Mill.	-	Better source of precosity, Ornamental tree
Malus domestica Borkh.	34, 51, 68	Single dominant gene (*Co*) identified for compact columnar habit, QTL identified for malic acid gene
Malus floribunda Siebold ex Van Houtte	34	Monogenic (V_f) resistance to apple scab, Semi-dwarfing rootstock
Malus angustifolia (Aiton) Michx.	34, 51, 68	Ornamental tree
Malus × atrosanguinea (hort. ex Spath) C. K. Schneid.	34	Resistance to fire blight

Contd...

Species	Chr. no. (2n)	Useful Characteristics
Malus baccata (L.) Borkh.	34, 68	Monogenic (V_b, V_{bj}) resistance to apple scab . source of monogenic resistance to powdery mildew, Semi-dwarfing rootstock
Malus coronaria (L.) Mill.	-	Better source of precosity, Ornamental tree
Malus domestica Borkh.	34, 51, 68	Single dominant gene (*Co*) identified for compact columnar habit, QTL identified for malic acid gene
Malus floribunda Siebold ex Van Houtte	34	Monogenic (V_f) resistance to apple scab, Semi-dwarfing rootstock
Malus fusca (Raf.) C. K. Schneid.	34	Resistance to fire blight
Malus halliana Koehne	34	Source of gene resistance to wholly apple aphid (*Eriosoma lanigerum*), Ornamental tree
Malus hupehensis (Pamp.) Rehder	51	Source of gene resistance to wolly apple aphid
Malus mandshurica (Maxim.) Kom. ex Skvortsov	34	Semi-dwarfing rootstock
Malus ×micromalus Makino	34	Monogenic (V_m) resistance to apple scab, Semi-dwarfing rootstock
Malus orientalis Uglitzk	34	Near dwarfing rootstock
Malus prunifolia (Willd.) Borkh.	34	Better source of precosity, cold hardiness, Semi-dwarfing
Malus pumila Mill.	34	Dwarf tree
Malus ×robusta (Carriere) Rehder	34	Resistance to WAA (*E. lanigerum* Hausm.) source of monogenic resistance to powdery mildew, Major gene for resistance to apple leaf curling midges, Resistance to Rosy apple aphid (*Dysaphis plantaginea*), Sd3 Resistance gene identified for Rosy curling aphid *(D. devecta),* Resistance to fire blight.
Malus sargentii Rehder	51, 68, 85	Polygenic resistance to apple scab, source of monogenic resistance to powdery mildew (*Podosphaera leucotricha*), Source of gene resistance to wholly apple aphid, Semi-vigorous rootstock.
Malus sieversii (Ledeb.) M. Roem.	34	Monogenic (V_r) resistance to apple scab, Better source of precosity
Malus sikkimensis (Wenz.) Koehne ex C. K. Schneid.	51	Ornamental tree, Semi-dwarfing rootstock
Malus sieboldii	34, 51, 68	its resistance to WAA and to *Phytophthora* rot, near dwarfing rootstock
Malus ×sublobata (Dippel) Rehder	34	Resistance to fire blight
Malus sylvestris (L.) Mill.	34	Polygenic resistance to apple scab

Contd...

Species	Chr. no. (2n)	Useful Characteristics
Malus tschonoskii (Maxim.) C. K. Schneid.	34	Source of gene resistance to wolly apple aphid (*Eriosoma lanigerum*)
Malus zumi (Matsum.) Rehder	34	Polygenic resistance to apple scab, source of monogenic resistance to powdery mildew, Major gene for resistance to apple leaf curling midges

Pear

The genus pyrus contains 24 primary species, six natural interspecific hybrids and at least three artificial hybrids (Bell *et al.*, 1996) and all the species are diploid and interfertile. Challice and Westwood (1973) reported that interspecific hybridization played an important role in pear domestication. The European pear or French pear (*Pyrus communis*) is a cultivated form derived mostly from *P. communis* var. *pyraster*, *P. caucasica* and *P. nivalis* while Rubtsov (1944) felt that modern cultivars of P. communis had characteristics derived from at least three species, *P. elaeagrifolia* and *P. syriaca*. These European pears have persistent calyx, fleshy pedicels, fruits mostly pyriform tapering base but sometimes apple shaped, highly susceptible to the fire blight disease. While the domestic Asian pears are derived mostly from *P. pyrifolia* (Burm) Nakai, known as Japanese sand pear from selection of *P. ussureinsis* Maxim. They have deciduous calyx, non–fleshy pedicels, shape like apple, resistant to fire blight disease. It bears poor quality fruit but are important for hybridization, propagation and other purposes. In the present cultivars of pear many are complex hybrids between the two species or with other Asian species. All species are self-sterile, cross-fertile and sexualdiploids (2n=34). Cultivars grown in northern China are probably belong to a hybrid complex involving *P. ussuriensis* and *P. pyrifolia* (Bell *et al.*, 1996).

Table 2: Useful horticultural traits carried by native pear species (from Bell 1991)

	Scientific name	Useful characteristics
European	*P. communis* L.	Resistance to fire blight, white spot, pear scab, pear decline, pear psylla; cold hardy
	P. nivalis Jacq.	Resistance to pear psylla; cold hardy
	P. cordata Desv.	P. cordata Desv. Resistance to codling moth; adaptation to warm winters, high pH, drought
Mediterranean	*P. amgydaliformis* Vill	Resistance to codling moth; adaption to warm winters, high pH, drought and clay soils
	P. elaeagrifolia Pall	Resistance to codling moth'sold hardy
	P. syriaca Boiss.	Adaptation to drought
	P. longipes Coss. & Dur.	Cold hardy'adaptation to wet and dry soils
	P. gharbiana Trab.	Adaptation to drought
	P. manorensis Trab	Adaptation to low pH

Contd...

	Scientific name	Useful characteristics
Mid Asian	*P. glabra* Boiss	Adaptation to drought
	P. salicifolia Pall	Adaptation to drought
	P. regelii Rehd.	Resistance to pear psylla
	P. pashia Buch .Ham. Ex D.Don	Adaptation to warm winters, low pH, clay soils
East Asian	*P. pyrifolia* (Burn,) Nak	Resistance to pear scab
	P. pseudopashia T.T. Yu	Resistance to codling moth; winter hardy
	P. ussuriensis Maxim	Resistance to fire blight, pear psylla; winter hardy
	P. calleryana Decne.	Resistance to fire blight, Fabraea leaf spot, pear psylla, codling moth; adaptation to warm winters, low pH, wet, dry and clay soils
	P. betulifolia Benge	Resistance to pear psylla, codling moth; adaptation to warm winters, low pH, wet, dry and clay soils
	P. fauriei C.K. Scheid.	Resistance to fire blight, pear psylla, codling moth; adaption to warm and cold winters, low pH, wet and clay soils
	P. hondoensis Kik. & Nak	Resistance to fire blight
	P. dimorphophylla Mak	Resistance to powdery mildew, pear psylla, codling moth; adaptation to low pH and wet soils
	P. kawakamii Hayata	Resistance to codling moth; adaptation to warm winters

Among four cultivated species, such as *P. communis, P. pyrifolia, P. ussuriensis* and *P. bretchneideri*, the systematic relationship is more distant and the gene basis is evidently different; interspecific hybridization makes additive effect; not only new fine pear varieties and selections can be bred by interspecific hybridization, but also the new bred pear varieties (selection) have productive, hardy, disease-resistant and dwarf characteristics respectively and at the same time, interspecific hybridization has very great potentialities for breeding new crisp-fleshed and flavorous pear varieties (Chengquen *et al.*, 1995).

During 1930 decades, breeders in North Americna first bred a new pear variety 'Le Conte' which is resistant to fire blight by (Ahmed and Tewfik, 1956). It has low chilling requirement, and can be grown in low and mid-hills and plain of Punjab and Haryana (Uppal *et al.*, 1993). Since then, many interspecific hybrids resistant to cold and disease such as 'Keiffer', 'Golden Spice', 'Garber', 'Harbin','Olia', 'Tioma' (Layne and Quamme, 1975), 'Mishirazu', 'Tietou', Chetou' (Chengquen *et al.*, 1995) *etc.* have been bred in success on by crossing between European and Oriental types.

Interspecific hybrid 'Jinxing' derived from *P. ussuriensis* and *P. communis* is more resistant to rot than *P. communis* cultivars (Pu Fushen and Fang Chengquan, 1990). Most pear variety of Oriental type is susceptible to pear scab, but cultivar of

P. communis are basically resistant. Interspecific hybrid 'Zaosu' and 'Wujiuxiang' (*P. bretschneideri* x *P. communis*), 'Jinxiang' and (*P. ussuriensis* x *P. communis*) *etc.* are markedly more resistant to scab then cultivar of *P. bretschneideri* or *P. ussuriensis* because they have the consanguinity of *P. communis*. *P. ussuriensis* which is originated in China is the most resistant to cold in *Pyrus* and is a ideal germplam material for breeding for cold hardiness. (Chengquen *et al.*, 1995).

Interspecific hybrid derived from *P. ussuriensis* x *P. communis* or *P. bretschneideri* x *P. communis* were more resistant to cold ('Jinxiang'), resistant to cold ('Zaosu' and 'Wujiuxiang') and very resistant to cold ('Ningmenghuang' and 'Xingcheng') respectively (Chen Changlan *et al.*, 1991). 'Aixiang' is a interspecific hybrid of *P. communis* and *P. ussuriensis* has dwarf tree and spreading tree posture (Pu Fushen and Jia Jingxian, 1985). *P. pyrifolia* and *P. bretschneideri* lack generally stronger flavour and aroma than that of *P. communis* and *P. ussuriensis*. At the same time *P. pyrifolia* and *P. ussuriensis* have more stone cells in flesh and have coarsertextured flesh than those of *P. communis* and *P. bretschneideri* (Chengquen *et al.*, 1995). Accoridng to Fu Fushen (1979) interspecific cross between the crispfleshed cultivar with a soft-fleshed cultivar it is possible in obtain both crispfleshed and stronger-flavorous varieties.

Low chilling cultivar Kieffeir cultivated extensively in the Kodaikanal area in South India (Singh, 1982) and Punjab and Haryana (Uppal *et al.*, 1993) is a cross between the French pear (*Pyrus communis*) and the Oriental pear (*Pyrus pyrifolia*). Further, an interspecific hybrid 'Baggugosha' reported by Dhatt and Grewal (1987) is most important variety of Kashmir valley and can also be grown successfully in submontanous areas. It bears soft fruit with better quality but it shows irregularity in bearing thus requires to be improved by adding some genes responsible for regular bearing (Chattopadhyay, 2009).

Reducing the chilling requirement of *P. communis* is a goal of some southern U.S.A. breeding programs to expand the range of cultivation. A number of inter specific *P. cammunis* x *P. pyrifolia* hybrids have been released with low chilling requirement by the University of Florida. Efforts are also been going on to reduce the chilling requirement of Asian pears at several centres in S.E. Asia, China and India.

Peach

Compared to those of other stone fruits, cultivars of peach (*Prunus persica* L.) are all self-compatible and have a relatively short juvenile stage (3–4 years). These traits make peach the ideal female parent for wide hybridization (Wen Liu *et al.*, 2007). P. persica L. has four types (Zhuang, 1993). Nectarine (*Prunus persica* var. *nectarina* Maxim) named Liguang peach is a mutant from. *P. persica* var. *densa* Makino which is used as the rootstock and germplasm for dwarf peach. Flat peach (*P. persica* var. *compressa* bean) also named as Liguang flat peach is a type of flat peach. *P. persica* var. *duplex* Rehd is used as an ornamental plant and *P. kansuensis* a red root type has a strong resistance to nematodes. It is thus a good rootstock for peach and plum. It is also a valuable breeding material for resistance to cold, drought and nematodes (Wang and Zhuang, 2001). *P. potaninii* is used as the rootstock for peach and has the advantage of more resistance to drought (Liu *et al.*, 2012). *P. davidiana* is mostly used

    Distant Hybridization in Horticultural Crops

as a rootstock for peach and plum (Wang and Zhuang, 2001). It has good resistance to cold, drought, saline-alkali, and green peach aphid, but it is intolerant to waterlogging. The production of winter-hardy peach varieties by means of hybridization with plum species has been attempted. Some success was achieved in the F from 2 backcrosses with myrobalan, but yield and fruit quality were not high, and were improved in the F by interhybrid crosses of the type (myrobalan 9-55 X AP1) X 3 (myrobalan Ashtarakskaya X peach) by Eremin (1980).

To overcome post-pollination incompatibility, embryo rescue technique was successfully used by Wen Liu *et al.* (2007) to produce interspecific hybrids by crossing peach (*Prunus persica*) as a female parent with apricot (*Prunus armeniaca*) and plum (*Prunus salicica*). In crosses that had 'Yuhualu' or 'Zhonghuashoutao' as female parents, hybrid embryos aborted from the 7th or 8th week after pollination mainly because of post-pollination incompatibility. An embryo rescue protocol was established to rescue such embryos and recover hybrid plants. Modified half-strength MS medium containing 4 mg l^{-1} 6-BA and 0.5 mg l^{-1} IBA produced up to 90% germination in the embryos. Modified MS medium with 1.0 mg l- 1 6-BA and 1.0 mg l- 1 IBA gave the highest bud induction and multiplication whereas modified MS medium containing 0.5 mg l- 1 IAA and 0.2 mg l- 1NAAgave the best rooting percentage.

Plum

The European plum (*P. domestica*) are hexaploids (2n=48), *Prunus spinosa* plum are tetraploids (2n=32) and the Japanese plums (*P. salicina*) are diploid (2n=16) as are *P. cerasifera, P. americana* and most other species. Since the two leading groups of commercial plums have different chromosome numbers, hybridization between them often gives poor results. Where both parents have same number of chromosomes, interspecific hybridization is generally successful. Many hybrids have been made, particularly with *P. cerasifera, P. salicina, P. simonii, P. besseyi, P. americana, P. angustifolia, P. hortulana, P. munsoniana,* and *P. nigra.* Hybrids have also been made with *P. japonica,* Chinese bush cherry. Hybrids between the first four species and apricot (*P. armeniaca* and *P. mume*) have also been successful, but many are not very productive.

Table 3: Important sources of germplasm for plum breeding

Species	Chr. No.	Useful characters
P. allegehaniensis Porter	16	Resistance to crown gall
P. americana Marsh.	16	Tough skin; very winter hardy
P. angustifolia Marsh.	16	Resistance to bacterial leaf spot; limited tolerance to plum leaf scald
P. besseyi Bailey	16	Late bloom; high heat threshold; very winter hardy;resistant to crown gall
P. cerasifera Ehrh.	16 (24,32,48)	Earliness; nematode resistance
P. domestica L.	48	High flavor and fruit quality
P. hortulana Bailey	16	Resistance to bacterial leaf spot

Contd...

Species	Chr. No.	Useful characters
P. maritime Marsh.	16	Late bloom; high heat threshold
P. mesicana S. Wats.	16	Large tree; low suckering
P. munsoniana Wight & Hedr	16	Good fruit; productive
P. nigra Ait.	16	Very winter hardy
P. salicina Lindl	16 (32)	Good size, color and attractiveness; exceptional firmness and keeping quality at high temperatures; very winter hardy
P. simonii Carr.	16	Firmness; upright tree
P. spinosa L.	16	Disease resistance
P. subcordata Benth	16	Drought tolerance; high chill requirement
P. umbellate Ell.	16	Resistance to crown gall

(Okie and Weinberger, 1996; Ramming and Cociu 1991.)

Winter-hardy myrobalan (*Prunus cerasifera*) varieties have been bred by crossing *Prunus cerasifera* (especially subsp. ponticavar. taurica) with diploid plums (notably *Prunus salicina*). Whereas, late-flowering myrobalans have been produced by crossing *Prunus cerasifera* with *Prunus americana* x *Prunus salicina* hybrids. Further, Crosses of myrobalan with cherry x plum hybrids such as Microcerasus [*Prunus*] *pumila* x *Prunus salicina* have given only a few highyielding hybrids. Some success has been achieved in improving characters of European plum (*Prunus domestica*) by crossing with diploid plum, apricot and small-fruited cherry species. Crosses of apricot with myrobalan have given hybrids with winter hardiness, resistance to fungal diseases, good flavour and high canning quality (Eremin, 1980). Success in breeding dwarfing rootstocks was achieved by sloe x plum crosses, notably hybrid 38-42, from a cross between the plum Finikovaya and *P. spinosa*. (Eremin, 1980).

A successful hybrid between blackthorn (*Prunus spinosa*) x plum (*Prunus domestica*) was greatly facilitated by using a pollen mixture of different plum varieties as compared with pollination by each plum variety separately. Form these crosses many hybrids combined the good winter-hardiness of blackthorn with the large fruits and good palatability of the plum, which are of commercial interest (Yenikeyev, 1965). Further he used one more variant of mixed pollination for interspecific hybridization; Sand cherry (*Cerasus besseyi*) (2n: 16) was pollinated by three distant stone fruit species; myrobalan plum (*Prunus cerasifera* Ehrh.), (2n: 16); blackthorn (*Prunus spinosa* L.) (2n: 32) and domastic plum (*Prunus domestica* L.) (2n: 48). Separate pollination produced hybrids between sand cherry and myrobalan plum and between sand cherry and blackthorn though yielded no results. However the cross of sand cherry (*Cerasus besseyi*) x plum (*Prunus domestica*) only succeeded when plum pollen was mixed with myrobalan pollen and a number of triple interspecific hybrids were obtained.

Burbank (1955) made crosses between Japanese plum (*Prunus salicina*) and apricot (*Prunus armeniaca*) and introduced term "plumcot". Later, the same term acquired a commercial importance and nowadays all the plum-apricot hybrids obtained between even different botanical species of plums and apricots, are given that name. In backcrossing of a plumcot with one of the parent species, other hybrids are obtained, known under the names of 'pluots' or 'apriums', depending on the larger genetic participation of the parent. By carrying out several series of isoenzymic analyses, Manganaris *et al.* (1999) developed methodological approaches for identification of plumcots with different level of participation of the parent species *P. salicina* and *P. armeniaca* with the aim of classifying the hybrids in concrete economic categories - plumcots, pluots and apriums. The apricot (*Prunus armeniaca*) cultivars 'Hungarian best' and 'Cacak's gold' crossed with the European plums (*P. domestica*) 'Pozegaca', 'Anna Spath', 'California blue' and 'Hall'. All hybrids obtained from such cross were sterile and the trees were ofvariable phenotypic characters (Paunovich, 1986). The first cultivars of plumcots, pluots and apriums were established in USAby the end of the 1980s. At present, a significant number of cultivars of the three new commercial groups have been established and distributed. For example 'Flavor Queen', 'Flavorich', 'Dapple Dandy', 'Flavor King', 'Flavor Supreme', 'Flavorella' (pluots) 'Flavor Delight' (apriums). 'Red Velvet', 'Royal Velvet', 'Rutland', 'Plum Parfaitp', 'Dapple Dandy', 'Spring Satin', 'Yuksa' as reported by Okie and Ramming 1999 and these new types of fruits are readily available and are being accepted by the consumers (Zhivondov, 2012). Dimitrova (2005) crossed *P. cerasifera* and *P. armeniaca* with the aim of establishing new rootstocks suitable for apricot. Georgiev (1972) also developed plumcot from *P. domestica* and *P. armeniaca*, 49 out of all the 67 hybrid plants died as early as the first vegetation and only one out of the 18 grown hybrids turned to be highly fertile.

The cultivar 'Standesto' was obtained by interspecific hybridization between *Prunus domestica* 'Stanley' *Prunus armeniaca* 'Modesto'. It is moderate in growth, showing an intermediate plum-apricot habitus. The long annual shoots are typically mixed, curved by the leaf nodes as in apricot. The leaves have an intermediate shape, but closer to the plum type. Fruits are oval-elongated in shape, asymmetrical. Fruit skin is finely fuzzy and dark violet-blue. Fruit flesh is golden yellow, moderately juicy, sweet, with slight acidity and a good mixed apricot-plum taste and mixed aroma. The stone resembles the plum type. 'Standesto' is a fully fertile cultivar and it bears fruits regularly and abundantly. It is tolerant to *Plum pox virus*. 'Standesto' and all the other plumcots obtained by interspecific combination of plum (*P. domestica*) x apricot (*P. armeniaca*) could reasonably be assigned to a new separate botanical species under the taxonomy name *Prunus* x *domestiaca* Zhiv. (Zhivondov, 2012). An another interspecific hybrid obtained from plum cultivar 'Stanley' and apricot cultivar 'Modesto' show increasing result against sharka virus (Zhivondov and Djouvinov, 2002). Several plum species have fruiting characteristics that would be valuable in a breeding program. Prunus salicina hybrids have good size, color, and attractiveness. Some have exceptional firmness and keeping quality. Prunus cerasifera plums transmit earliness. They produce progeny which are quite variable in hardiness, fruit form, and other characters even when selfed (Murawski, 1959). Tough skin is often carried by *P. americana* which can be used to improve shipping quality.

Yang *et al.* (2004) in distant hybridization of plum and apricot observed embryo abortion from three weeks after pollination. To overcome this problem an embryo rescue protocol was established to rescue such embryos and recover hybrid plants. The best germination differentiation medium was identified as MS + 6-BA2 mg l⁻¹ + IAA and 0.3 mg l⁻¹, and the rate to germination and differentiation reached up to 80 per cent, Bud induction and multiplication was found best with media MS + 6-BA1.5 mg l⁻¹ + IAA and 0.3 mg l⁻¹ and for rooting it was ½ MS + IAA 0.8 mg l⁻¹. Some hybrids were transplanted in field successfully.

Almond Vigorous, deep rooted trees are needed in almond orchards for maximum yields and to withstand the annual tree-shaking at harvest. Further, because of their large stature, root system of almond trees must also have adequate anchorage to reduce blow-over in the wet winter soils (Felipe, 1989). Currently, researchers are actively evaluating rootstock for almonds in field trials using various screening protocols. In recent year, emphasis on breeding for size controlling rootstock (dwarfing and semi-dwarfing) resistance to nematodes, precocity and higher fruit productivity has increased.

Almond generally performed well on almond rootstocks but peach rootstocks were also found satisfactory and mostly used for example Nemaguard (Prunus persica x Prunus devidiana). It is resistant to nematode. Almond roots are more resistant to lime induced chlorosis and sodium and boron toxicity than are of peach rootstocks (Chadha, 1995). Further, tree on peach are not as long-lived as those on almond (Grasselly and Crossa-Raynaud, 1980) because peach rootstocks have a mat-like root system as opposed to the vertical orientation of almond roots (Ledbetter and Sisterson, 2007). Thus, rootstocks that provide greater anchorage are needed for almonds. Peach and almond shows a very close genetic proximity.

In the areas where both species were cultivated natural interspecific hybrids, either almond x peach and peach x almond, vigorous and fully fertile. The fertility of peach　almond hybrids is the best evidence to show that both species are very closely related (Watkins, 1979) and that the two species could have very small differences between their genomes, based on the gene structure and order in the chromosomes (Arus *et al.*, 1999). This suggests allelism between most of the peach and almond genes. Peach x Almond hybrid G.F. 677 induces rapid growth and precocity (Monastra and Penbone, 1980). However, Peach-almond (PEAL) hybrids are becoming popular rootstock for almonds due to their enhanced vigor and anchorage as compared to almond propagated on peach rootstocks. Advanced generation peach-almond (PEAL) male sterile hybrids as seedlings rootstock for almond is better than seedling peaches rootstock such as Nemaguard. (Ledbetter and Sisterson, 2007).

Walnut

Juglans regia seedling is used as a rootstock in walnut. It is fully compatible, resistant to crown rot and tolerant to black line disease (McGranaham and Catlin, 1987). Juglans hindsii is tolerant to salts in irrigation water and resistant to oakroot fungus (Armillaria mellea). Hybrids of Juglans regia and Juglans hindsii called 'Paradox rootstocks' found suitable on poor hilly soil and found more resistant to crown rot and root lesion nematode (Chadha, 1995).

Cherry

There are more than 30 species of cherries, the main species are sweet cherry (*P. avium* L.), sour cherry (*P. cerasus* L.) and ground cherry (*P. fruticosa* Pall.). The Duke cherry, is presumed to have arisen from the pollination of sour cherry by unreduced pollen of sweet cherry. Tree and fruit type of Duke cherry are intermediate between these two species. Mazzard (*P. mahaleb* L.) is triploid hybrid between sour cherry and the diploid *P. canescens*.

Sour cherry (*Prunus cerasus* L.) is an allotetraploid species (2n=32) originated from natural hybridization between ground cherry (*Prunus fruiticosa* L.) (2n=32) and sweet cherry (*Prunus avium* L.) (2n=16), producing unreduced gametes (Dirlwanger *et al.*, 2007). Varieties of *Prunus fruticosa* and sour cherry (*Prunus cerasus*) (both 2n = 32) hybridized without difficulty and usually gave fertile progeny. Crossing these species with *Prunus avium* (2n = 16) was also easy but the hybrids (2n = 24) were usually sterile but some hybrids had 32 chromosomes and were fertile, some proved winter hardy. Further cultivar obtained from cross between sour and sweet cherry have improved quality traits such as higher sugar, dry matter and sugar to acid ratio. The variety Kent is derived from a cross between sweet and sour cherry (Kharitonova, 1970). For obtaining winter hardy and disease resistant hybrids Simagin (1987) suggested crossing *C. fruticosa* and *C. vulgaris* [*Prunus cerasus*] with Cerasus species (Simagin, 1987). Inter-specific hybridization with the tetraploid species *P. maackii* resulted in a cherry leaf spot resistant hybrid cultivar 'Almaz'. A seedling from this hybrid, named R1 (1) and the Gisela 6 *P. cerasus* x *P. canescens* hybrid were determined to be resistant to CLS when challenged with a wide range of CLS isolates (Wharton *et al.*, 2003).

Schmidt (1985) reported that the most widely used root stocks namely the triploid Gisea hybrids 5 and 6 have resulted from the inter-specific cross between the sour cherry cultivar 'Schattenmorelle' and *P. canescens*. Ochatt and Power (1989) successfully regenerated whole plants from mesophyll protoplast of 'Colt' and utilized protoplast cultures of 'Colt' to select for salt and drought tolerance.

Strawberry

Cultivated strawberry (*Fragaria ananassa* Duch.) is an octoploid species (2n = 8x = 56) belonging to the genus Fragaria of the family Rosaceae. It is hybrid of *F. chiloensis* × *F. virginiana*. Total of 23 species includes the recently discovered. *F.× bringhurstii* staudt, which is a pentaploid derived from interspecific hybridization between *F. vesca* and *F. chiloensis* (Hummer and Hancock, 2009). Intraspecific crosses of *F. ananassa* have been extensively used to develop new cultivars with improved agronomic traits but this species is susceptible to fungal diseases and bacterial pathogens. *F. vesca* species are valuable sources for disease resistance and stress tolerance (Marta *et al.*, 2004). Interspecific hybridization between *Fragaria* species resulted in a variety of gametophytic incompatibility and compatibility according to the crosscombination. In particular, interspecific hybridization between individuals with different ploidy level can hinder or prevent gene flow due to various internal and external barriers, and is often not successful (Marta *et al.*, 2004). Interspecific crosses were performed among accessions of the wild strawberry species F. Vesca and *F. nilgerrensis* and the cultivated strawberry, *F. ananassa*, following reciprocal crossing. Crosses between both *F.vesca* species 'Hawaii-30' and 'UC- 01' were successful regardless of the presence of reciprocal crossing, while crosses with 'Whiteberry' of the *F. nilgerrensis* species were not successful (Rae *et al.*, 2012).

REFERENCES

Ahmed, M.B. and Tewfik, M. 1956. *Indian Journal of Horticulture*, 13: 47-56.

Arus, P., Dirlewanger, E., Quarta, R., Tobutt, K.R., Ballester, J., Boskovic, R.,Dettori, M.T., de Vicente, M.C., Jauregui, B., Joobeur, T., Russell, K., Verde, I. andViruel, M.A. 1999. Location of 20 major genes of peach, almond and cherry on the Prunus linkage map. Plant & Animal GenomeVII , San Diego, 17–21 January.

Belfanti, E., Silfverberg-Dilworth, E., Tartarini, S., Patocchi, A., Barbieri, M.,Zhu, J., Vinatzer, B., Gianfranceschi, L., Gessler, C. and Sansavini, S. 2004. The HcrVf2 gene from a wild apple confers scab resistance to transgenic cultivated variety. *Proc Natl Acad Sci* USA101: 886-890.

Bell, R.L. 1991. Pears (Pyrus). *Acta Horticulture*, 290: 657-700.

Bell, R.L., Quamme, H.A., Layne, R.E.C. and Skirvin, R.M. 1996. Pears **In:** Janick, J.; Moore, J.N. (eds) Fruit breeding vol 1: Tree and Tropical Fruits, John Wiley and Sons, New York, pp: 441-514.

Boavida, L.C., Silva, J.P. and Feijo, J.A. 2001. Sexual reproduction in the cork oak (*Quercus suber* L). II. Crossing intra- and interspecific barriers. Sex Plant Reprod, 14:143-152.

Burbank, L. 1955. Izbranniye sochineniya (Selected Works), Moscow. (in Russian).

Bus, Vg., Alspach, P.A., Hofstee, M.E. and Brewer, L.R. 2002. Genetic variabilityand preliminary heritability estimates of resistance to scab (*Venturia inaequals*) in an apple genetics population. *NZ J Crop Hort Science*, 30: 83-92.

Chadha, T.R. 1995. In: Textbook of Temperate Fruits. Published by Dr. T.P.Trivedi, Project Director, Directorate of Knowledge Management in Agriculture,ICAR, Krishi Anusandhan Bhavan, Pusa, New Delhi.

Challlice, J. and Westwood, M.N. 1973. Numerical taxonomic studies of the genus *Pyrus* using both chemical and botanical characteristics. *Bot H Linn Soc*, 67: 121- 148.

Chattopadhyay, T.K. 2009. **In:** A Textbook on Pomology (Temperate fruits), Vol. 4. Kalyani publishers, New Delhi.

Chen Changlan, Jia Jingxian and Gong Xin. 1991. The first report of determination for cold hardiness of Pyrus. *North Hort.*, 1: 1-3.

Chen, J., Staub, J., Qian, Ch., Jiang, J., Luo, X. and Zhuang, F. 2003. Reproduction and cytogenetic characterization of interspecific hybrids derived from *Cucumis hystrix* Chakr. x *Cucumis sativus* L. *Theor Appl Genet*, 106: 688-695.

Cheng, F.S., Weeden, N.F., Brown, S.K., Aldwinckle, H.S., Gardiner, S.E. and Bus, V.G. 1998. Development of a DNA marker for V , a gene conferring resistance to apple scab. Genome, 41: 208-214. Chengquen, F. Fushen and P. Jingian, J. 1995. The effect of interspecifich ybridization in pear breeding. Cultivar improvement in Horticultural crops. *Acta Horticulture*, 403.

Chopra, H.R., Kanwar, J.S. and Gosal, S.S. 1994. Embryo culture as an effective aid for rescuing hybrid seedlings of early ripening peaches (Prunus persica Batsch). Journal of Research, Punjab Agricultural University, 31: 127-130.

Dhatt, A.S. and Grewal, G.P.S. 1987. *Punjab Hort. J.* 27: 157-160.

Dimitrova, Ì. 2005. Characteristics of interspecific hybrids (P. cerasifera Ehrh. X P. armeniaca L.). Scientific Works of the National Centre of Agricultural Sciences (in Bulgarian), 3: 88- 91.

Dirlwanger, E., Claverie, A. Wunsch, A. and Iezzoni, A.F. 2007. Cherry. In: Genome mapping an molecular breeding in plants. *Fruit and Nuts*, Vol 4. Kole C. (ed), pp. 103-118.

Eremin, G.V. 1980. Use of distant hybridization in breeding stone fruits. Doklady TSKhA, 261: 15-19.

Felipe, A.J. 1989. Rootstock for almond. Present situation. Options Mediterraneennes-Serie Seminaires, 5: 13-17.

Georgiev, V. 1972. Results of distant hybridization in *Prunus* genus. Some morphological and biological characteristics of hybrids between Prunus *domestica* L. and Armeniaca vulgaris Lam. 40th Anniversary of the Fruit-Growing Institute in Kyustendil, Sofia (in Bulgarian).

Gordillo, L.F., Jolley, V.D., Horrocks, R.D. and Stevevs, M.R. 2003. Interaction of BA, GA , NAA, and surfactant on interspecific hybridization of *Lycopersicon esculentum* x *L. chilense. Euphytica*, 131:15-23.

Gorshkova, T.A. 1947. Hybrids of apple and pear (In Russian). Sci. Rep. Michurin Centr. Genetic Lab.: 215-217.

Grasselly, C. and Crossa-Raynaud. 1980. G.P. Maisonneuve et Larose, Paris. Guillaumin, J., Pierson, J. and Grassely, C. 2003. The susceptibility to *Armillaria mellea* of different *Prunus* species used as stone fruit rootstocks. *Scientia Horticulture*, 46: 43-54.

Gygax, M., Gianfranceschi, L., Liebhard, R., Kellerhals, M., Gessler, C. And Patocchi, A. 2004. Molecular markers linked to the apple scab resistance gene Vbj derived from *Malus baccata* jackii. *Theor Appl Genet*, 109: 1702-1709.

Hormaza, J.I. 1998. Early selection in cherry combining RAPDs with embryoculture. *Scientia Horticulture*, 79:121-126.

Hummer, K.E. and Hancock, H. 2009. Strawberry genomics: botanical history, cultivation, traditional breeding, and new technologies. In: Folta, K.M., Gardiner, S.E. (Eds.), Plant Genetics and Genomics: Crops and Models. Springer Science + Business Media, UK, pp. 413-435.

Infante, R. and Gonzalez, J. 2002. Early maturing peach embryo rescue and *in vitro* survival at different fruit growth stages. *Acta Horticulture*, 592: 89-92.

Kato, J., Ishikawa, R. and Mii, M. 2001. Different genomic combinations in intersection hybrids obtained from the crosses between Primula sieboldii (Section Cortusoids) and P. obconica (Section Obconicolisteri) by the embryo rescue technique. *Theoretical and Applied Genetics,* 102: 1129-1135.

Kaushal, R., Malaviya, D.R., Roy, A.K., Kumar, B. and Tiwari, A. 2005. *Trifolium alexandrinum* x *T. resupinatum* - interspecific hybrids developed through embryo rescue. *Plant Cell Tissue Organ Culture,* 83: 137-144.

Kharitonova, E. N. 1970. Distant hybridization in cherry and sweet cherry breeding. In: Otdalennaya gibridiz. rast. i zhivotnykh, T.2. pp. 102-107

Khush, G.S. and Brar, D.S. 1992. Distant Hybridization of Crop Plants: Monographs on Theoretical and Applied Genetics, 16: 47-61.

Kuden, A.B., Tantiver, E. and Gulen, H. 1999. Embryo rescue of peach hybrids. *Acta Horticulture,* 484: 531-533.

Laurens, F. 1999. Review of the current apple breeding programs in the world: objectives for scion cultivar development. *Acta Horticulture,* 484: 163-170.

Ledbetter, C.A. and Sisterson, M.S. 2007. Advanced generation peach-almond hybrids as seedling rootstock for almond: first year growth and potential pollinizers for hybrid seed production. *Euphytica,* 160: 259-266.

Len'kov, M.F. 1990. Method of increasing the effectiveness of distant hybridization in plants. In: Dostizheniya nauki - v praktiku: Kratkie tezisy dokladov k predstoyashchei? nauchnoi? konferentsii: Puti uskoreniya selektsionnogo protsessa rastenii, pp. 25-26.

Liu, D., Zhang, H., Zhang, L., Yuan, Z., Hao, Y. and Zheng, Y. 2014. Distant Hybridization: ATool for Interspeci? c Manipulation of Chromosomes. A. Pratap and J. Kumar (eds.), Alien Gene Transfer in Crop Plants, Volume 1: Innovations, Methods and Risk Assessment, © Springer Science+Business Media New York 2014.

Liu, L., He, Y., Dong, B., Han, F., Wu, Y.X. and Tian, J.B. 2012. Review of the Peach Germplasm Resources and Breeding in China. Proc. XXVIIIth IHC – IS on the Challenge for a Sustainable Production, Protection and Consumption of Mediterranean Fruits and Nuts Eds.: A.M. D'Onghia *et al. Acta Horticulture:* 940.

Lobanov, G.A. 1960. Development of inheritance in pear seedlings from distant hybridization. *In: Agrobiologija (Agrobiology),* pp. 932-934.

Manganaris, A.G., Mainou, A., Goudaras, A. and Ledbetter, C. 1999. Identification of plum x apricot interspecific hybrids using isoenzyme polym prphism. *Acta Horticulture,* 488: 361-368.

Marta, A.E., Camadro, E.L., Diaz-Ricci, J.C. and Castagnaro, A.P. 2004. Breeding barriers between the cultivated strawberry, *Fragaria× ananassa,* and related wild

Murawski, H. 1959. Contributions to breeding research on plums. II. Further investigations on the breeding value of seedlings (in German). Zuchter, 29: 21-36.

Ochatt, S.J. and Power, J.B. 1989. Selection for salt and drought tolerance inn protoplast and explants-derived tissue cultures of Colt cherry (*Prunus avium* × *pseudocerasus*). *Tree Physiol*, 5: 259-266.

Okie, W.R. and Ramming, D.W. 1999. Plum breeding worldwide. *Hort Technology*, 9: 162-176.

Okie, W.R. and Weinberger, J.H. 1996. Plums. In: Janick J, Moore JN (eds) Fruit Breeding, vol. 1. Tree and tropical fruits. John Wiley and sons, Inc., New York.

Paunovich, S. 1986. Thirty five years of apricot breeding and clonal selections of apricot cvs. and rootstocks. *Acta Horticulture*, 192: 299-306.

Pu Fushen and Fang Chengquan. 1990. Anew dual purpose pear variety 'Jinxiang'. China Fruits, 4: 16-17.

Pu Fushen and Jia Jingxian. 1985. Some dwarfing germplasm recourses of pear. China Fruits, 1: 29-32

Rae, R., Yoon, J.H., Hyung, L. Ki L. and Choon-Hwan, L. 2012. Interspecific hybridization of diploids and octoploids in strawberry. *Scientia Horticulture*, 136: 46-52.

Ramming, D.W. and Cociu, V. 1991. Plums In: Moore JN, Ballington, JR (eds) Genetic resources of temperate fruit and nut crops. *Acta Horticulture*, 290: 233- 287.

Rubtsov, G.A. 1944. Geographical distribution of the genus Pyrus and trends and factors in its evolution. *Am Nat.*, 78: 358-366.

Rudenko, I.S. and Rudenko, I.I. 1994. Genotype variation in apple x quince progenies. In: Progress in Temperate Fruit Breeding. H. Schmidt and M. Kellerhals (Eds.). Kluwer Acedemic Publishers, pp: 229-223.

Schmidt, H. 1985. First results on a trial with new cherry hybrid rootstock candidates at Ahrensburg. *Acta Horticulture*, 169: 235-243.

Simagin, V. S. 1987. Results of distant hybridization in sour cherry and bird cherry in Novosibirsk. Problemy apomiksisa i otdalennoi? gibridizatsii, pp. 161-167

Singh, R. 1982. Fruits. National Book Trust. New Delhi, pp. 166-67.

Spiegel –Roy, P. and Alston, F.H. 1979. Chilling and post dormant heat requirement as selection criteria for late – flowering pears. *J Hortic Sci*, 54: 115- 120.

Suzuki, H., Ishikawa, T. and Hiura, I. 1988. Interspecific hybridization between sweet cherry (*P. avium* L.) and *P. apetala* Fr. Et Sav. Var. *Pilosa. J Yamagata Agric For Soc*, 45: 9-18.

Tolmachev, I.A. 1947. Hybridization of currant with gooseberry (In Russian). Sci. Rep. Michurin Centr. Genetic Lab.: 194-200.

Uppal, D.K., Chopra, S.K. and Chanana, Y.R. 1993. Improvement of Temperate fruits for Subtropical Climate. In: Advances in Horticulture Vol. 1- Fruit Crops Part 1. Eds: Chadha, K.L. and Pareek, O.P. Malhotra Publishing House, New Delhi, pp. 445-461.

Valk, P.V.D., Vries, S.D. and Verstappen, F. 1991. Pre- and post-fertilization barriers to backcrossing the interspecific hybrid between *Allium fistulosum* L. And *A. cepa* L. with *A. cepa. Euphytica*, 53: 201-209.

Wang, Z.H. and Zhuang, E.J. 2001. Chinese fruits records: peach volume: origin, cultivation and classification. Beijing, China, p.9-10, 17: 79-81.

Watkins R. 1979. Cherry, plum, peach, apricot and almond. *Prunus* spp. (Rosaceae). In: Simmonds, N.W., editor. Evolution of Crop Plants. Longman, London, pp. 242–247.

Wen Liu, Xuesen Chen, Guanjun Liu, Qing Liang, Tianming He and Jianrong Feng. 2007. Interspecific hybridization of *Prunus persica* with *P. armeniaca* and *P. salicina* using embryo rescue. *Plant Cell Tissue and Organ Culture*, 88: 289-299.

Wharton PS, lezzoni A, Jones AL (2003). Screening of cherry germplasm for resistance to leaf spot . *Plant Dis* 87: 471-477.

Yang H, Chen Y, Feng B, Liu H, Zheng Z (2004). Creating new germplasm by distant hybridization in stone fruits. *Scientia Agricultura Sinica* 37(8):1203-1207.

Yenikeyev, Kh. K. 1965. The method of pollination with a pollen mixture to obtain interspecific hybrids of plum and cherry. *Genetica*, 36: 301-306.

Zhivondov, A. 2012. 'Stendesto', the first Bulgarian Plumcot cultivar. Proc. 15th IS on Apricot Breeding and Culture Ed.: A. Avagyan. *Acta Horticulture*, 966.

Zhivondov, A. and Djouvinov, V. 2002. Some results of the plum breeding programme at the Fruit-Growing Research Institute in Plovdiv. Proc. 7th IS on Plum & Prune Genetics Eds. V. Djouvinov *et al.* Acta Horticulture, 577.

Zhuang, E.J. 1993. Peach. In: J. Shen (ed.), Agricultural Encyclopaedia in China: Fruits Volume. Beijing, China, p.328, 335-336.

Distant Hybridization in Horticultural Crops   *Pages* **45–52**
Editors: **M.R. Dinesh & M. Sankaran**
Published by: **ASTRAL INTERNATIONAL PVT. LTD., NEW DELHI**

4

Intergeneric Hybrdization in Papaya

K. Soorianathasundaram

Department of Fruit Crops,
HC &RI, Tamil Nadu Agricultural University,
Coimbatore-641003

INTRODUCTION

Papaya Ring Spot Virus (PRSV) causes devastating impact on papaya leading to severe yield reductions and huge economic losses to the growers. Since natural resistance is lacking in the cultivated papaya (Carica papaya), incorporating resistance from the allied genera Vasconcellea has been a possible conventional approach in India and elsewhere . Breeding efforts to enhance resistance in commercial cultivars is in progress using Vasconcellea cauliflora at both TNAU, Coimbatore and ICAR-IIHR, Bangalore. Partial success has been reported with V.quercifolia to PRSV strains of Philippines and Australlia . The challenges associated in such wide hybridization programmes , progress being made in breeding programmes in India and other countries and future approaches to introgress genes for resistance to PRSV from Vasconcellea to *Carica papaya* are highlighted in this chapter.

Papayas (*Carica papaya* L ; Caricaceae) are known for its attractive edible pulp having rich nutritive value, especially the carotene content and its proven pharmaceutical and industrial usages. Among the many problems faced by the papaya growers, Papaya Ring Spot Virus (PRSV-P) is considered as a major threat. Overcoming the devastating impact posed by this disease is a priority aspect of crop improvement across several countries where papayas are cultivated. Several approaches such as vector control and integrated plant protection measures are being used to manage the disease but with only limited success. Till date, commercial success has been achieved only through genetic engineering approach (Gonsalves,2006; Fitch *et al.*,1992). But public reservations and opinion against use of GE crop and as well as geographic variability of viral strains (Turner *et al.*, 2004) pose limitations in exploitation of such genetically engineered resistant crop varieties. The severity of economical losses hence requires due attention and development of papaya varieties with inherent horizontal resistance is essential to gain long term protection. Since there is no resistance in *C.papaya*, conventionally, use of wild relatives of papaya has been considered as one of the viable approaches to incorporate PRSV-P resistance, albeit various difficulties encountered in the process.

SOURCES OF RESISTANCE

The family Caricaceae comprises of six genera viz., Cylicomorpha, Carica, Jacaratia, Jarilla, Horovitzia and Vasconcella. Based on molecular evidences the cultivated Carica papaya is the only species placed within the genus Carica (Kim *et al.*, 2002). The species other than C. papaya referred in earlier years belonging to the genus Carica are now represented under the genus Vasconcella. While *Carica papaya* is highly susceptible to many of the biotic stresses, the genus Vasconcellea is reported to harbour resistance to PRSV especially in the species *viz.*, *V. cundinamarencis, V.cauliflora, V.quercifolia* and *V.stipulata* (Manshardt and Wenslaff, 1989; Drew *et al.*, 1998; Magdalita *et al.*, 1997, Mekako and Nakasone, 1975). Because of the severity of the PRSV, several attempts have been made in the past and as well as efforts are being continued to utilize these wild genetic resources.

GENETICS OF RESISTANCE

Monogenic and dominant nature of resistance was earlier attributed (Conover 1978; Mekako and Nakasone, 1975; Horovitz and Jimenez 1967) . Detailed studies on segregating progenies as well as back cross generations in Australia revealed possible involvement of more than one gene contributing to PRSV –P resistance. in V. quercifolia. In a recent report , Haireen and Drew (2014), confirmed and supported the hypothesis that genetic variations from resistance

to susceptibility was at determined by varying genetic sequences and more so due to structural differences of the serine / threonine protein kinase sequence (STK). An alternative splicing that occurs in *V. pubescens* mRNA and the presence of a peroxisomal targeting signal is hypothesized to be an important factor in contributing to the PRSVPresistance in *V. pubescence*. Alamary and Drew (2014) studied a homologous kinase gene in *V. quercifolia* and its mode of inheritance by manually inoculating the segregating populations and screening for PRSV-P resistance. Presence of the resistance phenotype among the majority of the hybrids indicated high stability of gene inheritance through successive backcrosses generations. Since the ratio of susceptible to resistant plants ranged from 4:1 to 1:3, it was not clear whether the resistant phenotype in *V. quercifolia* is controlled by a single dominant gene or multiple genes.

CHALLENGES FACED IN WIDE HYBRIDIZATION

In papaya, while conventional intervarietal hybridization is easy to achieve,post hybridization endosperm failure and seed sterility. Sterility is often are associated with differences in homology of chromosomes of wild and cultivated species although they do not differ in the chromosome numbers. It is important to understand the crossability barriers with other Vasconcella species and papaya and use of bridge species may be handy to incorporate genes from wild sources which do not readily cross with papaya. Attempt has been made in Australia suggesting the PRSV susceptible *V.parviflora* can be a bridge species and help to transfer PRSV-P resistance from *V. pubescens* (O'Brien and Drew, 2010) Smearing the stigmatic surface with 5 % sucrose solution was reported helpful to over come crossability barrier and in promoting fertilization and seed set when *V.cauliflora* was used as a male parent (Dinesh *et al.*, 2007). Similarly, 5 % Sucrose + 5 % boron and as well 5% sucrose +0.5% CaCl has been reported 2 to enhance fruit set and seed set in crosses involving cultivated papaya with *V.cauliflora* at TNAU (Jayavalli *et al.*, 2011).Refinement of tissue culture protocols has enabled the hybrid seedlings to be successfully developed by embryo rescue and ovule culture techniques as a means to overcome of intergeneric incompatibility (Magdalita *et al.* 1998; Manshardt and Wenslaff 1989; Drew *et al.* 1998, 2006). Successful somatic embryogenesis from immature embryos of *C. papaya V. cauliflora* crosses has been also reported 90 to 120 days old embryos can be used for embryo rescue with considerably high success (Magdalita *et al.*, 1996). Somatic embryogenesis protocols have been standardized for cv. CO.2 and CO.7 at TNAU using 90 -105 days old embryos post anthesis. The frequency of embryo formation was around 36 %. In another study by Azad *et al.*, (2012), successful recovery could be made in crosses involving *V.cauliflora* in 90-120 days old embryos. The number of embryos per fruit decreased at 120 days after pollination and these embryos were necrotic and deformed. No embryo was found at 150 days after pollination. Enhancing PRSV resistance through breeding programmes without hampering yield and edible quality is a major challenge as the fruit quality of the resistant wild sources is substandard in terms of consumer acceptance.

UTILIZATION OF *VASCONCELLEA CAULIFLORA* AT TNAU, COIMBATORE

The out break of PRSV in the late 1990s triggered research program in Tamil Nadu Agricultural University, Coimbatore with emphasis towards management of diseases including viruses. To develop PRSV resistance in cultivated varieties, intergeneric hybridization between ten *C. papaya* cultivars and *V. cauliflora* was initiated. From a total of 503 flowers pollinated, 283 fruits resulted. On extraction, these fruits yielded 222 seeds from crosses involving four varieties (CO 2, CO 5, CO 6 and CP 50). Rest of the crosses produced only parthenocarpic fruits. Out of 222 seeds, only six seeds germinated indicating necessity to undertake more number of crosses to generate variable population for selection when Vasconcella is one of the parent (Thirugnanavel, 2009).

In a follow up study, a large scale crossing programme was initiated using nine papaya varieties viz., CO 1, CO 2, CO 4, CO 5, CO 6, Pusa Dwarf, Pusa Nanha, CP 50 (all dioecious) and CO 7(a gynodioecious type) as females with *Vasconcellea cauliflora* as male parent (Jayavalli, 2010). Among the nine varieties used for crossing, only three cross combinations *viz.*, CO 7 x *Vasconcellea cauliflora*, Pusa Nanha x *Vasconcellea cauliflora* and CP 50 x *Vasconcellea cauliflora* produced viable seeds. Artificial screening for PRSV was carried out 27 days after sap inoculation on 419 seedlings derived from 1197 cross attempts.

Out of twenty nine F hybrid plants of CO 7 x *Vasconcellea cauliflora*, only 1 six plants were found free from PRSV symptoms. Similarly, out of fifty five F1 hybrid plants of Pusa Nanha x *Vasconcellea cauliflora* only twenty three were found free from the symptoms and seventy plants out of 335 plants of CP50 x *Vasconcellea cauliflora* were found free from PRSV symptoms. It can be presumed here that the dominant monogenic control of PRSV is absent. The hybridity of the selected F1 plants were also confirmed by use of six IISR primers. Physiological and biochemical characters were found to be higher in the male parent *Vasconcellea cauliflora* and also the cross combination involving Pusa Nanha *Vasconcellea cauliflora* recorded the highest chlorophyll content.

Activities of enzymes *viz.*, peroxidase and polyphenol oxidase were attributed towards tolerance to PRSV(Jayavalli *et al.*, 2012). In evaluations under field conditions with natural incidence of PRSV, the accession CP50 registered the highest fruit yield of 38.27 kg/ tree as compared to 18.98 kg in Pusa Nanha. Among the hybrids, the cross combination of Pusa Nanha x *Vasconcellea cauliflora* recorded the highest yield of 38.27 kg followed by CP 50 x *Vasconcellea cauliflora* registering 31.38 kg and the lowest tree yield of 13.99 kg was recorded by CO 7 x *Vasconcellea cauliflora* cross combination.

Based on the disease intensity score, reaction to the PRSV and performance, the cross combinations viz., CO 7 x *Vasconcellea cauliflora* (designated as CO7V3), Pusa Nanha x Vasconcellea cauliflora (designated as PNV9) and CP 50 x *Vasconcellea cauliflora* (designated as CPV23) were advanced to F generations. 2 Artificial screening for PRSV was again carried out 27 days after virus sap inoculation for all the F populations in the nursery stage .

Out of 138 plants of CO 7 x Vasconcellea cauliflora so generated, only 10 plants were found free from PRSV symptoms. Similarly, 45 plants out of 310 plants of Pusa Nanha x *Vasconcellea cauliflora* (F) and 33 plants out of 235 plants 2 of CP50 x *Vasconcellea cauliflora* (F) were observed to be comparatively free 2 from PRSV symptoms. In the putative resistant progenies, symptom expression was significantly delayed as compared to susceptible plants. Involvement of duplicate gene interaction in crosses of CO.7 with V. cauliflora with two dominant gene for susceptibility (15 :1 for susceptible to resistant) and inhibitory gene interaction with a ratio of 13: 3 (susceptible : resistant) in crosses of Pusa Nanha x *V. cauliflora* was postulated.

In F_2 evaluation, among the various characters studied, higher GCV, PCV,heritability and genetic advance as percent of mean were registered by the cross Pusa Nanha *V. cauliflora* for all the morphological characters and fruit yield parameters (Sudha *et al.*, 2013) Apparently healthy eighteen sibmating combinations (sibmating of selected, different male and female combination) from Pusa Nanha x *Vasconcellea cauliflora*, five sibmating combination from CP50 x *Vasconcellea cauliflora* and one selfed progeny from CO 7 x *Vasconcellea cauliflora* were forwarded to raise the F population (Sudha, 2011). Twenty plants 3 in each of the combination along with their parents were transplanted in the main field for performance evaluation. Estimates of genetical parameters revealed, preponderance of additive genetic action for most of the economic traits in the intergeneric crosses suggesting possibility of selection with resistance in later generations. Among the three cross combinations evaluated, Pusa Nanha x *Vasconcellea cauliflora* recorded the highest fruit yield of 30.32 kg per tree. Seven progenies (C1-10, C1-13, C1-21, C1-24, C1-38, C1-52 and C1-99) of Pusa Nanha x Vasconcellea cauliflora recorded significantly higher fruit yield per tree than Pusa Nanha. Similarly, C2-15, C2-16, C2-26 and C2-28 of the cross CP 50 x Vasconcellea cauliflora were found to be promising based on the disease intensity score, reaction to the papaya ring spot virus and mean performance for morphological, yield and quality attributes. These hybrid progenies were sibmated and have been evaluated in F4 generation. The combination of CO.7 x *V.cauliflora* failed to impress in F4 generation and hence was dropped from further advancement. Selected progenies from F population with high tolerance which was able to reasonably yield well 4 and were surviving for more than 14 months in the field have been raised for evaluation in F5 generation at present.

UTILIZATION OF *VASCONCELLEA CAULIFLORA* AT ICAR-IIHR BANGALORE

At the ICAR-Indian Institute of Horticultural Research, the intergeneric hybrids developed using Arka Surya X *V.cauliflora* has shown tolerance to 'PRSV. In the early stages of hybridization programme, smearing the stigmatic surface with 5 % sucrose solution was reported helpful in promoting fertilization and seed set when *V.cauliflora* was used as a male parent (Dinesh *et al.*, 2007). When red pulped fruit bearing parents as that of cv. Surya are involved in crossing with *V.cauliflora*, viz., castor type of leaf and as well as the yellow pulp colour inherited from *V. cauliflora* can be employed as morphological markers. Some of the hybrid progenies so developed are promising for yield and for disease tolerance under field conditions and these are presently being tested in advanced generation (Dinesh and Vasugi, 2013).

Utilization of Vasconcellea quercifolia at Philippines & Australia Vasconcellea quercifolia was known for its PRSV resistance but several earlier attempts have failed to introgress genes from V.quercifolia. Cytological observations by Drew *et al.*(2006) revealed variable chromosome numbers, presence of univalents, lagging chromosomes at anaphase and meiotic irregularities , non-functional gametes and infertility. Overcoming these difficulties, F hybrids have been produced by embryo 1 rescue 90 days post-pollination of female flowers of Carica papaya L. with pollen of the PRSV-P resistant species Vasconcellea quercifolia. Among the 300 progenies so raised, 75% were resistant to PRSV-P while 25% produced virus symptoms. Large backcross 1 (BC1) and backcross 2 (BC2) populations were produced in the Philippines and Australia when 3 of the male intergeneric hybrids were backcrossed to *C. papaya*. Trees produced commercially acceptable fruit containing high sugar content from *V. quercifolia* after 2 or 3 backcross generations (Drew *et al.*, 2006). First successful production of PRSV-P resistant backcross (BC) papaya plants following intergeneric hybridization between *C. papaya* and a *Vasconcellea* species was reported by Siar *et al.*, (2011).

According to them, from the PRSV-P resistant backcross 1 (BC1) male plant, 1465 plants [137 BC2, 546 SbC2 (BC2 sib-crosses), 147 BC3, 379 SbC3 and 256 BC4] were grown from seed and inoculated with PRSV-P and virus resistant BC3 and BC4 plants were selected from these generations. BC plants generally developed mild symptoms of PRSV-P after about 5 to 18 months in the field but many showed the ability to produce new growth free of symptoms. All control plants developed severe symptoms after 3 months in the field. According to Alamary and Drew (2014), PRSV-P resistance was also confirmed against Philippines and Australian PRSV-P virus strains in intergeneric hybrid derivative population involving *V.quercifolia* suggesting the possibility that the resistance gene in *V.quercifolia* could be effective to confer the resistance to virus strains in other countries.

MARKER ASSISTED BREEDING AND GENETIC MAPPING

Use of DNA markers in detecting association of genes related to susceptibility resistance would enable the breeder to screen a large population in the earlier stage and select only potential population for field screening. In special relevance to breeding for disease resistance, strategies can be refined by developing reliable genetic markers. Dillon *et al.* (2005) identified a potential functional sequenced characterized amplified region (SCAR) marker (Opa11 5r) for PRSV-P resistance which collocated with prsv-1, from a mapping population of (F) *V. pubescens* × *V. parviflora* combination. Using a bulked segregant 2 analysis approach, dominant randomly amplified DNAfingerprint (RAF) markers linked to prsv-1 were revealed in the resistant DNAbulk, derived from F_2 progeny of interspecific cross *V. parviflora* (susceptible) x *V. cundinamarcensis* (resistant). One marker, Opk4_1r, mapped adjacent to the prsv-1 locus at 5.4 cM, while a second, Opa11_5r, collocated with it. Sequence characterization of the Opk4_1r marker permitted its conversion into a codominant CAPS marker (PsiIk4), which was shown to correctly identify resistant genotypes 99% of the time when applied to interspecific segregating F_2 population (Drew *et al.* 2007). Later, O' Brien (2009) revealed that , the CAPS marker (Psilk4) was only partially reliable during screening of segregating progenies of intergeneric crosses involving *C.papaya* and *V. parviflora* and as well as in progenies of *C. papaya* and *V. quercifolia* as it was not able to resolve for heterozygous resistance condition.

FURTHER STRATEGIES

The attempts made in India and as well as elsewhere shows that there is a potential possibility to enhance resistance to PRSV-P through use of wild species. Bridge crosses especially use of *V. parviflora* to transfer genes from *V. pubescencs* to *C. papaya*, embryo rescue and refinement of screening methodology are crucial to accelerate such breeding programmes. Based on the overview of past and current efforts in conventional breeding, it can be seen that there is urgent necessity in India, to introduce species of Vasconcellea other than what is available at present for possible utilization in future breeding programmes to harness resistance not only to PRSV but also for other potentially dangerous biotic stresses such as mycoplasma or Papaya Leaf Curl virus or Lethal Yellowing Virus *etc.* Whenever the wild relative is used as one of the parents to impart resistance there is also immediate necessity to improve the quality in terms of flavor, taste and TSS to ensure consumer acceptability. Given the nature of variability being reported in the pathogenic population, it may be necessary to develop and build varieties and genotypes with horizontal and durable resistance. To avoid dilution of genes responsible for resistance, it is worthwhile to harness the in vitro techniques to recover and multiply useful recombinants. Standardization of protocols for mass multiplication is another important thrust area to counter possible dilution of resistance through seed propagation. Identification of genes conferring resistance in wild species and reliable marker development should assume priority.Recent forays being made in genomics of papaya and molecular approaches in manipulation of genes could trigger much more possibilities in utilization of the Vasconcellea and other allied genera for the improvement of papaya.

REFERENCES

Alamery, S. and Drew, R. 2014. Studies on the genetics of PRSV-P resistance genes in intergeneric hybrids between *Carica papaya* and *Vasconcellea quercifolia*. *Acta Hort.* (ISHS) 1022:55-61.

Azad, Md.A.K., Md. G. Rabbani and L. Amin . 2012. Plant regeneration and somatic embryogenesis from immature embryos derived through interspecific hybridization among different *Carica* species. *Int.J.Mol.Sci.*13: 17065-17076.

Conover RA and RE,Litz .1978. Progress in breeding papayas with tolernrence to papaya ringspot virus. *Proc. Flo.Sta.Hort.Soc.* 91. 182-184.

Dillon,S., C. Ramage, R. Drew, and S. Ashmore. 2005. Genetic mapping of a PRSV-P resistance gene in "highland papaya" based on inheritance of RAF markers," *Euphytica,* vol. 145, no. 1-2, pp. 11–23..

Dinesh. M. R., Rekha, A. , Ravishankar, K.V. , Praveen, K.S. and L.C. Santosh, 2007, Breaking the intergeneric crossing barrier in papaya. *Scientia Hort.* 114: 33–36.

Drew, R.A., C.M. O'Brien and P.M. Magdalita. 1998. Development of interspecific Carica hybrids. *Acta Hort.,* 461: 285-292.

Drew RA, Siar SV, O'Brien CM, Sajise AGC. 2006. Progress in backcrossing between *Carica papaya* and *Vasconcellea quercifolia* intergeneric hybrids and *C. papaya*. *Aust J Exp Agric* 46:419–424.

Drew RA, Siar SV, Dillon S, Ramage C, O'Brien C and AGC Sajise (2007). Intergeneric hybridization between *Carica papaya* and wild *Vasconcellea* species and identification of a PRSV-P resistance gene. *Acta Hort.*738, 165-169.

Fitch M, Manshardt R, Gonsalves D, Slightom J and Sanford J.1992. Virus resistant papaya plants derived from tissues bombarded with the coat protein gene of Papaya ringspot virus. *Bio/Technology* 10, 1466-1472.

Gonsalves, D. 2006. Transgenic papaya : development, release, impact and challenges. *Adv.Virus res.*67: 317-354

Haireen, M.R.R and R.A. Drew. 2014. Isolation and Characterisation of PRSV-P Resistance Genes in Carica and Vasconcellea. *International Journal of Genomics,* http://dx.doi.org/10.1155/2014/145403.

Horovitz S.and H. Jimenez .1967. Cruzamientos interspecificos intergenericos en caricaceas y sus implicaciones fitotecnicas. *Agron. Trop.* 17:323–343.

Jayavalli, R. 2010. Breeding for PRSV resistance in papaya (*Carica papaya* L.). Ph.D. (Hort.) Thesis, Tamil Nadu agricultural University, Coimbatore.

Jayavalli, R, T.N. Balamohan, N. Manivannan and M. Govindaraj. 2011. Breaking the intergeneric hybridization barrier in *Carica papaya* and *Vasconcellea cauliflora*. *Scientia Hort.*130 : 787-794

Jayavalli, R.; T.N. Balamohan, N and P.T.Selvi. 2012. Physiological and biochemical characterization of intergeneric hybrids of papaya. Madras Agric.J. 99 (4/6) pp. 161-165.

Kim, M. S., P. H. Moore, F. Zee, M. M. M. Fitch and D. Steiger. 2002. Genetic diversity of Carica papaya as revealed by AFLP markers. *Genome*, 45(3): 503- 512.

Magdalita, P.M., S.W. Adkins, I.D. Godwin and R.A. Drew. 1996. An efficient embryo rescue protocol for Carica interspecific hybrids. *Aus. J. Bot.*, 44(3): 343-353.

Magdalita, P.M., D.M. Persley, I.D. Godwin, R.A. Drew and S.W. Adkins. 1997. Screening *Carica papaya   C. cauliflora* hybrids for resistance to papaya ringspot virus-type *P. Plant Pathol.*, 46: 837–841.

Manshardt, R.M. and T.F. Wenslaff. 1989. Interspecific hybridization of papaya with other species. *J. Amer. Soc. Hort. Sci.*, 114(4): 684-689.

Mekako, H.V. and H.Y. Nakasone. 1975. Intergeneric hybridization among Carica species. J. Amer. Soc. Hort. Sci., 100(3): 237-242.

O'Brien, C. 2009. Marker assisted breeding for Papaya Ringspot resistancevirus resistance in Carica papaya , L. M.Phil Thesis submitted to GriffithUniversity, Brisbane, Australia.

O'Brien, C.M. and Drew, R.A. 2010. Marker-assisted hybridisation and backcrossing between *Vasconcellea* species and *Carica papaya* for prsv-p resistance . *Acta Hort.* (ISHS) 859:361-368.

Siar, S.V., G. A. Beligan, A. J. C. Sajise, V. N. Villegas, and R. A. Drew. 2011. Papaya ringspot virus resistance in *Carica papaya* via introgression from *Vasconcellea quercifolia. Euphytica*, vol. 181(2)159–16.

Sudha, R, 2011. Breeding for Breeding For Papaya Ringspot Virus Resistance In Papaya (*Carica papaya* L.). Ph.D. Thesis. Tamil Nadu Agricultural University, Coimbatore.

Sudha,R., T.N. Balamohan, K. Soorianathasundaram, N. Manivannan and R. Rabindran , 2013. Evaluation of F intergeneric population of papaya (Carica 2 papaya L.)for resistance to papaya ringspot virus (PRSV). *Scientia Hort.* 158 : 68-74.

Thirugnanavel, A. 2009. Breeding for PRSV resistance in papaya (*Carica papaya* L.) through germplasm screening and intergeneric hybridization. Ph.D. Thesis. Tamil Nadu Agricultural University , Coimbatore.

Turner, S.A., M.H. Ahmad and P. Tennant. 2004. Differential response of papaya cultivars to papaya ring spot virus isolates from Jamaica. *Trop. Agric.*, (Trinidad), 81: 223-228.

Distant Hybridization in Horticultural Crops　　　　　*Pages* **53–62**
Editors: **M.R. Dinesh & M. Sankaran**
Published by: **ASTRAL INTERNATIONAL PVT. LTD., NEW DELHI**

5

Interspecific Hybridization in Grapes

G.S.Karibasappa,

ICAR-Indian Institute of Horticultural Research,
Bengaluru 560089

INTRODUCTION

Global interest in tropical viticulture is gaining because table grapes are produced by early ripening ahead of temperate zones.Tropical grape production, has a narrow genetic base limited to a few *V.vinifera* varieties like Thompson seedless, Muscat Hamburg, Anab-E-Shahi and Kishmish Chernyi and their clones and few hybrid varieties grown for disease resistance like Isabella. Concerted efforts in varietal evaluations to exploit the genetic variability available elsewhere and local vine breeding are required to develop varieties more suited to tropical environments. Selection of table grape varieties, resistant to fungal diseases, more fruitful in basal buds, produce optimum bunch size, favour rapid shoot growth and maturity with fruits of likening taste and good colour as the predominant traits. Priority should also be given for development and selection of rootstocks, as they have potential to regulate vine vigor, fruitfulness, reduce problems associated with soil type and soil borne pests and improve vine longevity. Dogridge is commonly used for most of the table grape varieties for its drought and salt tolerance under tropical Indian conditions. Of late, Dogridge use is confronted with few problems, in scions exhibiting uneven bud burst after pruning, less fruitfulness and increased deadwood in the cordon as result of its inducing excessive vigor to scions. Hence rootstock breeding research should aim to develop new rootstocks that have vigour levels appropriate to end use (for table, raisin, juice and wine grapes) and combine risk reduction benefits with excellent cropping characteristics to give economic yields of optimum quality fruit. Table grape growers continue to seek vines with less variability in growth and yield between seasons. Hence, selected rootstocks must meet industry standards, tolerate root pests – phylloxera, nematodes, tolerate adverse soil conditions – salinity, high lime and pH, low nutrient status, poor water availability and possess desirable nursery and viticultural characteristics – high propagation rate, graft compatibility, appropriate growth, yield and fruit quality

A number of new interspecific hybrid varieties have been released from the grape breeding programs of USA and France and reported to have disease resistance and acceptable fruit quality even under warm, humid environments in Florida, Arkansas and NewYork State. Interspecific varieties such as Reliance Seedless, Mars Seedless, Venus, Marquis, Remaily Seedless, Suwannee, Conquistador and Orlando Seedless warrant testing in their own right and as parents in the tropics as they are resistant to anthracnose, oidium or powdery mildew and various bunch rots. Under tropical Indian conditions screening for anthracnose resistance indicated that cvs Gulabi, Phakdi, Bangalore Purple of *V. vinifera*, Catawba, Concord and Champion of *V.labrusca*, James of *V. rotundifolia*, Dog Ridge of *V. champini* and species; *V.aestivalis, V.lanata, V.arizonica, V. candicans, V. longii, V. palmata, V. riparia, V. rupestris* and *V. tilifolia* are tolerant (Patil *et al.*, 1990).

Screening under unsprayed field conditions at NRCG, Pune over a period of 3 years cv. Carolina Black Rose, 6 Seyve-Villard hybrids, Seibel 9836 and H533 were found immune to downy mildew (Table 1) and also moderately resistant to powdery mildew under Indian conditions.

Table 1. Evaluation of grape species/ varieties for downy mildew tolerance at NRCG, Pune

Sl.No.	Name of Variety/Accession	*In Vitro* Rating*	Field Rating
1	Thompson Seedless (California)	5.28	7.00
2	Kali Sahebi	5.14	9.00
3	Anab-E-Shahi	3.42	9.00
4	Delight	5.00	9.00
5	Gulabi(IIHR)	3.00	9.00
6	Gulabi (TN)	1.48	7.00
7	E -8/5	2.71	7.00
8	E-2/1	2.90	9.00
9	Alamwick	4.40	9.00
10	Carolina Black Rose	1.38	1.00
11	V.flexousavar.parviflora	3.57	5.00
12	H-533 (V.tilifolia x Gulabi)	5.50	3.00
13	SV-23501 (Poly-Vitis hybrid)	1.00	1.00
14	SV-12309 (Poly-Vitis hybrid)	1.86	1.00
15	SV-18315 (Poly-Vitis hybrid)	1.43	1.00

Contd...

Sl.No.	Name of Variety/Accession	*In Vitro* Rating*	Field Rating
16	SV-12364 (Poly-Vitis hybrid)	1.19	3.00
17	SV-12375 (Poly-Vitis hybrid)	1.00	1.00
18	Seibel 9813 (Poly-Vitis hybrid)	2.43	1.00
19	Seibel 9308 (Poly-Vitis hybrid)	1.00	1.00
20	Kanai Local (*V.lanata*)	4.14	5.00
21	Lake Emerald (*V.aestivalis*)	2.16	5.00
22	FruhroterVeltliner (*V.vinifera*)	3.00	5.00
23	SV-18402 (Poly-Vitis hybrid)	3.14	1.00
24	St.George (*V.rupestris*)	1.09	1.00
25	Bharat Ruba Black (*V.lanata*)	5.14	9.00
26	Flame seedless(*V.vinifera*)	5.70	9.00
27	Concord (*V.labrusca*)	1.19	5.00
28	Catawba (*V.labrusca*)	3.00	5.00

*Ratings 1- immune, 3-resistant, 5-moderately resistant, 7-moderately susceptible, 9-highly susceptible.

Screening for drought and salt Tolerance: Table grape varieties such as Marroo Seedless, Christmas Rose, Red Globe and Beauty Seedless and among wine varieties such as Cinsaut, Merlot, Maurvadre are drought hardy. At present 25 rootstock varieties are available at NRC for Grapes, Pune. Dogridge rootstock is widely recommended for table varieties under drought proned areas and saline soils. Other rootstocks such as 110R, 140 RU, 1103 P, Teleki 5BB and 99R are on going field evaluation for tolerance to moisture and salt stress conditions. Rootstocks such as Ramsey, Dogridge, deGrasset, SO- 4 and Freedom have been identified for salinity tolerance.

Utilization of germplasmin breeding programmes: There is an immediate need for developing seedless cultivars having tolerance to downy mildew under peninsular Indian conditions. Introgression of seedless character with downy mildew resistance, through conventional breeding with embryo rescue techniques at NRCG, Pune was taken up in collaboration with NCL Pune (Bharathy *et al.,* 2003 & 2005). The following crosses (Table 2) were effected and the plantlets of F1 progeny were raised by embryo rescue following Emershad and Ramming (1984). However very few F_1 progeny (49 mostly with Concord and Catawba as male, while Flame Seedless as female parent) have survived in the field for further evaluations.

CSIRO Merbein in Australia also have a breeding program for developing grapes for tropical environments involving complex species hybrids as parents along with *V. vinifera* table grape varieties as the recipient parents. A diverse range of genotypes with disease resistance e.g. *V. rotundifolia* (DR x 55), varieties of complex hybrid species, such as Chambourcin, Muscat St.Vallier, Villard Blanc, Aurelia, Carolina Black Rose,Lady Patricia,SV-12-309, SV-12-303, Illinois 271-1, and *V.labrusca* hybrids such as Isabella, Glenora, Mantey, Suffolk Red have been used in crosses with *V.vinifera* table grape varieties e.g. Muscat Hamburg, from the diverse germplasm maintained at Merbein (Possingham *et al.*, 1990). As well *In-ovulo* embryo culture technique is employed in the crosses ofseedless *V.vinifera* x *V.rotundifolia*for developing new breeding lines(Ramming *et al.*, 2000).

Table 2.Number of plantlets obtained through embryo rescue in F_1 progeny of various interspecific crosses

Sl.No.	Male Parents	Female Parents	
		Thompson Seedless (*V.vinifera*)	Flame Seedless (*V.vinifera*)
1.	Lake Emerald (*V.aestivalis*)	45	47
2.	S.V.18402(V.spp /polyvitis hybrid)	71	30
3.	*Vitistiliefolia*	5	27
4.	*V.candicans*	33	1
5.	Concord (*V.labrusca*)	76	60
6.	Catawba (*V.labrusca*)	65	135
7.	St.George(*V.rupestrisvar.*du Lot)	24	0
8.	FruhroterVeltliner (*V.vinifera*)	15	11
9.	Open Pollinated seedlings	5	3
	Total	339	314

Although there are various degrees of susceptibility between *V.vinifera* varieties which originated from Central Asian/European regions, there is less hope to breed disease resistant varieties without broadening the genetic base. There is considerable scope to increase resistance of *V.vinifera* grapes through interspecific breeding. The breeding programs have included routine screening for disease resistance by dual culture *in vitro e.g.* downy mildew (Stein *et.al.*1985; Barlass *et.al.* 1986) or Oidium (Stein *et al.*1985). Detailed information can be accessed by breeders to identify species and species combinations likely to be resistant to various diseases. For example cultivars of *V.labrusca* and hybrids between *V.vinifera* and *V.rupestris* were resistant to Oidium (Doster and Schnathorst, 1985). Some *V. labrusca* cultivars grown in tropical conditions are more tolerant to anthracnose than *V.vinifera*. Chinese have developed resistant quality table grapes from hybrids of *V.labrusca* and *V.vinifera* (*e.g.* Fortuna and Phoenix). At CSIRO, genes for resistance to downy mildew have been introgressed from complex hybridseg.,Marroo Seedless a cross of Carolina Black Rose and Ruby Seedless is resistant to downy mildew even under wet tropical conditions, the trait was attributed to one of its grandparent Villard Blanc. Marroo seedless is however susceptible to powdery mildew. In this regard there is potential to involve muscadine grapes (*V.rotundifolia*) which is resistant to most diseases as also some Asian species such as *V.amurensis* and *V.armata*.

Genotypes which are resistant to most of the parasites and ableto produce quality grapes by back crossing 5-6 times to *V.vinifera*using *V.rotundifolia* have been developed (Alain Bouquet at INRA, France, Ramming *et.al.*2000). They are still under experimentation to check the durability of the resistances, particularly against downy mildew.

Research at Division of Horticulture, CSIRO has been successfully employed the grapevine transformation technique with an experimental gene. The task lies ahead to introduce several other genes of economic importance, particularly protection from fungal diseases. Currently attention is being given to the response of plants to pathogenic infections. Many plants respond by synthesizing pathogenic response proteins (PR) which include hydrolytic enzymes to defend against pathogenic attack. Two such enzymes are chitinase which degrades chitin, the major cell wall component of oidium and -glucanase which degrades - glucans, constituents of cell walls of downy mildew hyphae.

In contrast to conventional breeding, where progeny are highly heterozygous and include varietal characteristics from both parents, a better approach might be to provide a selected variety with a gene for chitinase or -glucanase or both, by plant transformation methodology. The aim would be lyse the hyphae of the invading fungus. This would allow the conventional breeder to concentrate on the many pleitropic (multigenic) gene characters which are not suited to these new methodologiesi.e. yield, flavours *etc.*

ROOTSTOCK BREEDING FOR TROPICAL INDIAN VITICULTURE

Rootstocks were first used for grapes because of phylloxera, a root-feeding aphid. Phylloxera can kill grapevine varieties of the species *Vitis vinifera*. Many species of North American grapes are resistant to phylloxera. But only a few species were suited for direct use as rootstocks. The wild grapes *Vitis riparia* and *V. rupestris* were an important source of rootstock selections. These two species root easily from dormant cuttings and provide protection against phylloxera. Characterization of rootstocks based on their parental species is a helpful generalization, but only thorough testing determines the value of a given rootstock variety for particular pest, soil, or other horticultural conditions. (Peter Cousins, 2005). Breeders crossed *V. berlandieri* with *V. rupestris* and *V.riparia* and developed new families of rootstocks that combine adaptation to calcareous soils with ease of propagation. Since all three species provide protection against phylloxera, the hybrids among them also are recommended for use in phylloxera prone vineyards. The three groups of rootstocks formed by the hybridization of these species are the most important in viticulture today. The attributes of the rootstock varieties reflect the characteristics of their parental species. They tend to have deeper rooting than pure *V. riparia* rootstocks, which means improved drought tolerance. Hybrids of *V. berlandieri* and *V. riparia*, include important rootstocks such as SO4, Teleki 5C, Kober 5BB, and 420A Mgt. These rootstocks tend to be of lower to moderate vigor and are adapted to moister vineyard sites. *Vitis berlandieri* contributes adaptation to calcareous soils and both parents contribute to the excellent phylloxera protection provided by these rootstocks. Crosses between *V. berlandieri* and *V. rupestris* are adapted to deep, well-drained soils, such as hillsides. Like the *V. berlandieri / V. riparia* hybrids, they tolerate calcareous soils well and provide excellent protection against phylloxera. Many are noted for their high vigor. Rootstocks in this family include 110 R, 140 Ru, and 1103P.

Development and selection of rootstocks for tropical conditions is considered a top priority as they have potential to regulate excessive vine vigor, reduce problems associated with soil type and soil borne pests and improve vine longevity. Rootstock breeding program which aims to develop multispecies complexes suited to a range of soil types incorporating genes for nematode resistance from *V.champini*, phylloxera resistance from *V.berlandieri, V. rupestris* and *V.riparia*, drought resistance from *V.rupestris, V.champini* and tolerance of wetter conditions, *V. riparia* should be initiated for tropical Indian conditions. Other species distributed in more tropical regions i.e. *V.aestivalis, V.caribaea, V.longii, V.tilifolia,V.cinerea* and *V.rotundifolia* may also be included (Peter Cousins, 2005). For sandier soil types in particular drier regions, the nematode resistant rootstocks currently used for table grape production such as Dogridge, Ramsey and the lower vigorTeleki 5BB and other *V.riparia x V. rupestris* hybrid rootstocks may be included for breeding and selection program. (Possingham *et. al.* 1990).

AUSTRALIAN CSIRO'S PROGRAM ON GRAPEVINE ROOTSTOCKS

The introduction of nematode tolerant rootstocks by CSIRO in the 1950s made acritical contribution to the productivity of wineand dried grapes in Australia. The realisation that some of these rootstocks would also tolerate rootzone salinity led to the identificationof salt tolerant stocks. Rootstocks offer grape producers a risk avoidance strategy. While own-rooted vines will produce appropriate yields of high quality fruit,root systems of *Vitis viniferavines* run risks associated with rootzone pests, diseases and adverse conditions such as poor nutrient status or salinity. Rootstocks provide solutions to many of the risks present in the rootzone. However, problems of excessive yields of poor quality fruit can be associated with increased vigour imparted by the rootstock. CSIRO's rootstock breeding research aims to develop new rootstocks that have vigour levels appropriate to end use (for wine, drying or for table grapes) and combine risk reduction benefits with excellent cropping characteristics to give economic yields of optimum quality fruit (Possingham *et. al.*1990).

Vine rootstock breeding and genetics research began in the mid-1960s and aimed to developnew rootstocks which produce grafted vines ofmedium vigour that improve fruit quality, are tolerant of nematodes, phylloxera, and salinity and are water-use efficient.

A PROGRAM OUTLINE FORROOTSTOCK BREEDING UNDER TROPICAL INDIAN CONDITIONS (IIHR)

Rootstocks introduced from overseas *e.g.* Dogridge, 110 Richter have fulfilled many of our requirements, but high canopy vigour and negative effects on grape quality often restricts their use for some varieties of grapes. Further introductions from overseas breeding programs could be an answer to discovering improved rootstocks for local Indian conditions. These, however, would have been selected for conditions in the country where they were bred and may not be specifically adapted for tropical conditions.A local breeding program allows us to develop new rootstocks that meet our tropical table grape industryc specifications and are better adapted to our various climatic conditions and soils.

Rootstock breeding and selection is taking place at universities, research stations, and nurseries around the world and more and more new rootstocks are becoming available. We should know the species background and adaptation and testing history of a rootstock before we choose it for the vineyard and long before we plant it. This contributes to the selection of the appropriate rootstock and the development and maintenance of a healthy, productive vineyard. Important characteristics of rootstocks are summarized in Table 3.

i) Non-*vinifera* breeding parents are selected from the collection or from progenies of previous crosses, and crossed to produce new hybrid families. Different *Vitis* species are known for their ability to tolerate soil infestations of phylloxera and nematodes. In breeding new rootstocks, the aim is to combine resistance to these pestsalong with other key traits. In this, it is important to test hybrids for resistance to different racesor biotypes of the pests before they are selected for use as rootstocks in evaluation trials. Hybrids are screened for tolerance to different nematode biotypes and races of phylloxera.

ii) Hybrids are screened for rooting ability and graft compatibility. New rootstocks must strikewell as cuttings and produce strong root systems. They must also be compatible with major varieties. Glasshouse tests identify hybrids for further evaluation in the vineyard as grafted vines.

iii) The new hybrids are screened for key characteristics, after which they may be entered into short term viticultural performance trials. From these trials, elite selections are identified for release and commercial evaluation as new rootstock varieties.

iv) Cane production of own-rooted hybrids in the vineyard is a good indicator of the vigour they may impart to a scion variety. Replicated grafted trials serve to support data collected for own-rooted hybrids. Short-term performance trials help identify rootstock hybrids with good viticultural traits.

v) Rootstocks influence mineral uptake, chloride andsodium uptake are also issues, given that thereare limits for the levels ofthese ions in vines as well in fresh berries and wines. The bestsalt tolerant rootstocks arethose with high innate vigour and moderate to high chloride and sodium exclusion ability.Of particular interest for wine grapes is potassium uptake, where too much potassium in grape juice leads to high pH and a requirement for pH adjustment during wine making.New hybrids are therefore screened for key ion uptake characteristics.

Table.3. Characteristics of commonly used rootstocks for grapes.

Rootstock	Parentage	Relative scion vigor	Phylloxera protection	Rootknot nematode resistance	Dagger nematode resistance	Calcareous soil adaptation
St.George	*V. rupestris*	high	high	low	low	low
RepariaGloire	*V. riparia*	low	high	low	medium	low
1616C	*V.riparia x V.solonis*	low	high	high	medium	low
101-14 Mgt.	*V.riparia x V.rupestris*	medium	high	medium	Medium	low
3309C	*V.riparia x V.rupestris*	low to medium	high	low	low	Low
Schwarzmann	*V.riparia x V.rupestris*	medium	high	high	high	medium
Freedom	*V.champini, V.riparia, V.labrusca, V.vinifera, V.solonis complex hybrid*	high	medium to high	high	high	medium
Ramsey	*V.champini*	high	high	high	low	medium
Dogridge	*V.champini*	high	low	high	medium	medium
Teleki 5C	*V.berlandieri x V.riparia*	medium	high	medium	medium	medium
Teleki 5BB	*V.berlandieri x V.riparia*	medium	high	medium	low	medium
SO-4	*V.berlandieri x V.riparia*	medium	high	medium	medium	medium
110Richter	*V.berlandieri x V.rupestris*	medium	high	low	low	medium
1103 Paulsen	*V.berlandieri x V.rupestris*	high	high	medium	low	medium

Contd.

Rootstock	Parentage	Relative scion vigor	Phylloxera protection	Rootknot nematode resistance	Dagger nematode resistance	Calcareous soil adaptation
Boerner	*V.cinerea* x *V.riparia*	medium	high	high	high	low
Gravesac	*V.riparia, V.rupestris,* and *V.berlandieri* complex hybrid	low	high	low	information not available	low
41B Millardet	*V.vinifera* x *V.berlandieri*	medium	medium	low	low	very high
Fercal	*V.vinifera x V. berlandieri complex hybrid*	medium to high	medium	medium	information not available	very high
039-16	*V.vinifera* x *V.rotundifolia*	high	high	low	high	low

Expected Rootstocks therefore, must 1) meet industry standards 2) tolerate root pests – phylloxera, nematodes 3) tolerate adverse soil conditions – salinity, high lime and pH, low nutrient status, poor water availability 4) possess favourably nursery and viticultural characteristics – high propagation rate, graft compatibility, appropriate growth, yield and fruit quality.

REFERENCES

Barlass, M., Ramming, D.W., Davis, H.P. 1988. In-ovulo embryo culture: a breeding technique to rescue seedless x seedless table grape crosses. Austral.Grape Grower Winemaker, No.272: 123-125.

Bharathy,P.V., Karibasappa, G.S., Biradar, A.B., Kulkarni, D.D., Solanke, A.U., Patil, S.G. and Agrawal,D.C. 2003. Influence of Pre-bloom sprays of benzyladenine on in-vitro recovery of hybrid embryos from crosses of Thompson Seedless and 8 seeded varieties of grape (*Vitis* spp. L.).*Vitis* 42(4):199- 202.

Bharathy,P.V., Karibasappa, G.S., Patil, S.G. and Agrawal,D.C.2005. In ovulo rescue of hybrid embryos in Flame seedless grapes – Influence of pre-bloom sprays of benzyladenine. *Scientia Horticulturae,* 106(3) : 353-359.

Doster, M.A., Schnathorst,W.C. 1985. Comparative susceptibility of various grapevine cultivars to the powdery mildew fungus, Uncinulanecator. *Amer.J.Enol. Viticult.*36:101-104.

Patil, S.G., B.K.Honrao, V.G.Rao and V.P.Patil,1990. Screening of Grape (*Vitis* species) germplasm for resistance to three major fungal diseases. *Indian J.Agric.Sci.,* 60(12):836-38.

Peter Cousins. 2005. Evolution, Genetics, and Breeding: Viticultural Applications of the Origins of Our Rootstocks, in Proceedings of the Grapevine Rootstocks: Current Use, Research, and Application 2005 Rootstock Symposium OsageBeach, Missouri February 5, 2005 (Eds.) Peter Cousins & R. KeithStriegler,Published by Mid-America Viticulture and Enology Center Southwest Missouri State University Mountain Grove Campus 2005.

Possingham, J.V., Clingeleffer, P.R. and Kerridge, G.H. 1990. Breeding grapevine for Tropical Environments, pp.169-177, in Vitis Special Edition, Proc. 5th International Symposium on Grape Breeding, 12-16, September,1989. St. Martin/ Pfalz, Germany,

Ramming, D.W., Emershad, R.L. and Taralio, R. 2000. A stenospermocarpic seedless *Vitis vinifera* x *Vitis rotundifolia* hybrid developed by embryo rescue. *HortScience* 35(4): 732-734.

Stein, U., Heintz, C., Blaich, R. 1985. The *in-vitro* testing of grapevine cultivars for oidium and plasmopara resistance. Z.Pflanzenkrankh. *Pflanzensch.*92:355-359.

Distant Hybridization in Horticultural Crops   *Pages* **63–77**
Editors: **M.R. Dinesh & M. Sankaran**
Published by: **ASTRAL INTERNATIONAL PVT. LTD., NEW DELHI**

6

Improvement Possibilities Through Species Utilization in Banana

T.N.Balamohan and G.Karunakaran

*Dean(Hort.), Horticultural College & Research Institute,
Periyakulam-625 604, TamilNadu*

INTRODUCTION

Bananas and plantains, (*Musa* spp.) provide a staple food for many millions of people, are grown in more than 130 countries across the world in an area of 5.14 million ha producing 105.32 million tonnes of banana and plantain (FAO, 2012). India is the largest producer of banana in the world, producing 28.45 million tonnes from an area of 0.796 million hectares with a productivity of 35.7 MT/ha (NHB, 2012). They are important fruit crops in tropical and subtropical countries, and provide employment, nutrition, and food security. Banana fruit production is currently threatened by several diseases and pests, including bunchy top virus, burrowing nematodes (*Radopholus similes*), Moko disease (*Ralstonia solanacaearum*), black Sigatoka or black leaf streak (*Mycospharella fijiensis*), and Fusarium wilt (*Fusarium oxysporum* f sp. *cubense*). Nevertheless, conservation and characterization of genetic diversity, improvement of cultivars resistant to biotic and abiotic stresses, production technology for high productivity with good quality fruits and post-harvest technology needed more systematic research.

Most *Musa* cultivars believed to be one of the earliest plant species are largely sterile with parthenocarpic fruits (fruit development occurs in the absence of seeds) and derived from inter- and intraspecific hybridization between the species *Musa acuminata* and *Musa balbisiana*, designated with the diploid genomes AA or BB respectively (Heslop-Harrison and Schwarzacher, 2007). Most cultivars have arisen not through deliberate crossing by humans but through hybridisation events where the plants occur together, with farmers propagating parthenocarpic lines. Most, but not all, cultivars are triploid (2*n*=3x=33 chromosomes), typically with AAA, AAB or ABB genome constitutions. The word "banana" tends to describe fruits which are eaten fresh when ripe or used as sweet products; "plantains" and "cooking bananas" are usually cooked. Although there is no botanical distinction between the two, many cooking types are interspecific hybrids (AAB or ABB) while many sweet dessert types are AAA.

WORLD LEVEL BREEDING PROGRAMME

Targets for banana breeding, or indeed the breeding of any crop, can be divided into five interrelated groups: productivity characters, abiotic stress resistance, biotic stress resistance, post-harvest characters and those related to markets, environment and livelihoods. Resistance breeding in *Musa* has been attempted, but breeding in *Musa* is difficult because of sterility and polyploidy (Stover and Buddenhagen 1987). The objective in banana breeding was to develop new resistant hybrids akin to commercial cultivars (Dodds, 1943), but most commercial cultivars are either male- or female-sterile or both and this makes breeding difficult.

The systemic banana breeding programme was initiated during 1949 at the Central Banana Research Station, Aduthurai, Tamil Nadu, India. Later it was started at Tamil Nadu Agricultural University (TNAU), Coimbatore with the objective of evolving potential synthetic diploids and primary tetraploids to generate triploid hybrids having resistance or tolerance to pests and diseases (particularly *Fusarium* wilt and nematodes) coupled with good horticultural traits. In India NRC (Banana), Trichy, TNAU, Coimbatore and Banana Research Station KAU, Kannara, Kerala are engaged in banana crop improvement.

There are four hybridization classes:

Triploids which are the result of diploids X diploids crossings, with the original diploid male parental recombination only.

Tetraploids which are the result of triploids X diploids crossings, with the diploid male parental recombination only.

Tetraploids which are the result of crossings among tetraploids with segregation in the two parents.

Triploids which are the result of tetraploids X diploids crossings and segregation in the two parents.

Major breeding programmes that use conventional breeding methodologies are located at the Fundacion Hondurena de Investigacion Agricola (FHIA) in Honduras, the Centre de Coopération Internationale en Recherche Agronomique pour le Développement (CIRAD-FLHOR) in France and Guadeloupe, the International Institute of Tropical Agriculture (IITA) in Nigeria and Uganda, the Centre de Recherches Régionales sur Bananier et Plantain (CARBAP) in Cameroon, and the Empresa Brasiliera de Pesquisa Agropecuaria (EMBRAPA) in Brazil.

The first breeding strategy developed by FHIA, and now adopted by other programs, relies on the production of improved diploids as suggested by TNAU, Coimbatore. The diploids possess useful resistance characteristics introduced from wild sources in an improved genetic background, and are then used as male parents in crosses with the desired triploid as female parent. The improved diploids developed by FHIA are now being used for the improvement of many different types of bananas, including dessert varieties, plantains, cooking bananas, and East African highland bananas. The main constraints of this breeding strategy are: (a) the low fertility rate of the desired female parents, and (b) the possibility of residual fertility in the tetraploid progeny could lead to seed-bearing fruit under certain circumstances. To overcome this latter constraint, efforts are now being focused on the production of secondary triploids.

Widespread use of dwarf cultivars is a genetic contribution to this in banana, while cultivars with rapid field establishment, with genetic and agronomic contributions, have lead to shorter cropping cycles. It is critical that the plant can be propagated reliably at low cost; in banana, historically, this means that appropriate suckers are formed, but the ability to propagate efficiently by tissue culture is becoming increasingly important. For breeding new cultivars, some residual fertility is essential. 'Highgate' (AAA) – a mutant of 'Gros Michel' (AAA) – has been used as a female parent in the FHIA (*Fundacion Hondureña de Investigación Agricola*) breeding programme because after pollination it produces a few fruits with seeds. Cavendish (AAA), with no residual fertility, cannot be used as a parent. In the future, transformation may be used widely, and it is probable there will be genetic variation in transformability between lines, so cultivars where these procedures are reliable will be preferred.

BREEDING METHODS

Two different strategies were developed in breeding. The first aims at producing tetraploid hybrids by pollinating triploid varieties with wild or improved fertile male parents. Originating from female gametic restitution, the few selected tetraploid hybrids aim at combining the agronomic and commercial featured hybrids aim at combining the agronomic and commercial features of the variety with disease resistances contributed by the male parent. The second strategy developed by CIRAD, aims at producing triploid hybrids but starting from natural or improved diploid germplasm. After selection, diploid clones are doubled with a colchicine treatment. Thereafter, the auto- or allotetraploids are backcrossed with another diploid clone in order to create large segregating progenies of triploids. This last strategy offers the opportunity to benefit a larger natural variability than the first option.

GENE POOL AND UTILIZATION OF WILD SPECIES

Wild species and primitive cultivars of many crop harbor useful genes for resistant traits, land races for their quality characteristics and hybrids for superior agronomic characteristics. Banana breeders also largely rely on the introgression of genes from closely related wild plants to increase genetic variation in the cultivated varieties. Landraces, cultivable wild relatives and hybrids constitute the primary gene pool of banana breeding.

Table 1. Varietal holdings of banana field gene banks in India

Gene bank	State	No. of accessions
NRCB, Trichy	Tamil Nadu	690
TNAU, Coimbatore	Tamil Nadu	127
BRS, Kannara	Kerala	212
BRS, Kovvur	Andhra Pradesh	160
GAU, Gandevi	Karnataka	86
IIHR, Bangalore	Karnataka	320
AAU, Jorhat	Assam	96
AAU, Kahikuchi	Assam	87

Black sigatoka (*Mycospharella fijiensis*)

The most important source of resistance for black sigatoka is wild diploid banana *Musa acuminata spp.burmannicoides* widely known as **Calcutta 4** (Rowe and Rosales, 1996). Notable diploids SH 2829, SH 3437, SH 3142, SH 3217 were produced by crossing **Calcutta 4** with other bred diploids by FHIA (Fundacion Hondurena De Investigacion Agricola). Pisang lilin a parthenocarpic diploid accession (AA) in the *Musa acuminata spp.malaccensis* was also been used as resistance source for black sigatoka. The International Institute of Tropical Agriculture (IITA) used **Calcutta 4** and **Pisang lilin** as the main source for introducing host plant resistance to African plantains. A high yielding diploid accession Sanna chenkadali (AA) was found to be highly resistant to leaf diseases, infectious chlorosis and shade (Sathiamoorthy *et al.*, 2000).

Fusarium wilt (*Fusarium oxysporum* schlect.*F.sp. cubense* E.F.Smith)

There are four races of *Fusarium oxysporum* (Ploetz, 1994). The wild diploids of *Musa acuminata spp.malaccensis* and **Calcutta 4** and edible diploid **Pisang lilin** were reported to be the good sources for *Fusarium oxysporum* of race 1 and 2 (Shepherd *et al* 1994; Rowe and Rosales 1996). **Pisang lilin** was the first plant to be used as a source of resistance to develop Gros Michel like cultivar – Bodles Altafort (Jones 2000). Paka , an AA diploid from Zanzibar was also used as male parent. The FHIA diploids SH 3437, SH 3142, SH 3362 were found to be resistant to race 1 and 2 and SH 3362 was also found resistant to race 4. Also the genotypes **Yangambi 5** and **Kikundi** were found to be resistant to race 1 when tested at *Fusarium* infected field at Uganda (M.Pillay, unpubl.).

Bacterial wilt (*Xanthomonas campestris p. musacaerum*)

The disease causes total yield losses because the fruits are inedible in diseased plants. Vector transmission is perhaps the most important means of disease spread. Till now no banana genotype is found to be resistant to this bacterial wilt. Entry points for this virus appear to be the cushions of floral bracts where the initial symptom is observed. Plants that do not shed their floral bracts do not provide entry point for the disease to spread and appear to have escape the disease. Among the east African high land bananas the genotype Nakitembe did not shed the floral bracts (M.Pillay, unpubl.).

Viral diseases

Four virus namely banana bunchy top (BBTV), banana streak virus (BSV), cucumber mosaic virus (CMV) and banana bract mosaic virus (BBMV)are known to affect bananas (Peitersen and Thomas, 2001).The sources of resistance to different virus affecting banana are not known. Early identification and eradication of infected plants is the best means of controlling virus. Dahal *et al.* (2000) assessed natural incidence of BSV based symptom and virus indexing as well as relative concentration of BSV antigens in leaf tissue by ELISA. Some IITA and FHIA hybrids were found to have the higher concentration of BSV antigens in leaf tissue and regarded as field tolerant to BSV.

Weevils (*Cosmopolitus sordidus*)

Plantains (AAB) are found to be more susceptible to weevils than other genotypes (Fogain and Price,1994). Pavis (1991) demonstrated that cultivars from the Pisang awak group exhibit high levels of tolerance despite heavy tunneling by the weevil. While Yangambi Km5 was almost found to be free from the weevil attack. The other genotypes for weevil resistance include Bluggoe, Calcutta 4, Sanna chenkadali (AA),Sakkaki (ABB), Senkadali (AAA), Elacazha (BB), Njalipoovan (AB), Njaru, Muraru and *M.bulbisiana* (Kiggundu *et al.*, 2003.) A new method for resistance screening of banana genotypes for the banana weevils has been developed. It is possible to screen the resistance or susceptibility of the weevils within 7 months in a screen house than 2-3 years in field. It is likely that many of the Musa cultivars grown in Asia are resistant or partially resistant to banana weevil. Conventional breeding for pest resistance in banana is a long-term strategy; resistant clones may not satisfy consumer tastes. Therefore, there is increasing interest in breeding for banana weevil resistance by using biotechnology techniques. This would allow incorporation of resistant genes without changing other desirable properties in highland bananas and plantains.

Nematodes

The most important wide spread nematodes that are damaging banana are *Radopholus similis* (burrowing nematode), *Helicotylenchus multicinctus* (spiral nematode), *Pratylenchus coffeae* and *P.goodeyi* (lesion nematodes) and *Meloidogyne incognita* (root knot nematode). Calcutta 4 shows field resistance to *R.similis. Musa acuminata* spp. *malaccensis* has been used by FHIA breeding program as a source of nematode resistance. Some diploid cultivars like Pisang mas and Pisang lidi have moderate resistance whereas Pisang Jari bauya (AA)and Pisang bataua had resistance to *R.similis*(Fogain, 1996). A diploid clone Kunnan showed considerable resistance to *R.similis* and *P.coffeae* (Collingborn and Gowen, 1993). Yangambi Km5 (AAA) was reported to be resistant to *R.similis*(Pinochet *et al.*, 1998). Most accessions of *M.bulbisiana* have been found to be resistant to *R.similis* (Fogain 1996) while the ABB cultivar Pelipita was found to be moderately tolerant.

DIPLOID BREEDING

Normally the diploid breeding involved the use of Anaikomban, Matti, Namarai, Erachi vazhai, Pisang lilin and Tongat as male parent with Matti as female parent. Matti is resistant to leaf spot diseases while Anaikomban (AA) is resistant to nematodes. Namarai , Pisang lilin and Tongat were free from panama disease and nematodes. The diploid hybridization work resulted in occurrence of diploids, triploids and penta polyploids due to single and doubles restitution. Many of the synthetic diploids have been found to have good resistance to burrowing nematode, sigatoka diseases and Fusarium wilt. The hybrids namely, H.59, H.65, H.109 were found to have acceptable bunch characters with resistance to sigatoka and burrowing nematodes (Sathiamoorthy *et al*, 2000).

TRIPLOID – DIPLOID CROSSES

The initial steps for genetic improvement of plantains or bananas traditionally involve crossing these accessions to AA or BB diploid accessions that are disease-resistant. Most of the widely grown clones are triploid with low female fertility or male and female sterility. Meiosis in triploids can produce gametes with uneven chromosome numbers resulting in high sterility. Seed set per bunch in many clones is less than one seed and germination in soil is usually less than 1%. To overcome this problem and increase the number of progenies that can be recovered from 3x-2x crosses, embryo rescue techniques were adjoined to the breeding programme, and the number of progeny per pollination was multiplied from 3 to 10 times. Tissue culture therefore plays a pivotal role in Musa breeding. However, efforts are underway to improve pollinating procedures and methods of seed handling, including in vivo hybrid seed germination. The 3x-2x breeding scheme produces hybrids with varying ploidy levels and the most interesting selections have been (i) tetraploid hybrids that are high-yielding and resistant to black sigatoka, and (ii) diploid hybrids that have low yield potential but are resistant to black sigatoka. To date, more than 400 hybrids from 3x-2x crosses have been evaluated by IITA scientists and the most promising ones registered for public use (Vuylsteke *et al.* 1993a, b; Vuylsteke & Ortiz, 1995).

Breeding work at Tamil Nadu Agricultural University, Tamil Nadu, India, used a female-fertile clone ('Karpooravalli') and male-fertile diploid and triploid clones. The tetraploid hybrids were developed using 'Karpooravalli' ('Pisang Awak'), a triploid female-fertile/male-sterile clone belonging to the ABB genome group, as the female parent. The hybrids H-02-17 and H-02-18 developed from the diploid AA clone 'Pisang Lilin', H-02-19 from the diploid AA clone 'Eraichivazhai' and H-02-36 and the rest of the hybrids from the triploid clones 'Robusta' and 'Red Banana', respectively.

Triploid x diploid approach to breeding is a viable method to produce new hybrid combinations. Hybridization between triploid parents 'Karpooravalli', 'Red banana', 'Nendran' and Rasthali with already identified, potential male diploids / synthetic diploids. The hybrid H212, a diploid (AB), resembles Ney poovan (AB) and possesses resistance to nematodes and sigatoka leaf spot.

Table 2. Performance of triploid and tetraploid hybrids obtained from Karpooravalli crosses

Hybrid	H212 (AB)	96/7 (ABB)	H-02-34 (AABB)
Male parent	Pisang Lilin	H- 201	Red banana
Bunch wt (kg)	12.52	28.50	15.0
No. of fingers	160	202	116
Reaction to Nematodes	Tolerant	Tolerant	Resistant
Reaction to Fusarium wilt	Tolerant	Tolerant	Tolerant

Further, 3n x 2n breeding programme at TNAU resulted in identification of a promising hybrid designated as 96/7. This was evolved by crossing 'Karpooravalli' x H- 201 (3n x 2n) and its attributes are tabulated. The ploidy and genome were assessed to be ABB, similar to female parent.

Hybridization between triploid parent Poovan with already identified, potential male diploids / synthetic diploids. By this approach, new hybrid combinations named as H 531, H904, H914, H916 and H 923 were evolved. These hybrids possessed good horticultural traits coupled with resistance to nematodes and tolerance to *Fusarium* wilt.

Table 3. Performance of triploid and tetraploid hybrids obtained from Poovan crosses

Hybrid	H 531 (AAB)	H904 (AAB)	H914 (AAB)	H916 (AAAB)	H923 (AAA)
Male parent	Pisang Lilin	Rose	Ambalakadali	Erachi Vazhai	H 516
Bunch wt (kg)	17.50	17.85	28.35	35.87	21.85
No. of fingers	139	109	150	148	141
Reaction to Nematodes	Resistant	Resistant	Resistant	Resistant	Resistant
Reaction to *Fusarium* wilt	Tolerant	Tolerant	Tolerant	Tolerant	Tolerant

Table 4. Mean pollen output of diploids, triploids and tetraploids (Sathiamoorthy, 1987)

Clone	Genome	Pollen output	
		Per anther	Per flower
M. balbisiana	BB	47,142	2,35,710
M. acuminata	AA	40,119	2,00,595
Robusta	AAA	5,000	25,000
Virupakshi	AAB	5,000	25,000
Monthan	ABB	4,500	22,500
Boodles Altafort	AAAA	24,050	1,20,250
Klue teparod	ABBB	10,575	52,875

The pollen production was estimated using the haemocytometer method (Sathiamoorthy 1973). Pollen grains from a known number of anthers were suspended in a known quantity of water. The suspension was poured into the wells of the haemocytometer, the number of pollen grains per well was counted and the number of pollen grains per anther was calculated. The percentage of haploid, diploid and tetraploid spores was estimated based on the size of the pollen (Sathiamoorthy, 1987). Pollen size is an indicator of male fertility status of bananas, because the potential of a male parent depends on the proportion of fertile haploid spores it produces. The tetraploid hybrids derived from the male-sterile 'Karpooravalli' female parent produced a mixture of haploid, diploid and tetraploid spores. Similar results were obtained by Sathiamoorthy (1987). This may occur with normal meiosis (Sathiamoorthy and Balamohan, 1993). Production of a higher percentage of pollen grains with diploid spores offers more scope for using them as the male parent, especially where the female parent is a female-fertile diploid.

TETRAPLOID – DIPLOID CROSSES

The tetraploid hybrids are both female and male fertile and this often reduces fruit quality due to the presence of seeds in the pulp. However, this characteristic of the tetraploid hybrids facilitates the production of large number of seeds when they are crossed with other accessions. To restore seedlessness, crosses are made between the primary 4x and 2x hybrids to produce secondary 3x hybrids. Choice of crosses to make should take into account parental diversity and intrafamily variation to capitalize on heterosis and allow pyramiding of resistance genes from different diploid progenitors. It should be mentioned that 4x-2x crossing results in a vast array of gametes and genome associations, among which the most useful ones are triploids and diploids. However, unlike the 3x-2x scheme, it is expected that most gametes from 4x and 2x accessions will be balanced in terms of chromosome numbers, resulting in more viable zygotes.

The availability of large segregating populations has enabled IITA scientists to undertake genetic analyses aimed at understanding the inheritance of traits in Musa. Complex inheritances was reported for most growth and yield characteristics of Musa (Ortiz & Vuylsteke, 1996) and were attributed to the irregular meiotic behavior of the species and to unpredictable variation in genome size and structure both across and within generations. Therefore, parental performance may not accurately predict progeny performance for these traits. Thus, further genetic improvement will require control of Selection over prospective male and female parents through progeny testing to determine genetic parameters and assign accessions to compatible heterotic groups.

RECURRENT DIPLOID BREEDING

Inheritance studies in 3x-2x or 4x-2x cross-breeding suggest that most traits of economic importance are more predictably inherited from the diploid parents than from parents with a higher ploidy status (Tenkouano *et al.*,1998b and 1999). Furthermore, genetic analysis is easier in a diploid background due to disomic inheritance, which facilitates and accelerates breeding. This, in retrospect, provided a justification for past and current investment in the development of diploid breeding stocks as pursued by major programmes worldwide (Ortiz & Vuylsteke, 1996).

MUTANTS

Edible banana was presumed to be selected in prehistoric times from spontaneous mutations (Buddenhagen, 1987). One of the main difficulties in mutation induction is occurrence of high degree of chimerism in mutants and also time consuming. Till date only two banana accessions Novaria and Klue Hom Thong KU1 are the only registered improved mutant varieties. To overcome sterility problems in some of the diploid clones of potential breeding value *viz.,* Sanna Chenkadali (AA), Anaikomban (AA), Kunnan (AB), *in vitro* mutagenesis approach is being resorted (Sathiamoorthy *et al.,* 2000).

MARKER-ASSISTED SELECTION

DNA markers are also being sought for several characters of importance, including parthenocarpy, apical dominance, and resistance to black sigatoka, nematodes, and other pests and diseases. Fruit quality (colour, texture, ripening) are other candidate traits for DNA markers. Most of these traits are expressed only late in the life cycle of the plants or are difficult to screen for. Identification of markers linked to loci governing important traits will facilitate gene introgression and other marker assisted selection (MAS) applications. Similarly DNA markers will help to assess genetic stability (somaclonal variation), genetic relationships, including parental contributions to their offspring thereby facilitating parental selection for crosses. Accessing genes from various genomes, including the S (*M. schizocarpa*) and T (*M. textilis*) genomes, will increasingly become important to Musa breeding.

FUTURE METHODOLOGY

Genetic improvement presents a potentially cost-effective mechanism to address current constraints in smallholder production by providing high-performing varieties adaptable to diverse environments. The products of existing improvement programmes, drawing on sources of resistance from wild and edible genotypes, are not meeting several important criteria, such as widely-acceptable fruit-pulp quality, and only a fraction of the genetic diversity in diploid Musa is being used. Yet variation among wild and edible Musa species offers a wide spectrum of fruit and bunch qualities. For instance, the ecology of various wild species suggest that sources of resistance to abiotic stresses exist in Eumusa along the northern periphery of its distribution including mechanisms for tolerance to cold (*M. sikkimensis, M.basjoo, M.thomsonii*), water-logging (*M. itinerans*), and drought (*M. balbisiana, M. nagensium*). Recent collecting expeditions in northern India and Malaysia suggest that other poorly known or unexplored areas of diversity are likely to harbor other agronomically-interesting characteristics. In addition, the development of powerful molecular tools by initiatives such as the Global Musa Genomics Consortium provides an unprecedented opportunity to use more effectively the diversity available in wild and cultivated Musa.

GENOMICS

Research is needed to understand better the heterogeneity of the B genome, and the contribution of the A genome in activating virus integrants in advanced breeding lines. It is also crucial to identify the mechanism of silencing of BSV integrants in the genome. Research is also required to determine the geographical diversity of BSV vis-à-vis movement of germplasm, particularly with respect to epidemiological information and risk assessment. Virologists have also demonstrated the importance of developing resistance-screening methods, and also securing the supply of diagnostics.

SUPER DOMESTICATION

It can be argued that the application of genomic tools will enable banana breeding to advance rapidly. This process can be called super domestication (Vaughan *et al.*, 2007). Super domestication involves interactions of breeders and genomic scientists to design the characteristics required from a banana cultivar, and then considers how to produce this ideal cultivar – how to find and evaluate the genes responsible for particular traits and how to bring them together in a new cultivar. Super domestication requires definition of a cultivar with a suite of ideal characters ranging from biotic and abiotic stress resistance, through yield, to post-harvest ripening and packaging qualities – a few involving single genes, but many being quantitative trait loci with many genes involved and with changing patterns of expression depending on conditions. The diverse collections of *Musa* germplasm and the genes within the accessions underpin the search for desirable traits, now requiring work targeted to measuring, as well as finding and conserving, the biodiversity; novel approaches to understanding, conserving and using banana genetic diversity are now required urgently.

Increased understanding of the physiological basis of parthenocarpy, based on genetic knowledge (Ortíz and Vuylsteke, 1995) may even allow the character to be introduced into lines which previously were seeded. Exploitation of the reproduction of banana through seeds and the manipulation of gamete ploidy (Ortíz *et al.*, 2009; Sebuliba *et al.*, 2008) may allow new cultivars to be made with the planned combination of desirable characters.

FUTURE PROSPECTS

Genetic resources available in banana and plantains are huge and it is suspected that most of them are synonyms, which has to be identified and removed. The clear genetic status of all the collections is very essential for breeding programmes. To overcome these problems molecular characterization of these accessions is essential. The development of molecular markers and marker-assisted selection would enhance the efficacy on selection of improved cultivars for defined traits such as pests and diseases resistance, abiotic stress tolerance, and quality and its post harvest fruit characteristics.

The complete genomic knowledge in banana, which covers the complete sequence information, its genes, their expression, recombination and diversity are very important for the improvement of this crop. The genetic maps of banana will be useful to improve the selection of qualitative traits also allow better selection of parents for breeding programmes. The breeding in most commercially acceptable bananas is very difficult due to its parthenocarpic nature, triploidy and low fertility. Further, the hybridization is complicated by combination of different ploidy levels and by female restitution associated with the formation of unreduced female gametes.

Greater importance has to be given in evolving plants resistant to major biotic and abiotic stresses using molecular breeding and genetic engineering techniques incorporating resistant genes from wild *Musa* gene sources. In this regard, highest priority should be given for conservation of the available genetic wild material from the centers of diversity using molecular tools.

New generations of products are expected to contribute to an increased production of bananas in an environmentally friendly way. Both breeding programmes and recombinant DNA strategies require detailed knowledge of the genetics and genomics of the bananas. Therefore the quality of the products and their availability to farmers will depend on both the improvement strategy and the progress made on genetics and genomics.

FURTHER DEVELOPMENT

1. Cell and tissue culture

Production of somatic embryos should be initiated from shoot tip cultures, especially from virus-indexed *in vitro* plants, which are readily available throughout the year. Different factors, such as nitrogen sources, pH, *etc.*, affecting the growth behavior and regeneration capacity of somatic embryogenic cell suspension (ECS) cultures require more investigation.

2. Flow cytometry

Flow cytometric analysis should be carried out on proliferating meristem cultures and other intermediate steps leading to ECS. Ploidy of somatic embryogenic cell suspension cultures should be measured. Ploidy measurements should be made on local cultivars, wild types and *in vitro* multiplied banana plants in banana-producing countries.

3. Somaclonal variation

Somaclonal variation can become a serious problem in micro-propagated banana plantlets. It is desirable to refine the tissue culture protocol to minimize somaclonal variation in regenerated plants. Subculture number of differentiated shoots and somatic embryogenic cell cultures should be optimized to enhance genetic stability among regenerated plants. Molecular markers are needed for the early detection of somaclonal variants, such as dwarf types, erect and drooping leaves, *etc.* Aneuploids should be retained for genomic studies.

4. Genome analysis

A large number of simple sequence repeat (SSR) markers from *Musa acuminata* should be made readily available. The development of SSR markers for *Musa balbisiana* is needed. Stress-induced, developmentally regulated and tissue-specific promoters should be discovered in banana. An internationally accepted host differential set should be established (a group of cultivars with different resistance genes) for path typing *M. fijiensis* and *M. musicola* isolates.

5. Mutagenesis

The protocol developed for irradiation of ECS (embryogenic cell suspension) should be applied to a wide range of genotypes to prevent chimerism, and to improve the effectiveness of mutation induction. Fungal- and nematode-resistant bananas should be irradiated to knock out resistance, and identify resistance genes.

6. Nematodes

Biological, biochemical and molecular analyses should be carried out to study *Arabidopsis/Radopholus similis* and *Pratylenchus coffeae* interactions. The potential of lectins and lectin-related proteins for banana nematode control should be further investigated, both in transgenic model systems and in banana plants. Root-specific and wound-inducible promoters should be identified and tested for nematode control.

7. Genetic transformation

Non-antibiotic selectable markers should be tested (e.g. transposon *Ac* element in the isopentyltransferase (*ipt*) gene, or markers based on selection with sugar and amino acid analogues). Resistance genes (*R*-genes) should be identified in banana for controlling nematodes, fungi, *etc.* Pathogen-induced and tissue-specific promoters should be identified. Bananas should be transformed with genes and regulatory elements originating from the *Musa* genome. Collaborators should be identified for the evaluation of food safety for transgenic bananas.

CONCLUSIONS

The limited knowledge of the genetics of banana and the nature of the crop as a parthenocarpic, mostly triploid, sterile plant means that many aspects of breeding and selection that have been possible in other crops could not be applied in banana. The germplasm pool, as well as new mutations, includes a large amount of variability which new technologies will allow to be accessed and introduced into new and improved varieties in a designed manner. Superdomestication underpinned by genomic research with collection and assessment of the biodiversity in the genus are likely to ensure the future of the *Musa* crop.

In many countries banana has been a neglected crop in terms of research investment and scientists' effort. Key decision makers are beginning to realize the essential role of banana/plantain in food security, enhanced livelihoods, and resilient agricultural systems. The potential to breed superior hybrids has been demonstrated, and there are numerous opportunities for improving both the process and the product, and for realizing impact from already developed hybrids. The future for banana crop improvement looks promising.

H-916

H-926

H-902

REFERENCES

Buddenhagen, 1987. Disease susceptibility and genetics in relation to breeding bananas and plantains. In: Banana and plantain breeding strategies . *Proceedings No:m 21. Australia.* pp: 95-109.

Collingborn and Gowen, 1993. Screening of banana cultivars resistance to *Radophilus similis* and *Pratylenchus coffeae. Infomusa* 6:3.

Dodds KS. 1943. The genetic system of banana varieties in relation to banana breeding. *Empire Journal of Experimental Agriculture* 11:89–98.

Fogain, 1996. Screen house evaluation of *Musa* susceptibility to *Radophilus similis;* evaluation of plantains AAB and diploids AA,AB and BB. **In** : *Proceedings of the workshop: INIBAP, France* pp: 79-88.

Fogain and Price,1994. Varietal screening of some Musa genotypes for susceptibility to the banana weevil. *Fruits* 49 : 247-251.

Heslop-Harrison *JS* and Schwarzacher T. 2007. Domestication, genomics and the future for banana. *Annals of Botany* 100(5):1073-1084.

Jones 2000. Introduction to banana abaca and enset. In DR Jones (ed) Diseases of banana abaca and enset. CABI publishing, CAB international Wallingford, U.K.

Kiggundu A, Gold CS, Labuschange, M .T, Vuylsteke and D Schalk L.2003. Levels of host plant resistance to banana weevil. *Euphytica* 133 : 267-277.

Michael Pillay, unpubl. 2012. In book: Genetics genomics and diversity of bananas.

Ortiz, R. and Vuylsteke, D. 1996. Recent advances in *Musa* genetics, breeding and biotechnology. *Plant Breeding Abstracts* 66:1355-1363.

Pavis, 1991. Etude des relatiuons plante –insecte chez le charancon du bananier cosmopolites sordidus. Germar.In: Proceedings of a research coordination meeting on biological and integrated control of Highland banana and Plantain Pest and diseases.Cotonou, Benin IITA,Ibadan, Nigeria.pp:171-181.

Peitersen and Thomas, 2001. Overview of Musa virus diseases.In: Plant virology in sub Saharan Africa. http:// www.iita.org/info/virology/pdf_files /50-60.pdf.

Pinochet J, Jaizme MC, Fernandez C, Jaumot M and De waele D. 1998. Screening banana for rootknot and lesion nematode resistance for the Canary Islands. *Fund.Appl nematol* 21: 17-23.

Rowe and Rosales, 1996. Bananas and plantains.In: Fruit breeding Vol I:Tree and tropical fruits. John wiley and sons Inc. Newyork,USA. pp:167-211.

Sathiamoorthy S. 1973. Preliminary investigations on breeding potential of some banana clones [MSc dissertation]. Tamil Nadu Agricultural University, Coimbatore, Tamil Nadu, India.

Sathiamoorthy S. 1987. Studies on male breeding potential and certain aspects of breeding bananas [PhD dissertation]. Tamil Nadu Agricultural University, Coimbatore, Tamil Nadu, India.

Sathiamoorthy S, Balamohan T. N. 1993. Improvement of banana. In: Chadha K. L, Pareek, D.P, editors. *Advances in Horticulture, Vol. 1.* Fruit Crops, Part I. Malhotra Publishing House, New Delhi, India. pp. 303–335.

Sathiamoorthy S., Uma S., Valsalakumari PK.,Soorianathasundaram.K., Kumar,N ., Selvaraj N and Menon, R.2000. Banana improvement programme in India. Paper presented in PROMUSA meeting held in Bangkok, Nov. 6-8, 2000.

Shepherd K Dantas J.L.L De Oliviera e silva S. 1994.Breeding Prata and Maca cultivars in Brazil.In: In DR Jones (ed) Proceedings of the first global conference of the International Testing Program, INIBAP, France,pp: 57-168.

Stover R.H, Buddenhagen I.W. 1986. Banana breeding, polyploidy, disease resistance and productivity. *Fruits* 41:175–191.

Vuylsteke D, Ortiz R, Swennen R. 1993. Genetic improvement of plantains at IITA. In: Gamy J, editor. Breeding Banana and Plantain for Resistance to Diseases and Pests. INBAP, Montpellier, France. pp. 267–282.

Vuylsteke D and Ortiz R.1995. Plantain derived diploid hybrids with black sigatoka resistance. *Hortscience* 30: 147-149.

Distant Hybridization in Horticultural Crops   *Pages* **79–89**
Editors: **M.R. Dinesh & M. Sankaran**
Published by: **ASTRAL INTERNATIONAL PVT. LTD., NEW DELHI**

7

Intergeneric Hybridization of *Carica papaya* and *Vasconcellea Cauliflora*

R. Jayavalli[a*] *and T. N. Balamohan*[b]

[a]Assistant Professor (Horticulture), Horticultural Research Station,
Tamil Nadu Agricultural University, Thadiyankudisai, India
[b] Dean (HC&RI), TNAU, Periyakulam, Theni District, Tamil Nadu, India

INTRODUCTION

The present investigation was undertaken to develop PRSV resistant hybrids through intergeneric hybridization. Intergeneric hybridization was done involving nine *C. papaya* cultivars as female and *V. cauliflora* as male. To break intergeneric hybridization barrier, various nutrients were used. Among the nutrients used, sucrose 5 per cent, sucrose 5 per cent + boron 0.5 per cent and sucrose 5 per cent + $CaCl_2$ 0.5 per cent improved the fruit set and seed set percentage. A total number of 1197 flowers were pollinated resulted in 308 fruits. On extraction, 721 seeds were obtained from CO 7, Pusa Nanha and CP 50. Out of 721 F_0 seeds sown, 419 seeds germinated which accounted only 58 per cent. Artificial screening for papaya ringspot virus was carried out 27 days after sap inoculation. Out of twenty nine F_1 hybrid plants of CO 7 x *Vasconcellea cauliflora*, only six plants were found free from PRSV symptoms. Similarly, out of fifty five F_1 hybrid plants of Pusa Nanha x *Vasconcellea cauliflora* only twenty three were found free from the symptoms and seventy plants out of 335 plants of CP50 x *Vasconcellea cauliflora* were found free from PRSV symptoms. Molecular markers *viz.*, ISSR markers were used to check and verify the hybridity. Among the crosses, Pusa

Nanha x *Vasconcellea cauliflora* had higher yield under PRSV infected conditions, however, total soluble solids and total sugars were found lesser than the other two crosses. Based on the disease intensity score, reaction to the PRSV and performance, the cross combinations *viz.,* CO 7 x *Vasconcellea cauliflora* (CO7V3), Pusa Nanha x *Vasconcellea cauliflora* (PNV9) and CP 50 x *Vasconcellea cauliflora* (CPV23) were advanced to F_2 generations. Artificial screening for papaya ringspot virus was carried out 27 days after virus sap inoculation for all the F_2 populations. The F_2 populations of various combinations segregated for susceptibity and resistance in two different ratio. The cross combination CO 7 x *Vasconcellea cauliflora* (CO7V3) exhibited a 15:1 ratio explaining duplicate gene interaction and the crosses Pusa Nanha x *Vasconcellea cauliflora* (PNV9) and CP 50 x *Vasconcellea cauliflora* (CPV23) exhibited a 13:3 ratio explaining inhibitory gene interaction. The study revealed multi-genic inheritance of resistant to PRSV which uniquely exhibit inter-genic interactions.

Papaya (*Carica papaya* L.) is one of the most important fruits of tropical and subtropical regions of the world and belongs to family *Caricaceae*. It is believed to have originated in Central-America with South-Mexico and Costa Rica as origin (De Candolle, 1884). It was introduced to India during 16th century by Portuguese travelers. At present, it is cultivated throughout the world. India is the world largest producer of papaya with a cultivated area of about 129, 000 ha with that allow to produce 5190 million tonnes of fruits annually (NHBD, 2011). In India, it is commercially cultivated in Andhra Pradesh, Gujarat, Maharashtra, Karnataka, West Bengal, Assam, Orissa, Madhya Pradesh, Manipur, Tamil Nadu and Bihar and certain extent in Kerala. Papaya is affected by number of diseases caused by various pathogens and viruses. Nowdays the most destructive disease of *C. papaya* worldwide is papaya ring spot caused by papaya ring spot virus-type P Litz, (1984), Manshardt, (1992), a definitive potyvirus species in the *Potyviridae* (Shukla *et al.,* 1994). PRSV is grouped into two types, Type P (PRSV – P) infects cucurbits and papaya and type W (PRSV-W) infects cucurbits but not papaya (Gonsalves, 1998). Almost all cultivated varieties are highly susceptible. *Carica cauliflora* J., a wild species having non-edible fruits is known to be resistant for this viral disease (Jimenez and Horovitz, 1957). Now the species *cauliflora* has been grouped under the genera *Vasconcellea* (Vegas *et al.,* 2003).

Control measures to check the viral incidence against PRSV-P include cultural practices, cross-protection and planting of tolerant cultivars (Gonsalves, 1994). None of these has been very successful and the development of virus resistant cultivars through conventional breeding is the only reliable tool for long term control. None of the *Carica papaya* cultivars has natural-resistance to PRSV-P. Several related wild species of *Carica* have been reported as resistant to PRSV-P. Even though interspecific hybridization of *Carica papaya* with other species attempted, a very little work has been done using *Vasconcellea cauliflora* which has the desirable gene for PRSV resistance.Under these circumstances, an investigation entitled Intergeneric hybridization of *Carica papaya* and *Vasconcellea cauliflora* was carried out with the following objectives.

1. Breaking the barrier in intergeneric hybridization

2. To study the performance of intergeneric hybrids involving *Carica papaya* and *Vasconcellea cauliflora* against PRSV and genetic inheritance of F_2 populations

3. To find the best cross combinations for yield and quality along with PRSV resistance.

MATERIALS AND METHODS

Hybridization and Field Evaluation

Intergeneric hybridization of *C. papaya* with *V. cauliflora* was performed to transfer the desirable gene from *V. cauliflora* to C. papaya for PRSV resistance. The varieties CO 1, CO 2, CO 4, CO 5, CO 6, CO 7, Pusa Dwarf, Pusa Nanha andCP 50 were selected and used as female for crossing with *V. cauliflora* as male. Nutrients viz., sucrose, boron, calcium chloride and their combinations were smeared on the surface of the stigma before dusting of pollen to overcome the pollination barrier between *C. papaya* and *V. cauliflora* at the time of crossing. Cultivars and different nutrients solutions used are presented in Table 1. Female flowers which were about to open in the next day were bagged on the previous day evening. Pollen grains were collected from fully mature, unopened flowers of *V. cauliflora*. Pollination was done between 6.30 a.m. and 8.30 a.m. in every day. At the time of pollination, the bags on the female flowers were removed and the pollen were dusted on the stigma and bagged again. In gynodioecious varieties, andromonoecious flowers in tree were carefully emasculated on the previous day before dehiscence of anther and pollinated during the next day. Fruit set was observed 5–6 days after pollination, which were carried to maturity in 5–6 months. Matured fruits were harvested at colour break stage for seeds extraction. Collected seeds were treated with GA (Gibberellic acid) at 100 ppm for 2 h a day prior to sowing as it helps for better germination. Then the GA treated seeds were washed gently with running water to remove the traces of GA solution before sowing.

Seedlings thus raised were used for screening. All the seedlings were artificially inoculated with PRSV through artificial inoculation method (Manoranjitham *et al.*, 2008). The seedlings that are shown initial resistance (seedling resistance) alone were taken to field for further evaluation. One gram of infected leaves was ground in a pre-chilled mortar and pestle using 1 ml of 0.1 M chilled sodium phosphate buffer (pH 7.2) containing –mercapto ethanol and 0.01 M EDTA. The sap was rub inoculated using the pestle or glass rod on the young leaves of seedlings at 3 leaves stage which are previously dusted with carborundum powder 600meshes.After 5 min, the excess sap was washed off by distilled water.

Number of fruits borne on the tree during first crop was taken and expressed as number of fruits per tree. Mean fruit weight of five fruits was worked and expressed in kilograms (kg). The fruit yield per tree was arrived at by multiplying number of fruits harvested with mean fruit weight and expressed as kilograms. Total soluble solids of the fruit was determined by hand refractometer and expressed as Brix. Total sugars were estimated as per the method of Hedge and Horreiter

(1962) and expressed in percentage. Reducing sugars were estimated by the method suggested by Somogyi (1952) and expressed in percentage. Non-reducing sugars were estimated by subtracting the percentage of reducing sugars from the total sugars and expressed as percentage.

Molecular markers analysis

To confirm the hybridity of these intergeneric hybrids, ISSR marker analysis was carried out using six plants from CO 7 x *V. cauliflora*, twenty three from Pusa Nanha x *V. cauliflora* and seventy plants from CP50 x *V. cauliflora*. DNA from hybrid progenies and their parents' leaves was extracted following CTAB method (Doyle and Doyle, 1987). PCR reaction was performed using 6 (ISSR) primers. The details of the primers are presented in Appendix A. The reagents that required for performing PCR reaction are as follows (Benbouza *et al.*, 2006). PCR reaction was carried out in total volume of 10 l in 96 tubes PCR plates. Following were the master mix of solution for one reaction. For ISSR primers reagents of 10 Taq buffer + $MgCl_2$ (15 mM) on 1.0 l, dNTP (2 mM) on 1.0 l, Primers 10 M 1.0 l(0.5 l each for combination), Taq polymerase (3 IU/l) on 0.1 l, Sterile double distilled water on 4.9 l and Template DNA 10 ng/l on 2 l. Cycling profile – Touch down protocol was followed for all the primers. PCR cycles included initial denaturation at 94 ℃ for 3 min followed by 19 cycles of 30 s (0.5 ℃) denaturation at 94 ℃, annealing at 63 ℃ for 30 s and 1 min in extension at 72 ℃.Again 19 cycles of 15 s denaturation at 94 ℃, annealing at 55 ℃ for 30 s, 1 min in extension at 72 ℃, 10 min in final extension at 72 ℃ and infinitive final hold at 4 ℃. Electrophoresis was performed in 1.5% agarose with 120V for 2 h .

DAS ELISA of hybrids and parents

Apparently virus free plants from different hybrid combinations and parents were subjected for ELISA confirmation using PRSV specific antibodies. The later together with their positive samples were provided from DSMZ, Braunschweing, Germany. DAS-ELISA was performed for the detection of PRSV by following the manufacturer's instructions (DSMZ Gmbh, Braunschweig, Germany).

Raising F_2 progenies

Out of 99 F_1 hybrids taken to the main field, cross combinations involving CO 7 x *Vasconcellea cauliflora* (6 F_1 hybrid seedlings), Pusa Nanha x *Vasconcellea cauliflora* (23 F_1 hybrid seedlings) and CP 50 x *Vasconcellea cauliflora* (70 F_1 hybrid seedlings), 18 hybrid progenies were confirmed for hybridity through molecular markers. Seeds from these cross combinations were used for raising seedlings which served as F_2 population. F_2 populations were raised from S1 inbreds are screened and desired male and female plants are selected for further sibmating *i.e.* crossing between the female plant and male plant of the same cultivar.

Inheritance of susceptible and resistant plants in F_2 population

Chi Square ratio was worked out for susceptible and resistant plants for testing goodness of fit.

$$\chi^2 \quad = \quad (O - E)^2 / E \quad \text{with (n-1) df}$$

Where, O - is the Observed Frequency in each category

E - is the Expected Frequency in the corresponding category sum of

df - is the "degree of freedom" (n-1)

The test of significance of chi square (χ^2) was made referring to the table for chi square at 1 degrees of freedom at P= 0.05 and 0.01 level of significance.

RESULTS AND DISCUSSION

Intergeneric Hybridization Barrier

Intergeneric hybridization is a valuable tool for creating variability in plant breeding by broadening the germplasm base of the cultigens (Singh, 2003). Incorporation of resistant genes from wild relatives to the cultivated varieties is often hampered because of poor crossability, early embryo abortion, hybrid seed inviability, hybrid seedling lethality, and hybrid sterility due to incompatibility between the two genera. However, these post fertilization barriers in intergeneric hybridization could overcome by (1) the assemblage of diverse germplasm; (2) application of growth hormones to reduce embryo abortion; (3) improved culture conditions; (4) restoration of seed fertility by doubling the chromosomes of sterile F_1 hybrids; and (5) utilization of bridge crosses where direct crosses are not possible.

In the present study, nutrients *viz.*, sucrose, boron, calcium chloride and their combinations were used to overcome the intergeneric hybridization barrier between *C. papaya* and *V.cauliflora*. Most of the harvested fruits had 5-10 seeds, and many cases the seed did not have embryos. Lack of proper endosperm development was evident in a large number of seeds without embryos. Instead of endosperm, a clear mucous like substances, possibly degraded carbohydrate, was found in the seeds. Small translucent partly developed seeds were observed in the seed cavity in most of the F_1 fruits (Drew *et al.*, 1998; Thirugnanavel, 2010). The findings of present investigation are in close confirmity with these findings.

A total of 1197 crosses were affected from which 308 developed fruits were harvested. The result revealed that the seed set was increased considerably with the use of different combinations of nutrients. Among the nutrients combinations, application of sucrose 5 per cent on the stigmatic surface had produced 512 seeds, sucrose 5 per cent + boron 0.5 per cent had produced 200 seeds and sucrose 5 per cent + CaCl$_2$ 0.5 per cent had produced nine seeds (Fig.1and table 1). This may be due to the beneficial effect of sucrose which serves as the best carbohydrate source for pollen germination. This is in close confirmity with the findings of Dinesh *et al.* (2007), Balamohan *et al.* (2008) and Thirugnanavel (2010). They also reported that application of sucrose on the stigmatic surface of the female flowers may help to break the intergeneric hybridization barrier in papaya by increasing the pollen germination and pollen tube growth.

Screening of F₁ Progenies through Artificial Inoculation with PRSV Under Glass House Conditions

In a perennial crop like papaya, field screening for diseases is very difficult since, it requires a larger area for planting. Hence, screening in glass houses in the nursery stage proved quick and rapid method.

A total number of 29 seedlings in CO 7 x *Vasconcellea cauliflora*, 55 plants in Pusa Nanha x *Vasconcellea cauliflora* and 335 plants in CP50 x *Vasconcellea cauliflora* were artificially inoculated with papaya ringspot virus through sap inoculation method. Typical PRSV symptom of mottling of leaves and water soaked lesions on stems were observed in the susceptible parents and the hybrids. However, six out of 29 seedlings in CO 7 x *Vasconcellea cauliflora*, 23 out of 55 in Pusa Nanha x *Vasconcellea cauliflora* and 70 out of 335 in CP 50 x *Vasconcellea cauliflora* were found to be completely free from PRSV symptoms (Fig.2 and table 2). Regarding the female parents, all were found to exhibit the virus symptoms uniformly after sap inoculation. Symptom free F₁ hybrids were transplanted in the main field for further evaluation. The failures of PRSV symptoms to develop on the manually inoculated hybrid plants indicate the incorporation of genes resistant to PRSV. Further, the wild genus *V. cauliflora* was found to be completely resistant to the strain PRSV prevalent in Coimbatore area of Tamil Nadu, India (Manoranjitham *et al.*, 2008).

Hybridity Confirmation Using ISSR Markers

Three intergeneric hybrids from CO 7 x *V. cauliflora* crosses, eight hybrids from Pusa Nanha x *V. cauliflora* crosses and seven hybrids from CP 50 x *V. cauliflora* crosses were tested for hybridity through ISSR markers. The primer UBC – 856 produced unique banding patterns in *V. cauliflora* (male parent) in which five bands were prominent, out of which third and fifth were absent in female parent (Fig. 4) but these are present in CO 7 x *V. cauliflora* (CO7V3). The same primer produced distinguishable band between Pusa Nanha x *V. cauliflora* (PNV9) which was used for the identification of true hybrid (Fig. 5). In case of UBC – 807 primer, one prominent band was observed in male parent which was absent in female parent but present in CP 50 x *V.* cauliflora (CPV23) hybrid (Fig. 6). These primers were helpful to identify F1s in cross CO 7 x *V. cauliflora*, Pusa Nanha x *V. cauliflora* and CP 50 x *V. cauliflora*. Those intergeneric F1 hybrids which is confirmed the hybridity of plants were forwarded to F₂. Ruas *et al.* (2003) used

ISSR markers and successfully evaluated the genetic divergence among the eight *Coffea* species. Praveen (2005) also used ISSR markers to confirm the hybridity of intergeneric hybrids involving *C. papaya* x *V. cauliflora*.

Yield and Quality Performance

In the hybridization programme, yield is the prime character which decides the supremacy of any selection or hybrid for further advancement. In the present hybridization programme for disease resistance, among the female parents, the genotype CP 50 had the maximum fruit weight and yield / tree followed by Pusa Nanha and CO 7. However, among the cross combinations, Pusa Nanha x *Vasconcellea cauliflora* had more number of fruits per tree (35.00) and fruit yield

38.85 kg / tree (Fig. 3 and table 3). This cross combination has many other positive characters *viz.*, earliness, dwarf stature, maximum number of fruits and yield inspite of the virus disease very much prevalent in the experimental plot. Praveen (2005) also recorded higher fruit weight and yield in one of the interspecific hybrids evaluated for PRSV disease.

While involving a wild parent in the intergeneric hybridization, there are ample of possibilities for getting poor fruit qualities in the resultant hybrids (Table 4). In the present study also, the quality traits in all the hybrids were marginally affected due to intergeneric hybridization. However, among the cross combinations, the crosses involving CO7 x *Vasconcellea cauliflora* were found better in fruit qualities than the other two cross combinations. As a whole, the quality parameters were not drastically affected in the resultant hybrids due to the presence of wild male parent, *Vasconcellea cauliflora*. Praveen (2005) also reported that the crosses involving *Vasconcellea cauliflora* as male in the interspecific hybridization programme produced better quality fruits.

Study of F_2 Generations

Out of 99 F_1 hybrids taken to the main field, cross combinations involving CO 7 x *Vasconcellea cauliflora* (6 F_1 hybrid seedlings), Pusa Nanha x *Vasconcellea cauliflora* (23 F_1 hybrid seedlings) and CP 50 x *Vasconcellea cauliflora* (70 F_1 hybrid seedlings), 18 hybrid progenies were confirmed for hybridity through molecular markers. Seeds from these cross combinations were used for raising seedlings which served as F_2 population.

A total of 138 F_2 seedlings from CO 7 x *Vasconcellea cauliflora* (CO7V3), 310 F_2 seedlings from Pusa Nanha x *Vasconcellea cauliflora* (PNV9) and 235 F_2 seedlings from CP 50 x *Vasconcellea cauliflora* (CPV23) were raised and sap inoculated for screening of resistance to papaya ringspot virus. Five seedlings each of CO 7, Pusa Nanha, CP 50 and *Vasconcellea cauliflora* were also sap inoculated for PRSV symptoms. Among the parent, only *Vasconcellea cauliflora* seedlings alone did not produce PRSV symptoms, whereas all the female parents, CO 7, Pusa Nanha and CP 50 plants developed typical PRSV symptoms (Table 5).

It was observed that only 10 F_2 seedlings out of 138 from CO 7 x *Vasconcellea cauliflora* (CO7V3), 45 F_2 seedlings out of 310 from Pusa Nanha x *Vasconcellea cauliflora* (PNV9) and 33 F_2 seedlings out of 235 of CP 50 x *Vasconcellea cauliflora* (CPV23) did not exhibit PRSV disease symptoms. All the F_2 populations exhibited variation within the parental ranges which explained the resistance might be of oligogenic inheritance and not polygenic.

Serological test is more sensitive and convenient than back-inoculation tests when large numbers of plants have to be screened (Miller and Martin, 1988). Segregation can be studied only in the F_2 populations, which will throw light on the gene action .In the present study, in the cross combination, CO 7 x *Vasconcellea cauliflora* (CO7V3), 10 out of 138 F_2 seedlings did not show any PRSV symptoms when subjected to serological test. All of them except the male parent recorded OD values less than 0.300. Similarly, in case of F_2 progenies involving Pusa Nanha x *Vasconcellea cauliflora* (PNV9) all the F_2 progenies recorded OD values less than 0.300.

Similarly, in the case of CP 50 x *Vasconcellea cauliflora* (CPV23) all the F_2 progenies recorded OD values less than 0.300. However, susceptible parents showed OD values more than 0.962.

3.6. Frequency of Susceptible / Resistant Plants in F_2 Populations

Chi-square (χ^2) test of goodness of fit was employed to arrive the segregation ratio in F_2. Manshardt and Wenslaff (1987) studied the segregation of susceptible to resistant phenotypes in the F_2 and reported that single recessive gene or a group of tightly linked genes at a single locus was responsible for PRSV-P resistance in *V. cundinamarcensis* and *V. parviflora*. However in this study, none of the crosses produced 3:1 ratio and exhibited ratio explaining different di-genic interaction models. There might be couple of reasons for obtaining single locus inheritance ratio as explained followingly. If there were common alleles in other loci (or) only one locus between the parents is different (allelic) and the segregation would be recorded as a single gene inheritance. Moreover, study of small population in the F_2 generations might be another reason for recording it as a single gene inheritance.

In the present study, segregation ratio of 15:1 for susceptible to resistant was recorded in CO 7 x *Vasconcellea cauliflora* (CO7V3) (Table 6). The results indicated the presence of two dominant genes contributing the resistance for PRSV. Resistant to PRSV trait was observed to be governed by two recessive genes and the susceptibility was observed to be governed by two dominant gene interaction with the expected ratio of 15:1 explaining duplicate gene interaction determined by two completely dominant genes. These dominant genes produce the same phenotype susceptibility (S_1S_2) whether they are alone or together; the contrasting phenotype resistant was produced only when both the genes were in homozygous recessive state. The possible gene designations are furnished in Table 7.

Segregation ratio of 13:3 for susceptible to resistant phenotypes in Pusa Nanha x *Vasconcellea cauliflora* (PNV9) and CP 50 x *Vasconcellea cauliflora* (CPV23) were observed. The character, susceptibility to PRSV was observed to be governed by two dominant genes with inhibitory gene action (13:3) but the resistance to PRSV was governed by recessive gene. In inhibitory gene action, one of the two completely dominant genes produces the concerned phenotype or the character susceptibility while its recessive allele in homozygous state produces the contrasting phenotype. The second dominant gene has no effect of its own on the character in question; however, it has the ability to stop the expression of the dominant allele of the first gene. As a result, when the two dominant genes are present together, they produce the same phenotype as that produced by the recessive homozygote of the first gene. The recessive allele of the second gene does not affect the development of the character in anyway. Thus in inhibitory gene action, one dominant gene is capable of producing a character only if its expression is not prevented by another dominant gene known as inhibitory gene action and the F_2 ratio is modified to 13:3 ratio in this case. The possible genotype designations are presented in Table 10. In the present study, in the three crosses we have observed the genetic ratio which explained two different kinds of gene interaction. This is possible in the sense that the resistance to PRSV is governed by multiple loci which have independent and different types of interaction. The genetics of the trait, resistance to PRSV is explained with respect to individual cross and the two parents might be possessing different genes governing resistance mechanisms.

In the present study the three F_2 populations were raised from the seeds obtained from F_1 plants whose hybridity was confirmed using polymorphic ISSR markers. The F_2 populations of various combinations segregated for susceptibity and resistance in two different ratio. The cross combination CO 7 x *Vasconcellea cauliflora* (CO7V3) exhibited a 15:1 ratio explaining duplicate gene interaction and

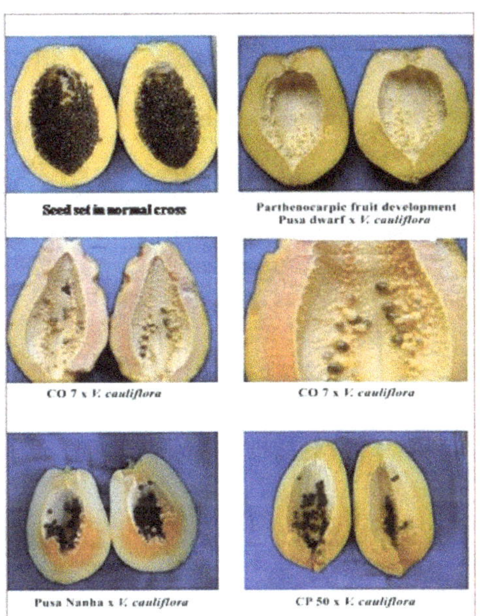

Fig.1 Seed set in intergeneric hybridization using different nutrient solutions.

Fig. 2 Mechnical inoculation of PRSV

Fig. 3 Intergeneric hybrids of *Carica papaya* **x** *V. cauliflora*

the crosses Pusa Nanha x *Vasconcellea cauliflora* (PNV9) and CP 50 x *Vasconcellea cauliflora* (CPV23) exhibited a 13:3 ratio explaining inhibitory gene interaction. The study revealed multi-genic inheritance of resistant to PRSV which uniquely exhibit inter-genic interactions.

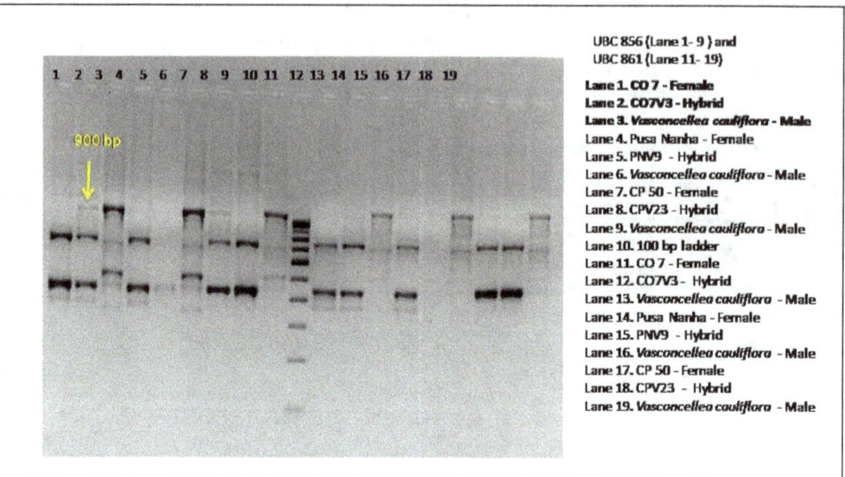

Fig. 4 ISSR markers profile detailing parents and intergeneric F1's using UBC 856 and UBC 861 primers.

REFERENCES

Balamohan, T.N., A. Thirugnanavel, J. Auxcilia and S.K. Manoranjitham. 2008. Breeding for PRSV Resistance in papaya through intergeneric hybridization. *Second International Symposium on Papaya*. 9-12 December, Madurai ,India S-II: Genetic resources and crop Improvement.

De Candolle, A. 1884. Origin of cultivated plants. John Wiley and Sons, Inc., New York, 281 p.

Dinesh, M.R., A. Rekha, K.V. Ravishankar, K.S. Praveen and L.C. Santosh. 2007. Breaking the intergeneric crossing barrier in papaya using sucrose treatment. *Scientia Hort.*, **114**: 33-36.

Doyle, J.J. and J.L.Doyle.1987. A rapid DNA isolation procedure for small quantities of fresh leaf tissue. *Phytoche. Bull.*, **19**: 11-15.

Drew, R.A., C.M. O'Brien and P.M. Magdalita. 1998. Development of interspecific *Carica* hybrids. *Acta Hort.*, **461**: 285-292.

Gonsalves, D. 1998. Control of papaya ringspot virus in papaya: a case study. *Ann. Rev. Phytopathol.*, **36**: 415-437.

Gonsalves, D. 1994. Papaya ringspot. **In:** Compendium of Tropical Fruit Diseases. (Ploetz, R.C, ed). MN, USA: APS Press.67.

Hedge, J.E. and B.T. Horreiter. 1962. **In:** Carbohydrate chemistry. 17 (Eds. Whisler, R. L. and J.N. Be Miller) Academic press, New York.

Jimenez, H. and S. Horovitz. 1957. Cruzabilidad entree species de Carica. *Agron. Trop.*,**7**:207-215.

Litz, R.E, 1984. Papaya. **In:** Evans DA Sharp WR, Ammirato PV & Yamada Y,eds. Handbook of Plant Cell Culture, Vol 2. New York, NY, USA: Macmillan, 349-368.

Manoranjitham, S.K., J. Auxcilia, T.N. Balamohan, A. Thirugnanavel and R.Rabindran. 2008. Confirmation of PRSV resistance in wild type *Vasconcellea cauliflora* through sap inoculation studies. S-II: Genetic resources and crop Improvement. *Second International Symposium on Papaya.* 9-12 December, Madurai India

Manshardt R.M. 1992. Papaya. **In:** Hammerschlag FA, Litz FA & Litz RE, eds. biotechnology of perennial fruit crops. Wallingford, UK; CAB International, 489-511.

Manshardt, R. and T.F. Wenslaff. 1987. PRSV resistant *Carica* interspecific hybrids backcrossed to papaya. *Hort. Sci.*, 22:160.

Miller, S.A. and R.A. Martin. 1988. Molecular diagnosis of plant disease. *Ann Review Phytopathol.*, **26:** 409-32.

National Horticulture Board Database. 2011. www. nhb.gov.in/**database2011**.pdf.

Praveen, K.S. 2005. Interspecific hybrid progeny evaluation in papaya (*Carica papaya* L.). **M.Sc. Thesis**. University of Agricultural Sciences, Bangalore.

Ruas, P.M., C.F. Raus, L. Rampim, V.P. Carvalho, E.A. Raus and T. Sera. 2003. Genetic relationship in *Coffea* species and parentage determination of interspecific hybrids using ISSR (Inter-simple sequence repeat) markers. *Genet. Mol. Biol.*, 26 (3):345-349.

Shukla, D.D, C.W. Ward and A.A. Brunt.1994. The polyviridae. Wallingford, UK: CAB International.

Somogyi, M. 1952. Notes on sugar determination. *J. Biol. Chem.*, 200: 245-247.

Vegas, A., G. Trujillo, Y. Sandrea and J. Mata. 2003. Obtention, regeneration and evaluation of intergeneric hybrids between *Carica papaya* and *Vasconcellea cauliflora. Interciencia.*, **28**(12): 710-714.

8

Distant Hybridization in Coffee Improvement in India

N.S. Prakash, Jeena Devasia,
V.B. Sureshkumar and Y. Raghuramulu

CRS (P.O.) - 577 117, Chikmagalur Dist., Karnataka, India

INTRODUCTION

Coffee (*Coffea* sp.) is one of the most economically important plantation crops grown in India. The genus *Coffea* belongs to the family Rubiaceae and comprises of over 100 species, but the commercial coffee production mainly depends on two species, *Coffea arabica* known as arabica coffee and *C. canephora* called as robusta. Arabica is suitable to high altitudes (900 to 2000 m MSL) while robusta coffee is well adapted to lowlands (600 to 900 m MSL). *Coffea arabica* is the only self-fertile, allotetraploid species ($2n = 4x = 44$) in the genus while other species including *C. canephora* are diploids ($2n = 22$) and generally self-incompatible. *C. arabica* is characterized by low genetic diversity while the diploid coffee species represent considerable variability useful for breeding. Arabica produces superior quality coffee but manifests susceptibility to major diseases and pests like leaf rust, Coffee berry disease, Coffee white stem borer and nematodes. Robusta is more tolerant to these diseases and pests, but the bean and liquor quality of robusta are inferior to arabica. Therefore, distant hybridization aimed at transfer of desirable genes in particular for disease resistance from diploid species into tetraploid arabica cultivars without affecting quality traits has been the main objective of arabica breeding. In case of robusta, bean and liquor quality improvement has been the main breeding focus. The difference in ploidy level between *C. arabica* and other diploid species is an important constraint for transfer of desirable genes to arabica through interspecific breeding. Nevertheless, occurrence of spontaneous hybrids between tetraploid arabica and other diploid species facilitated arabica coffee improvement through introgressive breeding not only in India but elsewhere in the world. Introgressive breeding has been proved to be the most successful strategy in arabica as out of 13 improved arabica cultivars released by the Central Coffee Research Institute, India, nine were the derivatives of the crosses involving spontaneous tetraploid inter-specific hybrids like S.26, Devamachy, Hibrido de Timor (HdeT). Two more cultivars were derived from inter-specific hybridization followed by back cross breeding. In case of diploid species, most of the diploid inter-specific hybrids failed to yield fertile lines except few cross combinations. In India, a compact inter-specific hybrid with improved bean and liquor quality was developed for commercial cultivation from the distant hybridization between two diploid species, *C. canephora* and *C. congensis*. Furthermore, the inter-generic crosses between *Coffea* and *Canthium* sp. as well as *Coffea* and *Psilanthus* a sub genera of *Coffea* were not be successful and resulted in maternal types. In the present paper, we highlight the accomplishments and disappointments of distant hybridization as well as future prospects of this strategy in coffee improvement.

COFFEE INDUSTRY – AN OVERVIEW

Popularly referred as 'Brown Gold', coffee is one of the economically important crops in tropics and sub-tropics and the most widely traded agricultural commodity. Globally, over 50 countries across Asia, Latin America and Oceania grow coffee on commercial scale and coffee exports constitute an important source of foreign exchange earnings for majority of these producing countries. The annual turnover of coffee trade is about US$ 70 billion and coffee sector provides livelihood to an estimated 100 million people in the areas of cultivation, processing and trade. *Coffea arabica* popularly known as Arabica coffee and *Coffea canephora* called as Robusta coffee are the two main species cultivated on commercial scale contributing 70% and 30% respectively, to the total Global coffee production.

In India, this beverage crop has a place of pride among the plantation crops grown and at present covers an extent of 4 lakh ha, mainly in the Southern states of Karnataka, Kerala and Tamil Nadu that accounts for 56%, 20% and 8% of the total area respectively. Coffee is also cultivated in about 58,000 ha in few other states like Andhra Pradesh, Odisha and North Eastern states which are considered non-traditional areas. The annual production is around 3.1 lakh MT and over 70% of the produce constitutes the exports. The coffee grower sector in the country is predominantly a small holder enterprise with 99% of coffee farmers own holdings of less than 10 ha. The two commercially cultivated coffee types, Arabica and Robusta share 49% and 51% of the total area under coffee. Arabica is suitable for cultivation in high altitudes (900 to 2000 m from MSL) while robusta is adaptable to low lands (600 to 900 m from MSL). Coffee cultivation in India is unique as it is grown under two tier shade canopy in simulated micro-climate, which leads to specific problems and varying prospects compared to cultivation in open conditions as practiced in South and Central America as well as Africa. The liquor quality of Indian arabica coffees is generally rated Good with good body, good acidity and fairly good flavour while high grown coffees are known for distinctive fragrance of liquor. The robusta coffee brew is often rated as neutral in taste with light to fair acidity with liquor quality rating as Fair Average Quality (FAQ) to Good and high-grown washed Robustas from India are well recognized world over.

COMMERCIAL COFFEE SPECIES AND BREEDING BEHAVIOUR

The highland forests of South west Ethiopia and Central Africa were considered to be the original habitats of *Coffea arabica* L. and *Coffea canephora* Pierre ex Froehner respectively. Coffee tree belongs to the genus *Coffea* of the family Rubiaceae, comprising of over 100 species. Recently, during 20th century, several new species of *Coffea* have been discovered from West Africa, Central Africa, Madagascar and East Africa (Charrier and Eskes, 2004). Inspite of the vast diversity at species level, only two species *Coffea arabica* L. (Arabica coffee) and *C. canephora* Pierre ex A. Forehner (Robusta coffee) are cultivated on commercial scale in India and elsewhere. Further, *Coffea arabica* L. is the only self-fertile, allotetraploid species (2n = 4x = 44) in the genus while all other species, including *C. canephora* are diploids (2n = 22) and generally self-incompatible (Charrier and Berthaud, 1985). Molecular–cytogenetic analysis established that *C. arabica* is an amphidiploid formed by natural hybridization between *C. eugenioides* and *C. canephora* or ecotypes related to these diploid species (Lashermes *et al.* 1999).

COFFEE GENETIC RESOURCES AND GENETIC VARIABILITY

Genetic variability and heritability are the important pre-requisites for genetic improvement of any crop. In coffee, the two commercial species, *Coffea arabica* L. popular as Arabica coffee and *Coffea canephora* Pierre ex Froehner referred as Robusta coffee, exhibit notable variation in terms of morphology, vegetative vigour, genetic diversity, yield potential, bean quality traits, breeding behaviour and also in genes conditioning resistance for major diseases and pests (Herrera *et al.* 2011). *C. arabica* is characterized by low genetic diversity (Lashermes *et al.* 1996a) which has been attributed to the allotetraploid origin, reproductive biology and evolution process of this species. In contrary, considerable variability was reported among diploid coffee species and some of these species form valuable gene reservoir, for different breeding purposes.

Genetic diversity of *C. arabica* cultivars and wild collections was anlysed using various DNA marker approaches *viz.*, RAPD, AFLP and SSR markers (Lashemes *et al.* 1996b, Anthony *et al.* 2001, Aga *et al.* 2003, Sera *et al.* 2003, Cristancho *et al.* 2004, Aggarwal *et al.* 2004) and all the studies established low genetic diversity within arabica varieties. Primary analysis of genetic diversity in Robusta coffee by enzymatic polymorphism defined the structure of the species into two main groups, the 'Guinean Group' and the 'Congolese group' including both wild and cultivated forms. Subsequent analysis using RFLP marker approach grouped robusta collections into five diversity groups (A,B,C,D,E) delineating the wild and cultivated accessions. Genetic diversity of 40 robusta accessions available in India was analysed in comparison with representative core collections of *C. canephora* and three accessions of *C. congensis* using AFLP and SSR markers (Prakash *et al.*, 2005). The studies revealed high amount of diversity in robusta collections compared to arabica and established that majority of the Indian robusta collections belong to `E` diversity group confirming the spread of cultivated robustas from Java, in early 19th century.

COFFEE GERMPLASM IN INDIA

Consequent to the establishment of Mysore Coffee Experiment Station near Balehonnur in Chikmagalur district of Karnataka, during 1925, systematic survey of existing plantations was taken up and collected the available variability in arabica and robusta populations. Thus, over 250 collections were established as indigenous collections in Germplasm bank at the Coffee Experimental Station (present Central Coffee Research Institute). These collections were evaluated for various traits of agronomic interest and exploited for developing early Indian cultivars. Subsequently, the coffee germplasm was further enriched with exotic collections from all the coffee growing countries with the support of international agencies. An FAO sponsored expedition of Ethiopia helped in introducing 80 wild collections from Ethiopia (Meyer *et al.* 1968). At present, the coffee germplasm at Central Coffee Research Institute comprises of about 300 surviving collections of arabica, 73 types of robusta and 17 different species of *Coffea*. In addition, few indigenous species of coffee have also been identified in India. These species earlier grouped under the genus *Coffea* have been regrouped under the sub genus *Psilanthus*. Out of these species, five species *P. bengalensis*, *P. travancorensis*, *P. wightianus*, *P. khasiana* and *P. bababudanii* were collected from their natural habitats in the forests of North East India and Tamil Nadu (South India) and established in the coffee gene banks at Central Coffee Research Institute and its Regional stations. Majority of the indigenous species are either caffeine free in nature or contain low levels of caffeine.

INITIAL FOCUS OF COFFEE IMPROVEMENT ACROSS THE WORLD

In arabica coffee, early breeding programmes undertaken between 1920 – 1940 focused on yield and quality improvement coupled with adaptability to local conditions. Simple selection or crossing within genetically homogeneous base populations was practiced to develop improved cultivars. Breeding for disease resistance was given low priority as there was no problem of diseases, except in South-East Asian countries like Srilanka, India and Indonesia, where coffee leaf rust epidemics (CLR) were noticed in late 1860s. Subsequently, gradual spread of dreaded CLR disease to other countries across the continents, and appearance of coffee berry disease (CBD) caused by *Colletotrichum kahawae* in highlands of Eastern and Central Africa, prompted a number of new breeding programmes (Van der Vossen, 2001). Thus, the main focus of arabica breeding was shifted to disease resistance coupled with productivity and quality. This phase of coffee improvement from 1950s to 1980s was critical and very productive, as several compact cultivars *viz.*, Catimor, Sarchimor, with high production potential and disease tolerance, suitable for high density planting were developed. The main strategy followed was distant hybridization of high yielding dwarf mutants *viz.*, `Caturra`, `San Ramon` and `Villasarch` and spontaneous tetraploid inter-specific hybrids like Hibrido de Timor (HdT) and Devamachy with high levels of disease resistance (Herrera *et al* 2011).

In order to overcome the epidemics of leaf rust disease on arabica coffee in South-East Asia, other tolerant *Coffea* species such as *Coffee canephora* (Robusta) and *C. liberica* (Tree coffee) were introduced in Indonesia and India, of which Robusta adapted well to lower altitudes. Robusta breeding programmes initiated in early 19th century were aimed at yield and bean quality improvement.

OBJECTIVES OF COFFEE IMPROVEMENT IN INDIA

The leaf rust epidemic on arabica coffee plantations in India during the later part of 18th century posed great threat for arabica coffee cultivation. Initial attempts made by some enterprising planters to cultivate disease tolerant materials by selection and introduction of rust resistant diploid species of *Coffea* like *C. liberica* and *C. canephora* provided temporary respite to tackle the disease on arabica. There was a felt need for systematic research on coffee and the Mysore Coffee Experiment Station was established during 1925 with the primary mandate of developing rust resistant varieties and to tackle the problems in coffee cultivation. Thus, unlike in other coffee growing countries, since beginning the main objective of arabica improvement in India has been on breeding for rust resistance coupled with high productivity, improved quality and wide adaptability.

BREEDING STRATEGIES ADOPTED FOR COFFEE IMPROVEMENT

C. arabica (Arabica coffee)

Genetic studies undertaken in Brazil (Krug and Carvalho, 1951) established that *C. arabica*, the only allotetraploid and self fertile species of the genus *Coffea*, exhibits diploid mode of inheritance for all the characters. Hence, breeding methods common to self pollinated diploid crops including the common strategy of pure line selection to complex distant hybridization techniques have been successfully employed.

The main limitation in arabica coffee breeding is the low genetic diversity which is a bottle neck to achieve a selection efficiency sufficient enough for quick progress besides maintaining adequate variability in subsequent generations. Nevertheless, *C. arabica* is a polymorphic species and limited variability has been observed with respect to various agronomic traits like plant size (dwarf, semi-dwarf, tall); branching habit (erect, spreading, drooping); fruit size (small, medium, bold); fruit ripening (early, late); quality traits and yield potential. In addition, a wide spectrum of resistance from complete susceptibility to moderate tolerance to specific races of leaf rust pathogen is also observed. Further, the tetraploid interspecific hybrids of spontaneous origin that manifest resistance to leaf rust pathogen due to introgression of diploid genomes provided additional variability that have been extensively used as donors in resistance breeding. By effectively managing the available variability, CCRI has successfully evolved 13 elite arabica selections for commercial cultivation. The salient accomplishments in coffee improvement and breeding methods used with special reference to distant hybridization are as follows.

Selection

In coffee, the major criterion for selection is fruit yield, bold beans and field tolerance to leaf rust. Yield and productivity of arabica coffee are largely influenced by the environment and upto 56% of variation in yield can be attributed to environmental factors (Carvalho, 1988). Thus, it is possible to select high yielding genotypes with desired level of linear response to environments on the basis of genotype – environment interactions. (Walyaro, 1983; Agwanda *et al.* 1999). In general low or moderate heritability for yield was observed in India (Srinivasan, 1982). Further, vegetative and reproductive characters such as stem girth, length and number of primary branches per plant, bush radius, internodal length, number of fruiting nodes per primary were correlated to yield and these criteria helped in developing selection indices to breed high yielding varieties (Srinivasan 1980, 1982 and Walyaro, 1983). However, the major constraint in dealing with improvement in quantitative characters like fruit yield is the role of non additive gene action in phenotypic expression, low heritability and limited permissible selection intensity and thus nature of genetic variability within the population is important for efficient selection. In general, low heritability makes selection of genotypes based on phenotypic expression very difficult.

Pure line selection: In the initial years, this strategy was applied wherein the elite individual plants with desired agronomical traits were identified in the populations. The selected individual plants were advanced by recurrent selfing to derive commercial lines. The early Indian coffee varieties, `Kents` & S. 288 were developed by this strategy. Subsequently, pure line selection strategy was used to exploit some Ethiopian Arabica collections of exotic gene bank, that manifest race specific resistance to leaf rust pathogen.

Hybridization

In coffee, hybridization strategies have been used to integrate or improve the agronomically desirable traits. Depending on the objective appropriate hybridization strategy has been followed. In the Indian context, as rust resistance is the primary objective of arabica improvement, distant hybridization techniques like introgressive breeding followed by pedigree selection and inter-specific hybridization, have been successfully exploited. Similarly, natural dwarf/semi-dwarf mutants were used in breeding towards development of high yielding varieties with compact bush types while Ethiopian land races have been used as donors in breeding for quality improvement.

A) *Breeding for rust resistance*

Resistance to coffee leaf rust is reported to be conditioned by atleast nine resistance genes, S_H1 to S_H9, either singly or in combination while the corresponding virulence factors in the pathogen are referred to as v1 to v9. Of these nine resistance factors, S_H1, S_H2, S_H4 and S_H5 were identified in the tetraploid *C. arabica* while S_H6, S_H7, S_H8 and S_H9 were introgressed to *C. arabica* from the diploid species *C. canephora* (Wagner and Bettencourt, 1965; Bettencourt and Rodrigues Jr. 1988) The S_H3 gene was introgressed from another diploid species *C. liberica*. In general, it was observed that the resistance genes identified in *C.arabica*, used either singly or in combination, have not provided durable resistance to most of the races of rust fungus. On the other hand, the genes introgressed from the diploid species such as the S_H3 gene from *C. liberica* as well as certain genes from *C.canephora* are found to be more durable and have provided long - lasting protection to arabica under field conditions (Srinivasan and Narasimhaswamy, 1975; Ramachandran and Srinivasan, 1979).

Hence, combined use of resistance genes of diploid species in arabica background is one of the promising approaches to achieve durable rust resistance.

i) Introgressive breeding followed by pedigree selection

Combined introgression of various genes governing resistance, into promising arabica cultivars through traditional breeding approaches is not a straight forward process because of the variation in ploidy levels of the recipient and donor species. Nevertheless, introgressive breeding strategy has been exploited to integrate the S_H genes wherein the spontaneous tetraploid inter-specific hybrids have been used as donor parents. The recipient parent is normally the proven genotype or possess different resistance genes than donor parent. The resultant hybrid progenies were subjected to recurrent selection by pedigree method. The famous and well known variety of India, S. 795 was developed during late 1940s, following this strategy. S.795 was developed from the cross between S.288, a derivative of the spontaneous *C.liberica* introgressed line and Kents. The variety became popular because of its good vegetative vigour, with consistent production potential of 1500 kg/ha coupled with superior bean quality traits and wide adaptability. At the time of release, this variety was resistant to leaf rust races I and II, prevalent in Indian coffee tracts (Ramachandran and Srinivasan, 1979). The rust resistance in S.795, is known to be governed by S_H3 resistance factor reported to be introgressed from *Coffea liberica* [Wagner and Bettencourt 1965; Prakash et al 2004]. Because of large scale expansion of S.795 variety by 1960s, resulted in selection pressure on the rust pathogen that lead to the development of new race VIII with gene combination (v2,3,5) and ability to overcome the resistance of S.795. Inspite of the development of compatible virulent race, the rust build up on S.795 variety was found manageable with prophylactic sprays due to durable and complex nature of resistance imparted by S_H3 gene.

Subsequently, 'Devamachy' a spontaneous tetraploid interspecific hybrid of *C.canephora* and *C. arabica* identified in an estate in Coorg region of Karnataka was exploited as source of resistance and two varieties Sln.5A and Sln.5B were developed. 'Devamachy collection' is characterized by intermediate plant phenotype to Arabica and Robusta with high vegetative vigour but low fruit set. Sln.5A was derived from the cross between Devamachy x S.881 (a wild Arabica collection from Rume Sudan). The hybrid plants shows vigorous vegetative growth with erect branching habit and manifest high field tolerance to leaf rust. It is a consistent yielder with production potential of 1200 to 1500 kg/ ha and adaptable to hot and humid climate and performs well across the locations in different elevations. Sln.5B was derived from the cross between S.333 x Devamachy and the plants of this line (S.2931) exhibit vigorous with semi drooping growth habit. The plants manifest resistance to leaf rust with the yield potential of 1500 kg/ha. Sln.5B is adaptable to different Arabica growing regions and performs well in medium to high altitudes.

Similarly, the other spontaneous tetraploid inter-specific hybrid used in arabica coffee breeding is Hybrido de Timor (HDT) spotted in Timor island of Indonesia during 1950s. HDT popularly known Timor hybrid is also natural hybrid between *C. arabica* and *C. canephora*, very much resembling arabica coffee, with high levels of resistance to major coffee diseases and pests. Hence, the Timor hybrid was extensively used in resistance breeding programmes world wide and still assumes significance for introgressive breeding. Introduced into India during 1961, HDT was used as donor parent and several hybrid populations were developed between different arabica genotypes and HDT. On detailed evaluation, the derivative of the cross between Tafarikela an Ethiopian collection known for horizontal resistance and liquor quality and HDT was released for commercial cultivation as Sln.9. The plants of Sln.9 are tall, vigorous with drooping branches and manifests high tolerance to leaf rust under field conditions. Sln.9 is an early ripener and drought hardy with yield potential of 1200 - 1600 kg/ha and widely adaptable to all the coffee growing regions. Fruits are bold with 60% to 65% `A` grade beans and liquor quality is excellent with strong body and possessing distinct flavour in cup.

Thus, introgressive breeding has proved to be the most successful strategy for arabica coffee improvement as it helped to a great extent in minimizing the time frame required for introgression of desired genes of diploid origin especially the disease and pest resistance compared to inter-specific hybridization as such.

ii) *Inter-specific hybridization*

The diploid species of *Coffea* are more heterogeneous because of their cross pollinating nature and provides additional variability for breeding purposes. Among several diploid species, *C. canephora* is the main source of resistance genes governing coffee leaf rust (*Hemileia vatatrix*), coffee berry disease (*Colletotrichum kahawae*) and root knot nematodes (*Meloidogyne* spp.). Similarly, other diploid species like *C. liberica* has been successfully used as source for rust resistance breeding in India (Srinivasan and Narasimhaswamy 1975) while *C. recemosa* was used as source of resistance to coffee leaf minor (Guerreiro Filho *et al.* 1999). Hence, inter-specific hybridization with an objective to transfer the desirable genes in particular for disease resistance from diploid species like *C. canephora* and *C. liberica* into tetraploid arabica cultivars without affecting quality traits has been the main objective of arabica coffee breeding (Van der Vossen, 2001). However, the ploidy difference between *C. arabica* (tetraploid) and other species (diploids) is the main limitation for developing fertile hybrids from direct crosses. Hence, two breeding strategies, tetraploid breeding and triploid breeding methods were followed to obtain fertile tetraploid inter-specific hybrids.

a) Tetraploid breeding method: In tetraploid breeding method followed in Brazil, the chromosome number of the diploid species (*C. canephora*) was doubled to maintain gametic balance and tetraploid robusta types were crossed with normal tetraploid arabicas. The resultant hybrids were back-crossed to arabicas and selection was exercised in the progeny for desirable types. The descendents of tetraploid breeding method tend to be very vigorous with characters resembling more towards robusta (bush stature, yield and resistance) but with bean quality traits of arabica. By using this strategy, a variety known as 'Icatu' was developed in Brazil for commercial cultivation.

b) Triploid breeding method: This method was followed in India and Colombia, wherein direct crosses were made between tetraploid *C. arabica* and diploid *C. canephora*. The resultant triploids were recurrently back crossed to arabica to develop a compact type of arabica variety with bearing characters and resistance of robusta. One of the arabica selections, Sln.6 (S.2828) a BC II line developed in

India by using triploid breeding strategy followed by back crossing to arabica, resemble arabica phenotype with good vigour, fruit cluster characters and resistance of robusta. The liquor quality of Sln.6 is similar to arabica as 'Kent' an old arabica variety of India known for intrinsic quality was used as recurrent arabica parent. However, both these strategies requires more time compared to introgressive breeding.

C) Diploid interspecific hybrids: Apart from tetraploid inter-specific hybrids, diploid interspecific hybrids have been successfully exploited for development of commercial lines of arabica and robusta. In case of arabica, spontaneous amphiploidy of a diploid inter-specific hybrid between *C. liberica* x *C. eugenioides*, gave rise to a fertile arabica line, Sln.11 (Narasimhaswamy and Vishveshwara, 1967). The bushes of Sln.11, resemble arabica with erect to semi erect branching habit and manifest resistance to leaf rust. It is a moderate yielder with yield potential of 1000 – 1200 kg/ha and adapt well to hot and humid conditions.

B) Breeding for high production

i) Mutation breeding – Exploitation of natural mutants

Inorder to exploit the potential of mutation breeding especially for arabica coffee with narrow genetic base, physical mutagens like gamma irradiation and chemical mutagens like EMS and DMS *etc.* were tried and no useful mutants were recovered. Nevertheless, exploitation of spontaneous mutants was successfully achieved in coffee. The dwarf (San Ramon) and semi-dwarf mutants (Caturra, Villasarchi) identified in Brazil were crossed with HDT and compact arabica varieties with high yield potential and disease resistance were developed. The Indian varieties such as Sln.7.3 (San Ramon hybrid), Sln.12 (Cauvery, a derivative of Catimor) Sln.13 (Chandragiri, a derivative of Sarchimor) belongs to this group where the mutants contributed for production potential while HDT was used as donor for disease/ pest resistance.

ii) Diploid interspecific hybrids – Robusta improvement

In robusta, inter-specific hybridization strategy was successfully used for improving the bean and liquor quality. Two diploid species *C. congensis* and *C. canephora* (robusta) were crossed and the F_1 hybrid was back crossed to robusta, and mass selection was followed in BC progeny. The compact bush stature was further stabilized by sib mating, which resulted in a highly fertile C x R hybrid variety. The hybrid plants are intermediate in size to both parents with compact and drooping secondary growth. The hybrid is gaining popularity among Robusta growers because of value addition in bean size and liquor quality, which is soft, neutral with light to fair acidity. This hybrid is suitable for high density planting (2.7 x 2.7m or 2.4 x 2.4m) compared to other *Robusta* varieties which requires a wider spacing of 3 x 3m.

Similarly, with an objective of reducing the ripening time and caffeine content in robusta, diploid inter-specific hybrids between *C. canephora* (robusta) and *C. recemosa* were developed (Sureshkumar *et al.* 2010). The fruit set in F_1 hybrids either on natural cross pollination or on controlled back crosses to both the parents is found to be very low. Hence, a small F_2 progeny has been raised by embryo-rescue method for further exploitation.

Apart from the above successful examples of diploid inter-specific hybrids, other cross combinations though yielded few F_1 hybrid plants, intermediate to both the parents, the fertility levels of these plants were observed to be low and failed to produce viable F_2 populations.

C) Breeding for low caffeine content:

i) Inter-generic hybridization:

Distant hybridization by inter generic crosses between *C. arabica* and *Canthium dicocum* were tried with an objective of reducing the fruit maturation time in arabica. Similarly, hybridization was attempted between *C. canephora* and indigenous species recently grouped under sub genus *Psilanthus* (*P. bengalensis, P. travancorensis, P. wightianus*) with a main objective of breeding for low caffeine. Embryo rescue technique was followed to generate F_1 population (Sreenath et al 1992). But, both the crosses gave rise to maternal types.

The current focus of coffee improvement in India is to maximize productivity through development of high yielding hybrids coupled with durable host resistance in Arabica and identification of drought tolerant robusta lines to cope with the changing climate. Accordingly, breeding programmes are being pursued with emphasis on development of heterotic F_1 hybrids in both arabica and robusta, pyramiding of the resistance genes in selected arabica cultivars through DNA marker assisted selection to achieve durability of rust resistance and search for potential sources of tolerance/resistance to white stem borer in coffee gene pool for breeding purposes. In this context, some of the spontaneous tetraploid inter-specific hybrids spotted in natural populations are showing promise in bio assays. Hence, distant hybridization has potential scope in arabica coffee improvement in general and for resistance breeding in particular.

REFERENCES

Aga, E., Bryngelsson, T., Bekele, E. and Salomon, B. (2003). Genetic diversity of forest arabica coffee (*Coffea arabica* L.) in Ethiopia as revealed by random amplified polymorphic DNA (RAPD) analysis. Hereditas 138: 36-38.

Aggarwal R. K., Rajkumar, R., Rajendrakumar, P., Hendre, P. S., Baruah, A., Phanindranath, R., Annapurna, V., Prakash, N. S., Santaram, A., Srinivasan, C. S. and Singh, L. (2004). Fingerprinting of Indian coffee selections and development of reference DNA polymorphism panels for creating molecular IDs for variety identification. *Proceedings of 20th International Scientific Colloquium on Coffee*, Bangalore, pp. 751-755. ASIC, Paris, France.

Agwanda, C. O., Baradat Ph., Cilas, C., Charrier, A., 1999. Prediction of yield stability in Arabica coffee based on stability of morphological components. In: XVIII International Scientific Colloquium on Coffee.pp.452-453, ASIC, Paris.

Anthony, F., Bertrand, B., Quiros, O., Wilches, A., Lashermes, P., Berthaud, J. and Charrier, A. (2001). Genetic diversity of wild coffee (*Coffea arabica* L.) using molecular markers. *Euphytica* 118: 53-65.

Bettencourt, A. J. and Rodrigues Jr C.J. (1988). Principles and practice of coffee breeding for resistance to rust and other disease. In: Clarke RJ, Macrae R (Eds) Coffee. Agronomy, vol 4. Elsevier Applied Science, London, pp 199–234.

Carvalho, A., 1988. Principles and practice of coffee plant breeding for productivity and quality factors: *Coffea arabica*. In: Clarke JR, Macrae R (eds) *Coffee: Agronomy*, Vol **4**, Elsevier Applied Science: London. pp 129–166.

Charrier, A., and Berthaud, J. (1985). Botanical classification of coffee. In "Coffee: Botany, Biochemistry and Production of Beans and Beverage" (M. N. Clifford and K. C. Wilson, Ed.), pp 13–47. Croom Helm, London.

Charrier, A., and Eskes, A. B. (2004). Botany and Genetic of Coffee. In "Coffee: growing, processing, sustainable production, a guidebook for growers, processors, traders, and researchers" (J. N. Wintgens, Ed.), pp. 25-56. Wiley-VCH, Weinhelm, Wiley.

Cristancho, M. A., Chaparro, A. P., Cortina, H. A. and Gaitan, A. L. (2004). Genetic variability of *C. arabica* accessions from Ethiopia evaluated with RAPDs. *Proceedings of 20th International Scientific Colloquium on Coffee*, Bangalore, pp 672-675. ASIC, Paris, France.

Guerreiro-Filho, O., Silvarolla, M. B. and Eskes, A. B., 1999. Expression and mode of inheritance of resistance to leaf miner. *Euphytica* 105:7-15.

Herrera, J. C., Cortina, H. A., Anthony. F., Prakash, N. S., Lashermes, P., Gaitan A. L., Cristancho, M., Acuna, R., and Lima, D. R. (2011). COFFEE` In "Genetic Resources, Chromosome Engineering and Crop improvement - Medicinal Crops" (Ram J. Singh Ed.), Vol 6, pp. 589-640. CRC press.

Krug CA, Carvalho A (1951). The genetics of *Coffea*. *Adv. Genet.* 4:127-158.

Lashermes, P., Combes, M-C., Trouslot, P., Anthony, F., and Charrier, A. (1996a). Molecular analysis of the origin and genetic diversity of *Coffea arabica* L. in "Implications for coffee improvement" Proceedings of EUCARPIA meeting on tropical plants. Montpellier, pp. 23–29.

Lashermes, P., Trouslot, P., Anthony, F., Combes, M-C. and Charrier, A. (1996b). Genetic diversity for RAPD markers between cultivated and wild accessions of *Coffea arabica*. *Euphytica* 87: 59-64.

Lashermes, P., Combes, M.C., Robert, J., Trouslot, P., D'Hont, A., Anthony, F., and Charrier, A. (1999). Molecular characterisation and origin of the *Coffea arabica* L. genome. *Mol. Gen. Genet.* 261:159–266.

Meyer, F. G., Fernie, L. M., Narasimhaswamy, R. L., Monaco, L. C., and Greathead, D. J. (1968). FAO Coffee Mission to Ethiopia, Food and Agricultural organisation of the United Nations, Rome.

Narasimhaswamy, R.L. and Vishveshwara, S. (1967). Progress on hybrids between diploid species of *Coffea*. *Turrialba* 17:11-17

Prakash, N. S., Combes, M. C., Dussert, S., Naveen, K. S. and Lashermes, P. (2005). Analysis of genetic diversity in Indian robusta coffee genepool (*Coffea canephora*) in comparison with a representative core collection using SSRs and AFLPs, *Genet. Resour. and Crop Evol.* 52: 333–343.

Ramachandran M. and Srinivasan C. S. 1979. Four generations of selection for resistance to race I of leaf rust in arabica cv. S.288 x 'Kents'. *Indian Coffee*, 43(6): 159-161.

Sera, T., Ruas, P. M., Ruas, C. F., Diniz, L. E. C., Carvalho, V. P., Rampim, L., Ruas, E. A. and Silveria, S. R. (2003). Genetic polymorphism among 14 elite *Coffea arabica* L. cultivars using RAPD markers associated with restriction digestion. *Genet. and Mol. Biol.* 26: 59-64.

Sreenath, H.L., Muniswamy, B., Naidu, M.M., Dharmaraj, P.S. and Ramiaiah, P.K. (1992). Embryoculture of three inter-specific crosses of Coffee. *Journal of Plantation crops* 20 (Supplement): 243-247.

Srinivasan, C. S.,1980. Accociation of some vegetative Characters with initial fruit yield in coffee (*Coffea arabica* L.), *Journal of Coffee Research*.10:21-27.

Srinivasan, C. S., 1982. Pre Selection for Yield in Coffee. *Indian Journal of Genetics*, **42**:15-19.

Srinivasan, K. H. and Narasimhaswamy, R. L., 1975. A review of coffee breeding work done at the Government Coffee Experiment Station, Balehonnur. *Indian Coffee* 39: 311-321.

Sureshkumar, V.B., Prakash, N.S. and Mohanan, K.V. (2010). A study of *Coffea recemosa* x *Coffea canephora* var. *Robusta* in relation to certain critically important characters. Int. J. Plant Breed. Genet. 4 (1):30-35.

Van der Vossen, H. A. M. 2001. Agronomy I: Coffee Breeding Practices. In: Clarke RJ, Vitzthum OG (eds) *Coffee Recent Developments*, Blackwell Science Ltd: London, UK. pp. 184–201.

Wagner, M and Bettencourt, A. J (1965). Inheritance of reaction to *Hemileia vastatrix* Berk & Br. in *Coffea arabica* L. Progress report 1960–1965, Coffee Rusts Research Center, Oeira, Portugal.

Walyaro, D. J. A. 1983. Considerations in breeding for improved yield and quality in Arabica coffee (*Coffea arabica* L.). Ph.D. Thesis, Wageningen Agricultural University.

Distant Hybridization in Horticultural Crops   *Pages* **103–109**
Editors: **M.R. Dinesh & M. Sankaran**
Published by: **ASTRAL INTERNATIONAL PVT. LTD., NEW DELHI**

9

Prospects of Interspecific Hybrids in Oil Palm

R.K. Mathur, K. Sunilkumar and P. Murugesan**

ICAR-Indian Institute of Oil Palm Research,
Pedavegi-534 450, West Godavari District, Andhra Pradesh
**IIOPR-Research Centre, Palode, Kerala*

INTRODUCTION

Oil palm is one of the most important crops in the world and by far out-yields other oil crops. There are more than 12 million hectares planted with oil palm in the world, with a production of over 45 million tons, which is expected to increase substantially due to growth in demand for oil by the food industry and the use of Biodiesel in Europe and in some South American countries (Fedepalma, 2010). Palm oil is the single largest vegetable oil produced, marketed and consumed globally. In recent past, oil palm (*Elaeis guineensis* Jacq.) has attained importance as the highest oil producing crop capable of ameliorating huge shortage of edible oil in India and hence, received increased attention of planners, researchers, entrepreneurs *etc.* In India, Oil Palm is being grown on substantial scale in different nontraditional agro climatic conditions like high temperature, low humidity, semi-drought and irrigated conditions predominantly in the states of Andhra Pradesh, Karnataka, Tamil Nadu, Kerala and other identified states. Since palm oil contains more saturated fats than oils made from canola, corn, linseed, soybeans, safflower, and sunflowers, it can withstand extreme deep-frying heat and resists oxidation. It contains no trans fat, and its use in food has increased. Palm oil is also used as biofuel. The development of oil palm varieties with a) high yield, b) compact canopy and dwarfness, c) high oil extraction ratio, and d) having tolerance to drought (varieties having high water use efficiency) have become very important.

Looking at the importance oil palm has assumed in terms of nutritional food security (as palm oil is rich in vitamin A and E) and as a major source of edible oil, the major thrust is production of high yielding varieties with better oil quality and dwarfness. Although, high yield is attained in cultivated species, oil quality, dwarfness, disease resistance are found available in less utilized species of the crop hence, interspecific hybridization has become imperative to combine desirable traits from two different species into the cultivated species. These necessitate elucidating the available genetic resources and their suitability for interspecific hybridization.

OIL PALM SPECIES RESOURCES

The success of any crop improvement programme necessitates the sound footing (assemblage)/ presence of wider spectrum of genetic variability (germplasm) in the species. This is all more important especially in respect to crops like Oil Palm, where narrow genetic base has been felt major constraint in research and development of Oil palm (Mathur *et al.*, 2005)

The genus *Elaeis* consists of two species viz. *E. guineensis* Jacq. and *E. oleifera* (HBK) Cortes (Uhl and Dransfield, 1987). Both species have 16 numbers of haploid chromosomes.

E. guineensis , the cultivated species and generally regarded as originated from West Africa (Chevalier, 1943 and Zeven, 1967) is commonly known as African oil palm. It is grown in the wild, semi wild and cultivated parts of the tropics within $10°$ latitude of the equator, Southeast Asia and South and Central America. However, it is endemic to the tropical lowlands of West and central Africa spreading from $16°N$ in Senegal to $15°S$ in Angola (Hartley, 1988). Although, this is the high yielding species, it has got demerits such as taller palms, susceptibility to diseases etc and the most of the research work is carried out focused on this species.

The *E. oleifera*, otherwise known as American oil palm, is native of South and Central America and is found scattered in wild palm groves in South and Central American countries like Brazil, Ecuador, El Salvador, Peru, Amazonian belt *etc.* (Rajanaidu, 1986). This species is under-utilized and is considered to have desirable traits such as dwarfness /slow vertical growth , high oil quality and tolerance to several biotic and abiotic stresses including lethal yellowing and spear rot (Hayati *et al.*, 2003). *Elaeis oleifera* oil is rich in carotene of about 4000ppm compared to 500-700ppm in guineensis. *E. oleifera* has got lower saturated fatty acid (33.5 ± 0.5) while cultivated oil palm has more saturation (47.3 ± 0.1). hence, *E. oleifera* as considered to as one of the precious genetic resources to overcome the problems faced by oil palm industry.

DEVELOPMENT OF INTER SPECIFIC HYBRIDS IN OIL PALM:

In spite of desirable qualities, cultivation of pure stand of oleifera is economically not viable due to its low yields (<1t oil per ha per year) compared t *E. guineensis* (4-5t oil per ha per year).

For widening the genetic base of oil palm, generation of pre-breeding populations and transfer of desired gene(s) from *Elaeis oleifera* into the *guineensis*, inter specific hybridization is a necessity. The basic chromosome number of two species (n) is 16 and show normal pairing of chromosomes on crossing, and hence are easily cross compatible. The two species are cross compatible and there is a wide scope of transgression of gene(s) responsible for dwarf and compact canopy, oil quality and biotic and abiotic stress tolerance from *oleifera* into *guineensis* through inter specific hybridization.

Inter-specific hybridization in oil palm is being followed for introgression of desirable traits such as dwarfness/compactness, high iodine value(for high proportion of liquid oil), high carotene content, high vitamin E content etc from *E. oleifera* into the cultivated species ie. *E.guineensis*. the progenies are then back crosses to its *E. guineensis* parent to improve yield. Genomic In Situ Hybridization (GISH) developed at Malaysia could be useful to ascertain the amount of introgressed parental genome into interspefic hybrids. The technique is able to show clear differentiation between the oleifera and guineensi genomes. Interspeciifc hybrids have thin shell, but alck fibre ring unlike that of DxP intraspecific hybrid of guineensis. Fruit detachment and build up of free fatty acids are slow in E. oleifera compared to guineensis. *E. oleifera* has marked tendency towars production of parthenocarpic fruits. Because of this, fruits per bunch in the hybrids tend to be more than guinneesis parent. This partly helps to compensate the lower percentage of oil in the mesocarp.

It is reported that OxG hybrids developed from Taisha (Ecuador) had the potential to produce high oil yields close to that of DxP seeds of *E. guineensis*. promising four back crossed families (sourced from south American countries) were developed from United Plantation Berhad, Malaysia, capable of frsh fruit bunch yield upto 35 tonnes per ha per year. Selection of ortets from these back crossed progenies are under progress with an objective to develop compact clones and seed progenies with 6.25 percent E. oleifera traits. Such hybrids are amenable to igh density planting of 200 palms per ha compared to conventional 136-160 palms per ha. Malaysia oil palm boards initiative resulted in production of high carotene contaianing planting material viz. PS4 by interspecific hybridization of high carotene *E. oleifera* in a full diallel crossing scheme. world over about 2.5 million interspecific hybrid seeds are produced every year from Ecuador (CIRAD partners-1 million), Columbia (la Cabana-0.3 million, Indu palma-0.2million) and Brazil (Embrapa-1 million) for commercial cultivation.

In interspecific hybrids, the refining efficiency is reportedly higher leading to more recovery of liquid oil compared to the traditional E. guineensis varieties. Inheritance of growth and yield parameters in interspecific hybrids was reported by Haradon (1969) and Hardon and Tan (1969). Sterling *et al.* (1987) reported the phenotypic characteristics of an exceptional segregant from *oleifera* X *guineensis* hybrid open pollinated with guineensis known as the Original Compact Palm (OCP). Commercial plantations are being developed in south American countries and the present interspecific sprout production is 2.5 million. The interspecific hybrids has gained importance due to the promising features like short palm height and the consequent advantages in plantation management and extending its economic life.

Inter specific hybrids exhibit intermediate traits compared to the parent species and show hybrid vigour for some traits of agronomic interest that offer great potential for improving the competitiveness and sustainability of the crop. From the production perspective, the advantages are slow vertical growth (20 cm year^{-1}) which means a longer economic life of oil palm plantations beyond normal 25 years; its partial resistance to the bud rot disease, which is perhaps the greatest threat to the oil palm industry in the America (Torres *et al.*, 2010); delayed degradation of the fruit and lower formation of free fatty acids, which translates into a better stability of the fruit after harvest and lower percentage (less than 2%) of acidity (Corley and Tinker, 2003); high productivity (28-33 tonnes of Fresh fruit bunches ha^{-1} year^{-1} with an oil extraction rate between 18% and 19%), which ensures high incomes (Torres *et al.*, 2004); and, the best oil composition in terms of iodine index (68-70) and content of antioxidants such as carotenes, tocopherols and tocotrienols (Corley and Tinker, 2003). Therefore, the planting of OxG hybrid materials has been intense.

In recent years, due to the presence of diseases, plantations of the OG interspecific hybrid, a cross between the American palm and the African palm, have increased substantially in Latin America because of its apparent partial resistance to the bud rot disease caused by *Phytophthora palmivora* (Torres *et al.*, 2010). O G hybrid, while having a lower oil-yield than the ténera, does have the advantage of resistance to the disease known as "lethal bud rot", which in the past years had decimated plantations along the Atlantic seaboard. This increase in resistance is inherited from the mother strain *oleifera* which is a native of those latitudes. The oil yield of this hybrid varies between 12 and 18 per cent of the fruit weight.

ACHIEVEMENTS IN INDIA

Crosses developed: a no of interspecific crosses developed using available accessions of oleifera with Eo as female and Eg as male parent as well as reciprocals (Table 1) and evaluated with DxP at IIOPR.

Table 1.Details of oil palm interspecific hybrids developed

Sl.No	Cross ID*
1	12Eo x 82Eg
2	19Eo x 81Eg
3	16Eo x 18Eg
4	360Eg x 13Eo
5	361Eg x 11Eo
6	15Eo x 18Eg
7	16Eo x 81Eg
8	DxP control (*E. guineensis*)

*Eo- *Elaeis oleifera*; Eg- *Elaeis guineensis*

Evaluation of two crosses 360*Eg* x 13*Eo* (ISH-1) and 361 *Eg* x 11 *Eo* (ISH-2) at IIOPR-Pedavegi showed that there was significant variation for vegetative growth and yield parameters between as well as within the hybrids.

Of the growth parameters studied, significant variation was observed with respect to height of palm, annual height increment, leaf dry weight, leaf area, specific leaf weight, vegetative dry matter and total dry matter. Girth of trunk, number of leaves, bunch dry matter or bunch index did not show any significant difference between hybrids. With respect to bunch characters, mean bunch weight, number of fruits and average fruit weight showed significant variation among the hybrids. Percent of partthenocarpic fruits was slightly higher for interspecific hybrids. The character was more prominent in oleifera and could have been transmitted from that parent. In oleifera and its hybrids the bract surrounding the inflorescence persist to cover the inflorescence to a large extent and this seemed to make them less accessible for airborn pollen and results in more parthenocarpic fruits. The reduced pollen fertility of some interspecific hybrids (Sunilkumar *et al*, 2012) could also be attributed to higher proportion of parthenocarpic fruits. However, there was no significant variation in   case of oil to bunch ratio. The mesocarp to fruit content varied from 43.76 to 77.26 percent for interspecific hybrids. Some of the palms recorded mesocarp content on par with commercial DxP material. Corely and Tinker (2003) reported that interspecific crosses involving Dura (*E. guineensis* parent) had mesocarp to fruit percent of 40 -50 % and crosses involving Tenera/Pisifera, 58-74%.

Six out of seven palm shortlisted based on low stem height viz. 2, 3, 10, 11, 20 and 73 belonged to SH-1 and only one palm viz.43 belonged to ISH-2. Imposing of multiple trait based selection index resulted in were shortlisting of eight palms; five from ISH-1 and three from ISH-2. Of the five promising palms from ISH-1,   four were having annual height increment between 30 and 33 cm. Number of leaves produced by interspecific hybrid is lower compared to *E. guineensis* and this was in conformity with an earlier report by Hartley (1988). Palm number 28 recorded a bunch yield of 148.25 kg, height increment of 33.1 cm and had a bunch index of 0.29 which is on par with normal D x P. Though ranked low, palm numbers 2 and 3 were promising with respect to height increment and had moderate yield.

Palm 28 of ISH-1 showed potential of the maximum oil yield of 4.8 tonnes per ha resulting from the high mesocarp content and oil to bunch ratio. This was followed by 3.8 tonnes per ha in palm 25, also from ISH-1. Among promising palms from ISH-2, palm 34 recorded the highest oil yield of 3.4 tonnes per ha. Hardon (1969) and Meunier (1975) quoted low oil to bunch ratio in interspecific hybrids compared to *E. guineensis* (Tenera) material. The kernel size was the maximum for palm number 34, which in turn would be useful in the development of genotypes for high kernel oil. The shortlisted palms would be used for back crossing with better parent for development of dwarf genotypes. Back crosses are generally reported to have better fruit set and oil/bunch ratio with oil composition close to that of *E. guineensis* (Corley and Tinker, 2003).

Evaluation of pollen viability in inter-specific crosses indicated that significant variability exist among different crosses as well as palms within each cross. In the 360Egx13Eo cross the pollen viability varied from 43.52 to 98.33 % with a mean of 91.15% . In case of 361Eg x 11Eo, reduced pollen viability was noticed with a mean of 66.39%, whereas the mean *in vitro* germination   was 62.75%. This could be one of the reasons for slightly higher percent of parthenocarpic fruits in interspecific hybrids compared to DxP control. This has got practical implication in selecting the parents for back crosses in an effort to develop dwarf/ compact palms as well introgress other traits. As a result, inter-specific hybrids with compatible parents having sufficient pollen viability and fruit set in the progenies could be selected and advancement of the population could be achieved.

More palms were shortlisted in F-1considering the chance of segregation in subsequent generation and these palms would be used for back crossing with recurrent parent for development of dwarf varieties with better oil yield.

REFERENCES

Chavelier, A.1943. Taxonomie biogeography et selection dis palmiers due genere Elaeis. *Rev. Bot.etd' Agric. trops*, 23:295.

Corely, R.H.V. and Tinker, P.B. 2003. *The oil palm*. Fourth edn, Blackwell Publishing company, 541 p.

Corley, R.V.H.; Tinker, P.B. 2003. The Oil Palm. 4. ed. Blackwell Science, Oxford, UK.

Federación Nacional de Cultivadores de Palma de Aceite [Fedepalma]. 2010. Statistical Yearbook 2010. Fedepalma, Bogotá, Colombia.

Hardon, J.J. and Tan, G.Y.1969. Interspecific hybrids in the genus *Elaeis*. I. Crossability, cytogenetics and fertility of F_1 hybrids *Elaeis guineensis* x *E. oleifera* . *Euphytica*,**18**: 372-379.

Hardon, J.J. 1969. Interspecific hybrids in the genus *Elaeis*. II. Vegetative growth and yield of F_1 hybrids *Elaeis guineensis* x *E. oleifera*. *Euphytica*,**18**: 380-388.

Hartley , C.W.S.,1988. *The Oil Palm*, 3rd Edn, Longman, London. 806p.

Hayati, A., Wickneswari, R., Maizura, I. and Rajanaidu, N., 2003. Genetic diversity of oil palm (*Elaeis guineensis* Jacq.) germplasm collections from Africa: implications for improvement and conservation of genetic resources. *Theoretical and Applied Research* 108: 1274-1284.

Mathur, R.K.; Murugesan, P.; and Pillai, R.S.N. 2005. Oil Palm genetic resources-introduction, utilization and future needs. *Indian J. Plant Genet. Resour.*18 (1): 112-113.

Meunier, J., Vallejo, G. and Boutin,D. 1976. l'hybride *E. melanococca* x *E. guineensis* et son amelioration. Un nouvel avenir pour le palmier a huile. *Oleagineux* **31**:519-528.

Rajanaidu, N., 1986. Variation in the natural population of oil palm (*Elaeis guineensis* Jacq.), Ph.D. Thesis, University of Birmingham, UK.

Sterling, F., Richardson, D.L., and Chaves, C. 1987. Some phenotypic characteristics of descendants of QB049 , an exceptional hybrid of oil palm. *Proc. oil palm- palm oil conference: Progress and prospects*. PORIM, pp134-146

Sunilkumar, K. Mathur, R.K. and Sparjan Babu, D.S. and Reddy , A.G.K. 2012. Pollen viability and *in vitro* germination in *E. guineensis* x *E. oleifera* interspecific hybrids of oil palm. In: *Global conference on Horticulture for food, nutrition and livelihood options*, (Eds) Rajan, S., Garg N, Singh, B., Bajpai, A. And Kumar, M., OUAT and ASM Fundation, May 28-31, 2012, Bhubaneshwar, Odisha, India, p-122.

Torres, G.A.; Sarria, G.A.; Varon, F.; Coffey, M.D.; Elliot, M.; Martinez, G. 2010. First report of bud rot caused by *Phytophthora palmivora on* African oil palm in Colombia. Plant Disease 94: 1163.

Torres, G.A.; Sarria, G.A.; Varon. 2004. First Report of Bud Rot Caused by *Phytophthora palmivora* on African Oil Palm in Colombia. *Plant Disease*. 94(9): 1,163.1.

Uhl, N.W. and Dransfield, J. 1987. Genera Palmarum: A classification of palms based on the work of Harold E. Moore, Jr. Lawrence, Kansas, Allan Press.

Zeven, A.C. 1964. On the origin of the oil palm. *Grana Palynol.*, 5, 50.

Distant Hybridization in Horticultural Crops   *Pages* **111–162**
Editors: **M.R. Dinesh & M. Sankaran**
Published by: **ASTRAL INTERNATIONAL PVT. LTD., NEW DELHI**

10

Wild Species Utilization in Vegetable Crop Improvement

Dutta O. P., Sadashiva A. T., Madhavi Reddy K. and Pitchaimuthu, M.

Division of Vegetable Crops, ICAR-IIHR, Hesaraghatta-560 089

INTRODUCTION

Use of wild relatives of vegetables in breeding is likely to intensify as crop adaptation to climate change becomes more pressing .Presently there is a wide gap between the existing mosaic of genes in wild relatives and their utilization in countering biotic and abiotic stress parameters. The advances made in embryo rescue, marker assisted selections and mapping genes conditioning biotic and abiotic stresses will further assist a rapid gene introgression into the cultivated species. Another approach to counter the biotic and abiotic stress is through grafting using wild relatives of vegetables as root stock. Now a days, grafting is regarded as a rapid alternative tool to the relatively slow breeding methodology, aimed at increasing environmental stress tolerance of fruit vegetables. It provides an excellent resistance / tolerance to soil borne diseases such as fusarium wilt, bacterial wilt, verticillum wilt and root knot nematodes in crops like tomato, eggplant, watermelon, muskmelon and cucumber. It also provides tolerance to abiotic stresses such as thermal stress, hydric stress and organic pollutants. Commercial grafting is done in watermelon, muskmelon, cucumber, eggplant and tomato. In Japan and Korea 97% of eggplant, cucumber and watermelon are grafted, in Spain 85 % of watermelon are grafted. In Turkey 90 million grafted seedlings of watermelon,eggplant and tomato are being cultivated. In China , more than 300 grafted seedling nurseries are operative and in Mexico, more than 1250 acres of area is under grafted tomato.(Lee., 1994 , 2003).

Wild relatives of Tomato : Wild tomato species are important sources of genes that could improve adaptation to biotic and abiotic stresses. Table 1 gives an over view of diseases and pest resistance genes present in wild releatives of tomato.

Table 1. Tomato diseases and wild tomato species as sources for resistance genes

	Diseases	Pathogen	Resistance sources	Resistance gene	Chromosomal location	Reference
Soil borne disease	Alterneria Stem canker	*Alterneria Spp.*	*S.pennellii*	Asc	3	Scott and Gardner (2007)
	Fusarium wilt	*Fusarium oxysporum f.sp.lycopersici*	*S.pimpinellifolium* *S.pennellii*	1 1-2 1-3	7,11	Scott & Gardner 2007,Grattiige & O Brien (1982) Scott & jones (1990),Huang &Lindhout (1997)
	Fusarium crown and root rot	*Fusarium oxysporum &sp.radicis-lycopersici*	*S.peruvianum* *S.pennelli* *S.lycopersicum var.Cerasiforme*	Frl	9	Scott and Jones(1990)
	Verticillium wilt	*Verticillium dahliae*	Peru wild "VEDA"(Eriopersicon)	Ve	9	Diwan *et al.*, (1999), Stamova(2005)
Foliar diseases	Cladosporium leaf mold	*Cladosporium fulvum*	*S. pmpinellifolium* *(22 genes)* *S.habrochaites* *S.peruvianum*	Cf(26 genes) 1 1	- - -	Scott and Gardner (2007) Foolad *et al.*, (2000)
	Early blight	*Alterneria solani*	*S.habrochaites* *S. pmpinellifolium*	- -	- -	
	Grey leaf spot diseases	*Stemphyllium spp.*	*S.pimpinellifolium*	Sm	11	Scott and Gardner (2007); Parlevliet(2002)
	Gray mold	*Botrytis cinerea*	*S.peruvianum* *S.habrochaites* *S.pimpinellifolium*	R b e q 1,2,4	1,2,4	Egashira *et al.*, (2000);Nicot *et.al.*,(2002),Finkers *et.al .*,(2007)

Contd.

Diseases	Pathogen	Resistance sources	Resistance gene	Chromosomal location	Reference
Late blight	*Phytophthora infestans*	*S.pimpinellifolium*	Ph-1	7	Merk *et al.,* (2012);
		S.habrochaites	Ph-2	10	Foolad *et al.,* (2008)
		S.pennelli	Ph-3	9	
		S.habrochaites	S e v e r a l QTLs	All12	Brower *et al.,* (2004)
		S.pennelli			Li *et al.,* (2011)
		S.pimpinellifolium	2QTIs	1,10	Merk *et al.,* (2012)
Powdery mildew	Odium spp	*S. habrochites*	01-1 -01-3	6	Huang *et al.,* (2000)
		S.esculentum var. cerasiforme	01-2	4	Huang *et al.,* (2000)
			3QTLs		
		S. parviflorum		6,12	Bai *et al.,* (2003)
		S. peruvianum	01-04	6	Bai (2004)
		S.habrochites	01-05	6	Bai (2004)
Phytophthora root and crown rot	*Phytophthora parasitica*	*S. lycoperssicum var. cerasiforme*	QTLs		Blaker and Hewitt (1987)
	P.capsici				Scott and Jones (1990)
Southern blight	*Slcerotium rolfsi*	*S.piminellifolium*	QTL		Scott , (1993)
Bacterial Canker	*Clavibacter michigonesis*	*S.peruvianum*	3 QTIs	5,7,5	Sandbrink *et al.,* (1995)
	Sub sp. michigonesis				Van Heusden *et al.,* (1990)
Bacterial wilt	*Rolstonia solanaerum*	*S.piminellifolium*	BWr-6		Hanson *et al.,* (1998), Wong *et al.,* (2013 a)
		S. lycoperssicum var. cerasiforme	Bwr-12	6, 12	

Contd...

Diseases	Pathogen	Resistance sources	Resistance gene	Chromosomal location	Reference
Bacterial speck	Pseudomonas syringae pv.tomato	*S.pimpinellifolium*	PTO	-	Pitblado &Kerr (1980)
Bacterial spot	Xanthomonas species	*S.pimpinellifolium* *S.pennelli*	Rx1-3	-	Yu *et al.,*(1995);yang & Fransis (2007)
Viral pathogens Tomato yellow leaf curl	Various Begomo virus species	*S.pimpinellifolium*	QTL	-	Perez de Castro *et.al.,* (2007) Scott(2007)
		S.peruvianum	5 recessive genes	-	Pilowsky &Cohen (1990)
		S.Chilense	Ty-1 Ty-3 Ty-4	6,3	Zamir *et.al.,* (1994),Ji *et.al.,* (2008)
		S. habrochaites	Ty-2	11	Vidavsky & Czosnek (1998), Hanson *et.al.,* (2006)
		S.peruvianum	Ty-5	4	Anbinder *et al.,* (2009)
		S.Chilense	TyLCV1 TyLCV2 TyLCV Ty6-1	6,10	Agrama & Scott(2006) Kadirvel *etal.,* (2013)
		S. cheemaniae	Ty10-1		Hassan *et al.,* (1984)
Tomato Spotted wilt Virus	Tospo virus species	*S.Cheesmaniae*	Sw-5	-	Hassan *etal.,*(1984)
		S.peruvianum	Sw-6	-	Stevens *etal.,*(1995) Rosello *etal,*(1998) Gordillo *etal.,*(2008) Stevens *et al.,* (2007)

Cont'l

Diseases	Pathogen	Resistance sources	Resistance gene	Chromosomal location	Reference
		S. chilense	Sw-7	-	
		S.peruvianum			
		S.pimpinellifolium			
		S. habrochaites	-	-	Dianese *et al .,* (2011)
		S.Corneliomulleri			
		S. arcanum	-		
				-	
	Bemisia argentibolii	*S.pennellii*	4 QTLs	2,3,6,8,9,10,11	Heinz and Zalon (1995),
		S.peruvianum			Snyder *et al.,*(1998),Nombela *et al.,* (2000),Kennedy (2007), Momotaz *et al.,* (2010)
		S.habrochaita			
	Trialeurodes vaporarioum	*S.habrochaita*	2QTLs (reduced	1,12	Maliepaard *et al.,* (1995)
		S.pennellii	ovi position); 3QTLs (trichomes)	1,5,9	Berlinger *et al,*(1991), Kennedy (2007)
Insect Pests mites and nematodes				-	
Potato aphid	*Macrosiphum euphorbiae*	*S.pennellii*	-	-	Kennedy (2007)
	M.persicae				

Contd...

Diseases	Pathogen	Resistance sources	Resistance gene	Chromosomal location	Reference
Root Knot nematode W	*Meloidogyne incognit*	*S.peruvianum*	Mi	6	Kennedy (2007)
	M.javanica				Smith (1944)
	M.arenaria	*S.peruvianum*	*Mi-2* to *Mi-9*	-	Williamson (19940
					Ammiraju *et al* ., (2003)
					Pereira –Carvalho *et al*.,2010 ,Wang *et al.*, (2013a, 2013b)
White files	*Bemisia tabacci*	*S.pennellii*	4QTLS	2,3,6,8,9,10,11	Heinz and Zalom (1995)
		S.peruvianum			Snyder *et al.*, (1998)
		S.habrochaites			Nombela *et al.*, (2000)
					Kenneydy *et al.*, (2007)
					Memotaz *et al.*, (2010)
Flower Thrips	*Frankliniella occidentalis*	*S.habrochaites*	-	-	Kumar *et al.*, (1995a)
		S.pennellii			
		S.chilense			
Spider mites	*Tetranychus urtiae*	*S. lycopersicum var. cerasiforme*			Fernandez-MunoZ *et al.*, (2000)
	T.cinnabarinus	*S.habrochaites*			Kennady (2007)
		S.pimpinellifolium			

Sources : Andreas Roland (2015)

Among the foliar diseases of tomato, late blight caused by *phytopthora infestans* is most destructive worldwide. Genetic resistance has been reported in *S.pimpinellifolium ,S.habrochaites,* and *S.Pennellii. In S.pimpinellifolium* accessions,three late blight resistance genes (ph-1,ph-2 and ph-3)have been identified (Merk *et al.,*2012). Ph-1 confers resistance to tomato race of *Phytophthora infestans,* where as, Ph-2 provides partial resistance to some isolates of *P.infestans* whereas, Ph-3 provides strong resistance to late blight .Presence of sexual reproduction of *P.infestans* leads to the formation of new aggressive *P.infestans* isolates, thus overcoming the resistance of Ph-3 (Foolad *et al.,*2008.). Apart from this quantitative resistance of late blight has been observed in *S.habrochaites* accessions LA 2099 and LA 1777 (Li *et al.,*2011). Late blight resistance QTLs were detected on all 12 chromosomes (Brower *et al.,* 2004). Severe linkage drag and the large number of QTLs involved in *S.habrochaites* and *S.pimpinellifolium* hinders the breeder to use them in their breeding programme. However, one *S.pimpinellifolium* accession PI 270443 available in *S.pimpinellifolium* core collection of AVRDC exhibited strong resistance to seven *P.infestans* isolates and was selected for inheritance studies (Merk *et al.,*2012). Another *pimpinellifolium* accession PI 270443 having resistance to late blight is controlled by two genes and is highly heritable and is being used in tomato breeding programme (Merk *et al.,* 2012).

Bacterial wilt caused by *Rolstonia solanacearum* is another serious threat for tomato production under tropical and subtropical conditions. PI 127805A and PI 129080 both *S.pimpinellifolium* accession and CRA66 a *S.lycopersicum var cerasiforme* accession contributed largely in the development of tomato cultivars resistant to bacterial wilt (Hanson *et al.,*1998).In addition of this several QTLs located on chromosome 6 played a dominant role in conferring bacterial wilt resistance Wang *et al.,* 2013a confirmed the polygenic nature of resistance to bacterial wilt in tomato and identified two major QTLs(BWr-12 and BWr-6) conferring stable bacterial wilt resistance in the tomato cultivar Hawaii7996,most likely derived from PI 127805, a *solanum pimpinellifolium* accession (Hanson *et al.,*1998).

Presence of large variation of the pathogen,lack of consistency of high resistance level over locations and the linkage of resistance QTLs to small fruit size hinders the substantial progress in resistance breeding to bacterial wilt(Yang and Francis,2007)

Among the viral diseases of tomato, The tomato yellow leaf curl viruses (TYLCV) transmitted by white fly are the most wide spread and currently rank 3rd among the economically and scientifically most important viruses worldwide (Scholthof,KBG *et al.,* 2011). Wild tomato relatives such as *Solanum pimpinellifolium, S.peruvianum ,S.chilense ,S habrochaites* and *S.cheesmaniae* exhibit high degree of disease resistance (Pico *et al.,* 1996,Ji *et al.,* 2007b ,Scott, 2007). A total of six TYLCV genes have been identified which are named as Ty-1,Ty-2,Ty-3,Ty-4,Ty-5 and Ty-6 (Hanson *et al.,* 2000,2006, Ji *et al* .,2007a,2009b, Hutton *et al.,*2012). Sources of TYLCV genes are summarised in table -2

Table:2 Sources TYLCV Resistance (Ty genes) in Tomato

Sources	Species	Gene/alleles	Chrosomes	Reference
LA1069	*Solanum chilense*	Ty-1	6	Zimir *et al.*,1994
B6013	*S. habrochiates*	Ty-2	11	Hanson *et al.*, 2000
LA1932 LA 2779	*S. chilense*	Ty-3, Ty-3a, Ty-3b	6	Ji *et al.*, 2007a
LA 1932	*S. chilense*	Ty-4	3	Ji *et al.*, 2009
PI 126926				Anbinder *et al.*, 2009
PI 126930 PI 390681	*S. peruvianum*	Ty-5	4	
Ty king	*S. lycopersicum*	ty-5	4	Hutton *et al.*,2012

Hanson *et al* ., 2013

There are 14 begomoviruses causing leaf curl diseases in tomato in India. Out of these 14 *begomoviruses*, the tomato leaf curl New Delhi virus (ToLCNDV) and tomato leaf curl Palampur virus (ToLPalV) have bipartite genome, DNA A and DNA B, whereas, tomato leaf curl Gujarat virus (ToLCGuV) exists as monopartite in some states and as bipartite in Uttar Pradesh .The other eleven viruses are monopartite. At present, the monopartite *begomoviruses* like tomato leaf curl Karnataka virus (ToLCKaV) is present throughout India. Similarly the bipartite ToLCNDV is not restricted to northern region but is very much present in southern India in cucurbitaceous hosts. Mixed infection is common. Spread of ToLCNDV and ToLCKaV is all over the country (Malathi , V.G. 2013). Among the major Ty genes available to the breeder, Ty-1 and Ty-3 are allelic (Verlan MG *et al.*, 2013). Ty-1 and Ty-2 are both dominant and provide high level of resistance whereas, other genes provide partial resistance and are additive to partially dominant or recessive as with ty-5. Two or Three gene combination of Ty-3 Ty-4 and Ty-5 in semi isogenic background provided high level of resistance to Tomato yellow leaf curl virus(TYLCV).Tomato mottle virus (ToMoV) trait in Florida using the same material was severe but indicated that Ty-3,and Ty-4 had small but significant effects in lowering ToMoV, Ty-5 had no significant effects on resistance. Combining Ty-6 and Ty-5 homozygously provided a high level of resistant to TYLCV, while the combination of homozygous ty-5 with heterozygous Ty-6 was also good but not as resistance in the former (Hutton *et al.*, 2015) . Further it was observed that Ty-3 is the strongest candidate gene against Tomato mottle virus (ToMoV) . Pyramiding Ty-3,Ty-4 and Ty-5 causes a significant reduction of disease symptoms. There is a significant yield reduction of Ty-6 and Ty-5 introgression in tomato (Shekasteband *et al.*, 2014). In almost all TYLCV resistant material viral replication occurs (Naresegowda *et al.*,2003). This also holds true for Ty -1/Ty-3, where the donors was *S.chilense* as well as in commercial line with Ty-1 introgression (3761 AB seed) The TYLC was replicating and was detectable, (Peroz de castra *et al.*, 1971) although the level did not exceed more than 10 % of that in susceptible tomato cultivar (Fargette *et al.*,1996) and Prasanna *et al.*, (2014) were successful in pyramiding Ty-2 and Ty-3 genes for resistance to monopartite and bipartite tomato leaf curl virus of India at IIVR Varanasi and could develop five stable breeding lines with good horticultural qualities and yield.

At IIHR, Bengaluru, Sadashiva (2014) able to introgress Ty-2 gene from *S.haborochaites* into bacterial wilt and early blight resistance lines through marker assisted selection and was successful in developing two tomato hybrids Arka Rakshak and Arka Samrat with triple diseases resistance to TyLCKav, bacterial wilt and early blight. In Cuba, Olimpia Gomez consuegra *et al.*, (2015) were successful in pyramiding Ty-1 gene from *Soloanum chilense* (LA 1969) and Ty-3 gene from *S.chilense* (LA 2779) and spotted wilt virus resistance gene Sw-5 from *S.peruvianum* with the help of early and simultaneous molecular screening for resistance and productivity evaluation under field conditions.

Among the wild relatives of tomato, *S .peruvianum* is considered as the most variable . Many genes have been introgressed into tomato. Such as Sw-5 gene that confers resistance to tomato spotted wilt virus (*TSWV*) (stevens *et al.*, 1992). *Pyl* gene confering resistance to the fungus *Pyrenochaeta lycopersici*,(laterrot 1978). Tm-2 (Laterrot and Pecant , 1969) and Tm-2 gene (Hall 1980) confering resistance to tomato mosaic virus (ToMoV). The mi gene is associated to resistance to the main species of nematodes of the genes Meloidogyne (Gilbert, 1958) . Ty-5 gene confers resistance to the tomato Yellow leaf curl virus (Anbinder *et al.*, 2009). The accession *S. peruvianum* (PI-126944) in particular has been described as resistant to TWSV (Paterrson *et al.*, 1989) , Tobacco mosaic virus (TMV)(Yamakawa and Nagata 1975). Tomato leaf curl virus (ToLCV) (Muniappa *et al.*, 1991) and to *Fusarium oxysporum* f.sp. radicis lycopersici (Rowe and Farley 1981). In order to better exploit the potential breeding for resistance of PI 126 944, Jubon, *et al.*, (2013) were able to constract a set of introgression lines with combined resistance to TLCV and Tswv in the genetic back ground of tomato.

Thermal, Hydral and Salt Stress Tolerance in Tomato_: Stress tolerance is a stage specific phenomen . Tolerance at one stage of plant development is not necessarily correlated with tolerance to at other stages (Foolad and Lin 1997,Foolad , 2007).Screening for stress tolerance should be done at specific ontogenetic stages such as seed germination seeding survival ,initial plant growth, vegetative growth and reproduction. *Solanum chilense* offered good prospects for raising level of heat tolerance in tomato (de la pene *et al.*,2011) . Sources of salt tolerance have been reported in several wild *solanum species* such as *S. pimpinellifolium, S.peruvianum, S.chilense,S.cheesmaniae ,S habrochites ,S.Chmielewskii,S.esculentum var. cerasiforme,S pimpinellifolium* and *S.pennellii* (Tal *et al.*, 1979,Asins *et al.*, 1993, Foolad and lin 1997,Foolad,2004, Robertson and Labate,2007,Rao *et.al* ., 2013). *Solanum Pimpinellifolium* having good capacity to form lateral roots, readily cross compatible with *S.lycopersium* and with beneficial salt tolerance traits is being utilized for salt tolerance in cultivated tomato (Foolad 2004).

Drought tolerance has been reported in *S.cheesamaniae, S. chilense, S.pennellii,S. pimpinellifolium* and *S.lycoprsium ver.cerasiforme* (Foolad 2007) . Drought tolerance at the seed germination stage is a quantitive trait and four QTLs have been identified on chromosomes 1,8,9 and 12. Potential sources of drought tolerance during vegetative growth and reproduction has been identified in *S.Chilenese* and *S.pennelli* (Rick 1982). *S.Chilenese* is able to grow in the desert because of its long primary roots and extensive secondary root system.

Flood tolerance have been reported in *S. esulentum var. cerasiforme,S. juglandifolium* (Rick 1982) and *S.ochronthum* (Robertson and Labate,2007). Tolerance against chilling injury has been reported in *S.habrachaites, S.chilense* and *S.lycopersicoides* (Robertson and Labate,2007).

Wild Relatives of Capsicum: Among the 31 Capsicum species, five are domesticated species which includes

Botanical name	:	Common Name
Capsicum annuum L. Var. *annuum*	:	Chilli pepper
		Bell pepper
		Paprika
Capsicum baccatum L. Var. pendulum (wild) Eshaugh	:	Peruvian pepper
Capsicum chinense Jacq	:	Bonnet pepper
		Hobanero pepper
C. frutescens L.	:	Tabasco pepper
		Bird pepper
C. pubescens Ruiz & Pavon	:	Rocoto
		Apple chilli

Source: Heiser and Pickessgill (1969)

Other species with economic importance are

Scientific name	:	Commen name
Capsicum annuum L. var. *globriusculum* (Dunal)	:	American Bird Pepper.,
C. baccatum L. var. *baccatum*	:	Locoto
C. cardenessi Heises and P.G. Sm	:	Utupica
C. eximium (Hunz)	:	Ulapuca

These four species as well as *C. baccatum* L. var. *praetermissum* (Heises and P.G. Sm)

C. chacoense (Hunz) and *C. tovarii* (Estibaugh *et al.*,) are considered semi domesticated

(Ribeiro et al 2008) Remaining 20 species are considered as wild species

In Capsicum there are three complexes which include

1. *C. annuum*   (*C. chacoense, C. chinense, C. frutesense and C. galapagoense*)

2. *C.baccatum*   (*C. practermissuman, C. tovarii*)

3. *C. pubesecens*   (*C. eximium* & *C. cardanasii*)

Djian – caporalino *et al.*, 2007

Most interspecific crosses with in the same complex produce partially fertile hybrids without embryo rescue. Cross ability between two species is not always reciprocally successful (Djian – caporalino *et al.*, 2007)

Successful interspefic crosses within the secondary pool have been summarized in Fig.1   by Djian – caporalino *et al* ., 2007

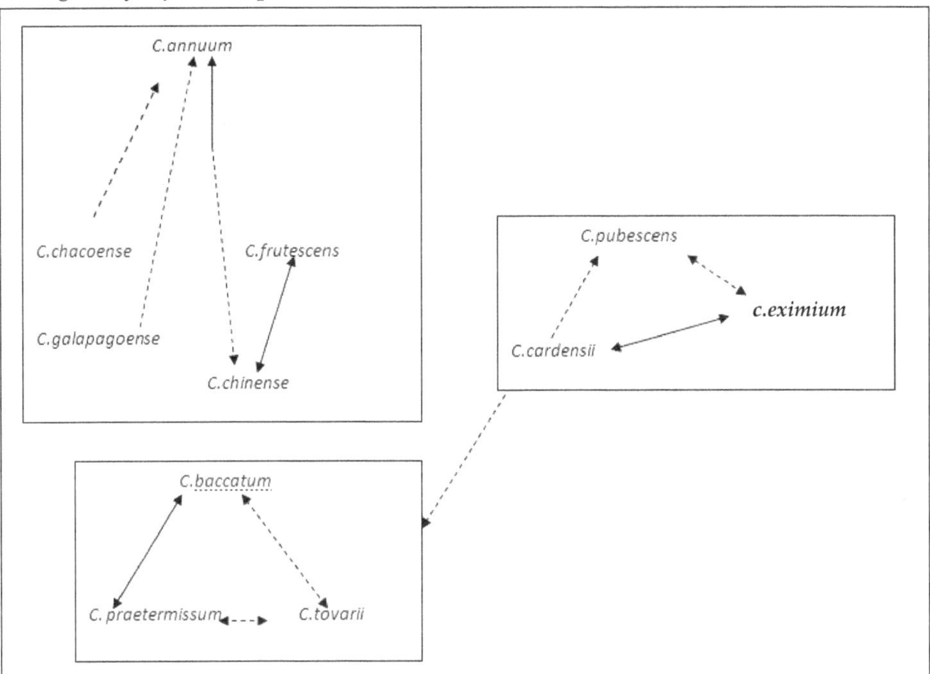

Fig.1. Gene flow between & within gene pools in the Capsicum genes(Arrow points female parent, solid lines indicate relatively successful fertile hybrids, broken lines indicate very few hybrids with low fertility. Adapted from Djian-caporalino *et al.*, (2007)).

When crossing *C. annuum* with *C. chacoense* or *C. galapagoense*, or crossing *C. pubesences* with *C. cardenasii*, it is recommended that *C. annuum* and *C. pubescens* should be the female parents to gain more success.

By using *C. annuum* as female parent to hybridize with *C. baccatum*, embryo rescue technique was adopted to overcome embryo abortion (post fertilization barrier). However, in cases where post fertilization embryo abortion occurred before globular stage, tri species bridge crosses were successful. *C. chinense* was used in the bridge parent to accomplish the hybridization between *C. annuum* & *C. baccatum* (Yoon and park, 2005).

WILD RELATIVES OF CAPSICUM

Disease Resistance

a) Anthracnose

Pepper anthracnose is caused by several *Colletotrichum* spp. such as *C.acutatum, c.gloeosporioides, C.capsici* and *C.coccodes* (Park & kim 1992). *C.acutatum* & *C.gloeosporoides* are the most destructive and are widely distributed (Sarath Babu *et al.*, 2011 & Voorrips *et al.*,2004). The pathogen attacks pepper fruit at both the green and the red fruit stage and can cause lesion on Pepper leaves and stem. The main sources of resistance to anthracnose have been identified in *Capsicum baccatum* and *C. chilense* at AVRDC in 1990, which have been used for studying the inheritance of anthracnose resistance by various researches (Kim *et al.*,2010, Lee *et al.*, 2010, Lin *et al.*, 2002, Pakeleevaraporn *et al.*,2005 and Voorrips *et al* 2004). Genetic analysis of segregating populations showed that the resistance inheritance patten varied depending on the *Colletotrichum* species and isolates, the resistance source and also the fruit maturation stage.The genetics of anthracnose resistance is *Colletotrichun* spp. is summarized in table-3.

Table-3: Genetics of anthracnose resistance in *Colletotrichum* spp.

Pathogens	Capsicum spp	Resistance	Fruit maturation stage	References
Colletotrichum acutatum	*Capsicum chinense* (PBC 932)	2 complemenry dominant genes	Green fruit	Lin *et al.*, (2007)
		2 recessive genes	Red fruit	
	C. chinense (AR)	Single recessive gene	Green fruit	Kim *et al.*, (2008)
	C. baccatum (P159413)	Monogenic	Green fruit	Kim et al., (2008 a)
	C. baccatum (PBC 80)	Monogenic Recessive	Green fruit	Mahasuk *et al.*, (2009 b)
		Monogenic dominant	Red fruit	
Colletotrichum capsici	*Capsicum chinense* (PBC-232)	Single recessive gene	Green fruit	Kim *et al.*, (2008b)
				Mahasuk *et al.*, (2009
			Red fruit	Pakdleevaraporn *et al.*,2005
		Single recessive with epistatic effect		Cheema *et al.*, (1984)
Colletotrichum capsici	*C. annuum* 83-168 Chung Young	Dominant recessive		Park *et al* ., (1990 a) , (1990 b) Lin *et al* 2002)

Genetic mapping and quantitative trait loci (QTL) analysis further specified the quantitative & polygenetic or oligo genic nature of resistance in *C. chienensis* line PBC932 and *C. baccatum* line PBC-80.(Kim *et al.*, 2010, Lee *et al.*, 2010and Veerips *et al.*,2004) Resistance to anthracnose *(Colletotrichum acutatum)* of Capsicum is controlled by one major dominant QTL on chromosome P_5 at red mature fruit stage, where as, in green fruit stage the resistance is dependent on major P_5 QTL plus minor QTLS (Chunying sun *et al.*, 2015). Different resistant genes are expressed at different stages of fruit maturity was also observed by Taylor *et al.*, (2007). A similar result was observed by Mahasuk *et al.*, (2009) who studied the inheresistance of resistance to *C.capsici* at different stages of fruit ripening (unripe & ripe) from crosses between *C.annuum* (Susceptible) X *C.chinese* (Resistant). They verified the involvement of tow different genes responsible for resistance in ripe and unripe fruit, suggesting that the change in fruit maturation may have triggered the expression of different genes at different stages.

b) *Phytophthora* spp.

Resistance to *Phytophthora capsici* in capsicum has been reviewed by O.Mongkolporn and P.W.J.Taylor (2011). *Phytophthora* isolates collected from blight affected leaf tissues of hot pepper in India have been identified as *P. boehmeriae* and *P.Capsici ,P.boehmeriae* isolates were highly aggressive (Madhur *et.al.*, 2015). *P.Capsici* is a soil borne pathogen and can infect different parts of plant during its growth causing sudden wilt of the plant .Two wild accessions of *C.annuum*, CM 334 and *Ac2258* derived from P120134 possess high level of resistance to *P.capsici* (Ares *et.al* 2005).The resistance in *CM334* is controlled by two genes with dominant and recessive epistasis (Reifchneider et.al., 1992). Resistance in AC 2258 is controlled by two dominant genes without additive effects (Smith et.al.,1967). In fact *P. capsici* causes separate disease syndromes for foliage blight, fruit rot, root rot and stem blight and the capsicum species have evolved independent resistance mechanism to protect against these various disease syndromes (walker and Bosland 1999,Sy & Bosland 2005).Thus genes for resistance to phytobthora root rot do not provide the host with protection against phytopthora foiler blight and stem blight.

c) Powdery mildew in capsicum

Powdery mildew is caused by *Leveillula taurica* in capsicum. Many *C. chinense* and *C. baccatum* types seem to be either immune or poor hosts for the pathogen. Resistance has been described in some *C.annuum* sources (Deshpande *et al.*, 1985). Resistance in *C. chinense* has been transfered into a jalapeno background with extremely low heritability suggesting multiple recessive genes or incomplete penetrance.

d) Pepper viruses

The main virus groups infecting pepper are transmitted by aphids, white fly and thrips. There are 68 different virus species infecting peppers (Pernezny, Robert, Murphy and Goldberg 2003). Out of these 68 species, 20 virus species have been reported to cause severe damage to pepper crops (Moury and verdin 2012). The main viruses infecting pepper in India belongs to *potyvirus, cucumovirus, begomovirus and tospo virus* (Reddy and Reddy, 2010 , Lawrence *et al.*, 2014, Reddy *M.K et al.*, 2014) and are summarised in table-4.

Table-4: Important viruses infecting capsicum spp in India

Virus group	Virus	Vector
Poty virus	Chilli Veinal mottle virus (chivmv)	Aphids
Cucumo virus	Cucumber mosic virus (CMV)	Aphids
Begomo virus	Chilli leaf curl virus(chiLcv)	White fly
	Pepper leaf curl varanasi virus (Pep LCVaV)	White fly
	Tomato leaf curl joydepur virus (ToLCJaV)	White fly
	Tomato leaf curl Karanataka virus (ToLCKv)	White fly
	Tomato leaf curl New Delhi virus (ToLCNDV)	White fly
Tospo virus	Tomato spotted walt virus (TSWV)	Thrips
	Groundnut bud necrosis virus (GBNV)	Thrips
	Peanut bud necrosis virus (PBNV)	Thrips

Many monogenic and polygenic virus resistance have been detected in *C.annuum,C.frutesens* and *C.chinense,* some of them which have been exploited commercially, are summarised in table-5 (Lawrance Kenyon *et al.*,2014).

Wild Relatives of Eggplant-(*Solanum melogena*)

Assessment of crossability of *Solanum melengena* with other *solanum* species has been studied by several workers (Nishio,T.*et al.*, 1984, Attavian,BN *et al.*, 1983, Mohammad Ali *et al.*, 1989,1990).

Table-5 Virus resistance identified in *Capsicum* spp.

Virus group	Gene Q=QTL	Virus	Donor Species	Donor cultivar or accession	Chromos-omal location	Type of gene action	Reference
Begomo virus	2 recessive	Pep GMV PHYVV	*C.chinense*	BG-3821	?	Virus replication & movement	Garcia-Neria & Rivera-Bustomante (2011)
	1 recessive	PepLCVaV	*C.frutescens x C.chinense*	Bhut jolokia	?	-	Rai *et.al.,* (2014)
Cucumo virus	Cmr I	CMV-FNV	*C. annuum*	Bukang	P2	M o v e m e n t epidermal cell to mesophyll	Kang *et.al.*(2010)
	CmV.hb-4.1(2) / CmV.hb 8.2(Q)	CMV-HB	*C. annuum*	BJ0747	P5(LG4)/ P11(LG8)	-	Yao ,Li,wang and ye (2013)
	CmV11.1(Q) / CmV 13.1(Q)	CMV1 Israel	*C. annuum*	Perennial	P11/?	-	Ben- chaim, Grube,lapidot, Jahn and paran 2001
	C m V 4 . 1 Q / CMV6.1(Q)				P4/P6		
Tospovirus	Cac	CaCV	*C.Chinense*	P1 290972	?	-	Persley ,McGroth,Sharman and walker (2010)
	1 recessive	TNRV	*C.annuum*	PY-6423 PY-6424	?	-	Puangmalai,Potapohn, Akarapisam-Cheewachaiwit, & Insuan (2013) & Puangmalai potapohn, Akarapisan and pascha (2013)

Contd...

Virus group	Gene Q=QTL	Virus	Donor Species	Donor cultivar or accession	Chromos-omal location	Type of gene action	Reference
	Tsw	TSWV	*C.chinense*	P1 152225 P1 159236	P10 (Pvr4, Pvr7)	HR-like localization NBS-LRR	Jahn *et.al*,(2000)
Poty Virus	Both Pvr1[1] and Pvr1[2] alleles / prv5	PVY	*C. annuum* *C. annuum*	P4			Jeong *et al* .,(2012) Parrella *et al.*, (2002)
	Pvr 6 In combination with Pvr1[2]	PVMV					Caranta *et al.*, (1996)
	Natural variants of Pvr6	Chi VMV PVMV				-	Rubio *et.al* (2009)
	Single recessive gene Pvr3	PepMov					Murphy *et.al.*, (1998)
	Dominant Gene Pvr4	PVY PepMoV					Janzac *et.al* (2009)
	Dominant Gene Pvr7 tightly linked to Pvr4	Pep Mov					Chaine-Dogimont *et al.*, (1996)

Source: Lawrence Kenyon *et.al* (2014)

Contd...

Table: 6. Sources of resistance in wild relatives of Egg plans

Solanum Species	Resistance	Reference
Solanum incanum	Fruit & shoot borer,Verticillium wilt, Fusarium wilt;Phomopris fruit rot, Bacterial wilt	Datar & Ashtaputra(1988),Yamakawa K(1979)Stravato *et.al.,* (1993), Yamakowa *et.al*(1979), F.Boyaci *et.al* (2010)
Solanum macrocarpon	Spider mite,Fruit and shoot borer Bacterial wilt,fusarium wilt	N.K. Krishna Kumar and Sadashiva(1996)
Solanum forvum	Fusarium wilt,Bacterial wilt Phomopis fruit rot,nematodes	Gousset *et.al* (2005) Hebert (1985) Datar and Ashtaputra(1988)
Solanum auriculatum	Immune to little leaf pathagon	
Solanum Integrifolium	Hypersensitive reaction to little leaf pathogen, Fusarium wilt,Bacterial wilt	Gausset et.al,(2005),F.Boyaci *et al.,* (2010),A.K.Chakrabarti and B.Chaudhari (1975)
Solanum gilo	Hypersensitive reaction to little leal pathogen	A.k.Chakrabarti and B.Chaudhari (1975)
Solanum indicum	Fusarium wilt and Bacterial wilt	Gausset et.al,(2005),Rizza et *al.,*(2002),Stravato *et al* (1993),Yamakawa & Mochizuki (1979)
Solanum sisymbriifolium	Verticillium wilt, Bacterial wilt Root knot nematode, Fusarium wilt	Do
Solanum aethiopicum	Fusarium wilt, Bacterial wilt	G. Ano *et al.,* (1991), Laura Toppine *et al.,* (2008)

Potential sources of resistance to various soil borne diseases, root and shoot borer, spidermites,phomopsis fruit rot and little leaf pathogen have been identified in various *solanum* species and are compailed in the table-6.

Compatibility in *solanum* species

Nishio et.al (1984) classified *solenum* species into 3 groups to study the compatibility in it,

Group A: Comprises-	*Solanum melongena*
	Solanum incanum
	Solanum macrocarpon
Group B: Comprises-	*Solanum integrifolium*
	Solanum gilo
	Solanum nodifl orum
Group C: Comprises-	*Solanum indicum*
	Solanum mammosum
	Solanum torvum
	Solanum sisymbriifolium

Crosses were compatible within and between group A and B but were otherwise incompatible. *Solanum indicum* when used as a male or female parent provided mostly unfilled seeds in eggplant however, it provided 50% filled seeds when crossed with cultivar Uttara. This will help in transferring *fusarium* wilt resistance of *solanum indicum* into *S.melongena* (Rajshaker 1960).

Successful hybrids were developed in *S.melongena* and *S.torvum* through embryo rescue technique (Kumchai et.al,2013) *Solanum Integrifolium* and *S.gilo* showed resistance due to hypersensitive reaction to little leaf pathogen(A.K.Chakrabarti and B.chaudhari (1975) *Solnum aethiopicum* and S.macrocarpon are cultivated for their large glabrous leaves which are consumed as a vegetable . *Solanum incanum* is more closely related to *Solanum melongena* and can be used as a good source of resistance to fruit and shoot borer, fusarium wilt, bacterial wilt and phomopses fruit rot. *Solanum incanum* is also a rich source of phenolic acid (chlorogenic acid) a potent antioxidant. Phenolic acid content is controlled by numerous genes. Breeder can select plants that produce marketable fruit with enhanced antioxidant content (Prohen *et.al* 2013).

Wild Relatives of Cucumis species

In the genus *Cucumis*, the melons and most other species with basic chromosome number n = 12 are referred to as the African group. *Cucumis sativus* and *Cucumis sativus* var.hardwickii with basic chromosome number n = 7 are referred to as Asian group. *Cucumis hystrix* is the only cucumis species with 2n = 24 and is native to Asia. It bears a morphological resemblance and biochemical affinity to *C. sativus* while its chromosome number is the same as that of *C.melo* (Chen *et al.*, 1997a). In general, crosses among wild species are frequently possible but all attempts to cross any of these wild species with the two cultivated species : *C. sativus* and *C. melo* failed (Deakin *et al.*, 1971). Cross compatibility among the wild relatives as well as with cultivated species have been summarized in **Fig.2.** ,

Table 7. Wide – cross attempted between cultivated and wild cucumis species

Cross	Result	Source
C. sagitatus × *C. melo*	Embryos only	Deakin *et al.*, 1971
C. metuliferus × *C. melo*	Embryos only	Fassuliotis, 1977
C. sativus × *C. melo*	Globular stage Embryos only	Niamirowiez-Szezytt and Kuhicki, 1979
C. metulliferus × *C. melo*	Fertile F_1	Norton & Granberry, 1980
C. propheterum × *C. melo*	Fruit with non viable seeds	Singh and Yadav, 1984b
C. zeheri × *C. sativus*	Fruit with nonviable seeds	Custers and Nijs 1990
C. sativus × *C. metulliferus*	Embryos only	Franken *et al.*, 1988
C. melo × *C. metulliferus*	Embryos only	Soric *et al.*, 1990
C. sativus × *C. hystrix*	Sterile plants (2n+4n)	Chen *et al.*, 1998

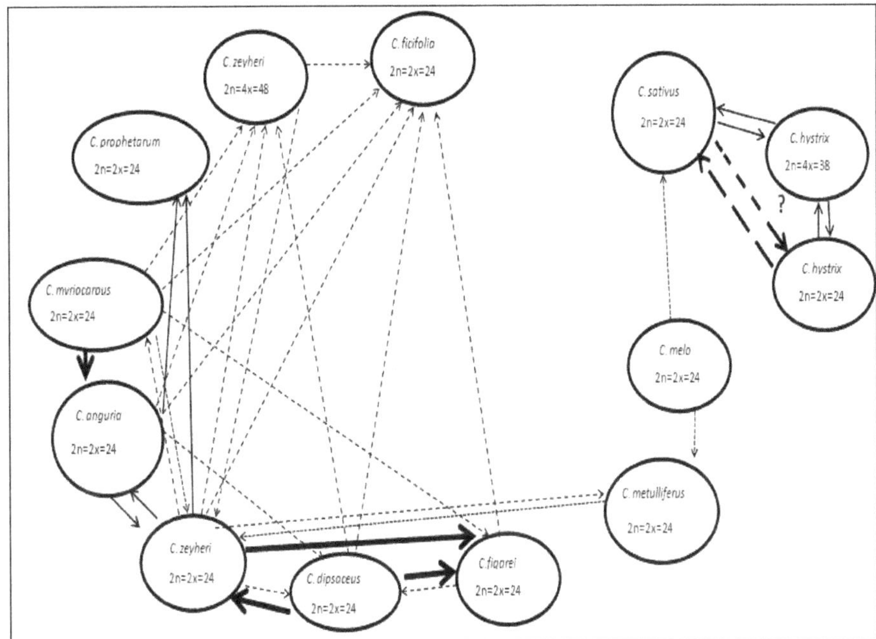

Fig. 2: Polygon of crossability in cucumber species (modified by Sehen et al., (2000) from Nijs & Vissar, 1985). Arrow points to the female parent. 1) Moderately to strong self-fertile and cross fertile hybrids (Thick solid lines). 2) Sparingly self-fertile and moderately cross fertile hybrids (thin solid lines). 3) Self-fertile, usually not cross fertile hybrids (thick dashed lines). 4) Self-fertile and cross fertile hybrids (long dashed lines). 5) Obsense of a line indicates that seeded fruits were not obtained. 6) Question mark means that the information needs to be confirmed.

Some of the successful crosses between cultivated and wild releatives have been reported by various authors and are summarized in table-7 However, in practice , most of these results were not repeatable and did not result in fertile hybrids.

Conspecificity in Cucumis species

Conspecific means belonging to the same species. List of conspecific species which are cross compatible with each other and produce viable F_1 hybrids are summarised in **Table 8.**

Table 8. Cross Compatible conspecific species

Conspecific spp.	Reference
Cucumis anguria and *Cucumis longipes*	Pearl – Treves et al., 1985
Cucumis dinteri and *Cucumis sagittatus*	Dane **F.** *et al.,* 1980, Deakin **J. R.** *et al.,* 1971, Kroon *et al.,* 1979, Kho *et al.,* 1980
Cucumis sativus and *Cucumis hardwickii*	Pearl – Traves *et al.,* 1985, Dane *et al.,* 1980, Deakin *et al.,* 1971
Cucumis myriocarpus and *Cucumis leptodermis*	Deakin *et al.,* 1971, Pearl – Traves *et al.,* 1985,

The incompatibility between cultivated cucumis species and their wild relatives is mainly characterized by 1) delayed growth of pollen tube 2)arrested pollen tube growth on the stigma 3) inability of the pollen tube to reach the ovule (Kishi and Fukishite, 1969) 4) lack of cell division of the zygote and abortion of endosperm (Kishi and Fukishite, 1970). Approaches to overcome compatibility barriers include 1) growth regulator (Behera and Cohn, 1994) 2) use of menter pollen (Kho et al., 1980) 3)bud pollination (Chatterjiee and More, 1991) and 4) embryo rescue (Chen *et al.*, **1997b**).

A cross between *C. hystrix* (2n = 24) and *C. sativus* (2n = 14) was made that resulted into a sterile F_1 hybrid **(2n = 19)** rescued by embryo rescue technique (Chen *et al.*, **1997b**). The amphidiploids were developed (2n = 38) through colchiploidy, which produced fertile flower and fruit set with viable seeds and was designated by a new species namely *C. hytivus* (Chen et al., 2000). These amphidiploids were cross compatible with *C. sativus* enabling gene transfer from *C. hystrix* to *C. sativus* (Chen *et al.*, 2003a).

Disease resistance in Cucumis species

Sources of resistance to various fungal and viral diseases along with resistance to various pests in different cucumis species have been summerised in Table 9.

Table 9. Sources of resistance to *Cucumis* spp.

Cucumis spp.	Resistance	Reference
Cucumis metulliferus	Nematode, CGMMV, WMV, SqMV	Nijs and Custers (1990)
		Wehner *et al.*, (1990)
Cucumis anguria	Nematodes, spider mites, CGMMV	Laboda (1996,1999, 2007)
Cucumis africans	Nematode, spider mites, CGMMV	MeCarthy *et al.*, (2001)
Cucumis frigonus	Nematodes and CGMMV	
Cucumis myriocarpus	Mites	
Cucumis zeyheri	Mites	
Cucumis hardwickii	CGMMV and Nematodes	
Cucumis callosus	Fruit fly	
Cucumis melo (Mp-1)	Downy mildew	

Accessions of *C. anguria, C. metuliferus* and *C. zeyheri* have high level of resistance to root knot Nematode (Nirs & custers, 1990 , Wehner *et al.*, 1990) high level of resistance to squash mosaic virus and watermelon mosaic virus in C. *metulliferus* (MeCarthy *et al.*, 2001). *C. melo* line MR-1 has some resistance to cucumber downy mildew (Labeda *et al.*, 1996, 1999, 2007).

Wild Releative of Cucurbita species

The domesticated cucurbita species such as *C. pepo, C. moschata, C. maxima, C. argyrosperma* (formerly called *C. mixta*) and *C. ficifolia,* do not hybridize readily with each other. Based upon species crossability, Whitekar and Davis (1962) concluded that *Cucurbita moschata* occupies the central position and can be crossed with *C. maxima*, *C. pepo* and *C. mixta* producing partially fertile hybrids. The interspecific crossing is highly cultivar dependent, usually produces few viable seeds as shown in the Fig 3 (Robinson and Deaker-walter, 1999).

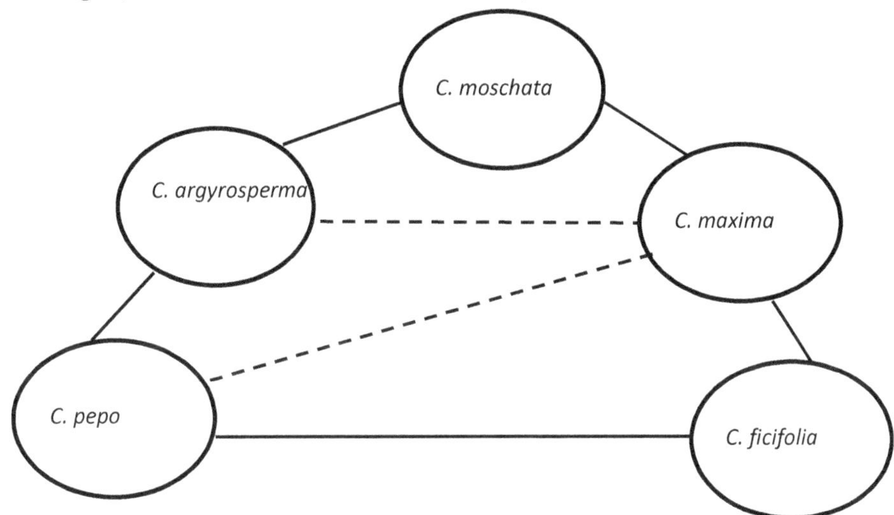

Fig . 3. Crossability among domasticated Cucurbits species

:——————— Partially fertile hybrid
:------------------- F_1 viable but self sterile
Sources: Robinson and Deaker-walter, (1999)

To obtain fruits and fertile seeds from F_1 plants of interspecific crosses, additional techniques like repeated pollination, bud pollination, mixed pollen pollination, embryo culture, amphidiploidy and adjustment of environment of flowering are adopted (Bemis 1973, Cheng *et al.*, 2002, Hiroshi 1963, Shifriss 1947, Rakho *et al.*, 2012). However, these techniques could not releave the male sterility/ incompatibility problems in subsequent generations (Stebbins, 1956).

Cucurbita lundeliana, when used as a female parent is cross compatible with nearly all cultivated species and is used as a bridge to facilitate gene flow. In a trispecific cross *C. pepo* (*C. lundelliana C. moschata*) the bush habit of *C. pepo* can be transferred to *C. moschata* (Rhodes, 1959).

The perennial species *C. okeechobeensis* and *C. martinezii* are cross compatible with *C. moschata* (Whitaker and Bemis, 1965, Hurd *et al.*, 1971). *Cucurbita argyrosperma* also has been used as genetic bridge to transfer gene in less compatible cucurbita species (MeCondless, 1998, Wessel-Beaver *et al.*, 2004). Successful attempts have been made by Qizhang *et al.* (2012), in the development of interspecific bridge lines among *C. pepo, C. moschata* and *C. maxima* with normal sexual compatibility by varietal recombination and successive selection towards increasing compatibility and interspecific crosses and the successively derived generations, as shown in Fig, 4

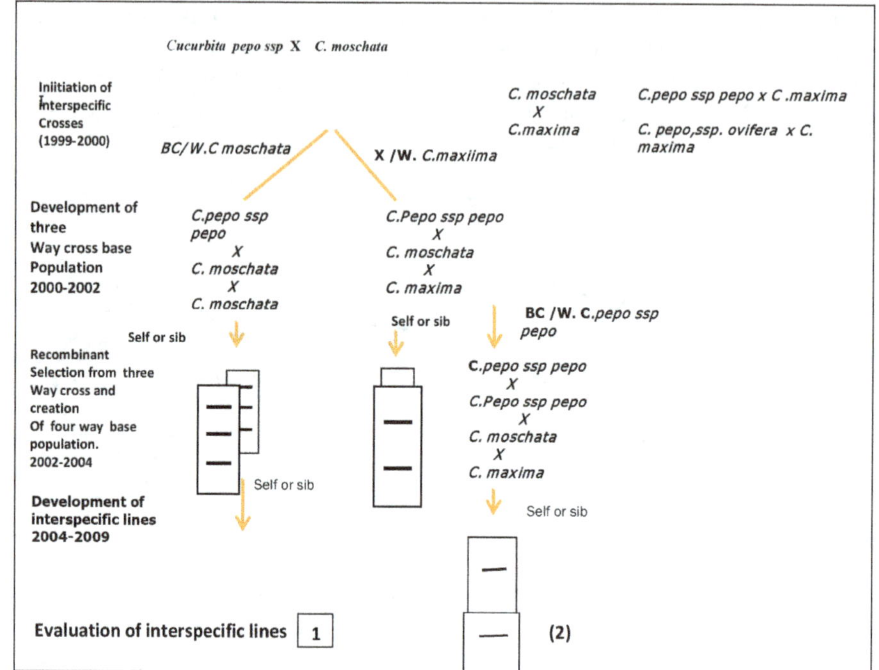

Fig 4: Development of advanced Interspecific bridge lines in among *Cucurbita* species

Source: Qizhang *et al.*, (2012)

This important breakthrough has created a powerful platform for breeders to transfer the favourable characteristics among the species freely, without introgression of unfavourable trait from wild species during the breeding process.

Sources of multiple virus resistance have been explored by Provvidenti (1980, 1990) and are summarized in table 10 below.

Table.10. Multiple viral resistance in Cucurbita species

Cucurbita species	Multiple virus resistance
Cucurbita ecudorensis	CMV, TRSV, WMV-1 or PRSV, WMV-2
C. foetidissima	CMV, TRSV, WMV-1, WMV-2
C. lundelliana	CMV
C. martinezii	CMV, TRSV
C. maxima	TRSV, BYMV, PRSV-W
C. moschata	TRSV, PRSV-W, ZYMV
C. pepo	TRSV
C. ficifolia	WMV-1, WMV-2, CMV, TRSV, BYMV-S, TmRSV

The only important virus of cucurbita spp for which high level of resistance was not found was squash mosaic virus. All species become infected, however *C. ecuadorensis, C. okeechobeensis* and C. *martinezii* could recovered from initial systemic symptoms (Provvidenti 1990). *C. ecuadorensis* because of its good resistance to several viruses and its compatibility with *C. maxima* is an excellent source of resistance.

Herrington *et al.*, (1991) were able to transfer resistance of WMV-2, PRV-W, ZYMV and CMV of *C. ecuadorensis* to *C. maxima* to produce cultivar Redlands Traiblazer *Cucurbita foetidissima* is equally good in virus resistaance but it is difficult to hybridize it with other member with genus cucurbita. *Cucurbita martinezii* on the other hand is not resistant to many viruses, but can be used to transfer resistance to CMV and powdery mildew to *C. moschata* and the F1 can be used as a bridge to transfer these resistant to *C. pepo* (Munger 1976).

Wild relatives of Citrullus species

The cultivated watermelon *Citrullus lanatus* includes two botanical varieties - *Citrullus lanatus* var. *lanatus* and *Citrullus lanatus* var. *citroides* also called citron or **tsama**. Other citrullus species are

Citrullus rehmii (annual)

Citrullus colocynthis (perennial)

Citrullus ecirrhosus (perennial)

Citrullus naudinianus (perennial)

All these citrullus species are cross compatible with each other, Sources of resistance to various viral and fungal diseases located in different citrullus species have been summarised in Table 11.

Table: 11 Sources of resistance to viral and fungal diseases in *Citrullus* species

***Cuccurbita* species**	**Resisatance sources**	**Pathogen**	**References**
Citrullus lanatus var. *lanatus*	P1 595203	ZYMV	*Guner and Wehner, 2008*
		PRSV	Stranger *et al.*, 2002
		ZYMV-FL	Boyhan *et al.*, 1992
		ZYMV-ch	Xu *et al.*, 2004
		WMV	Xu *et al.*, 2004
	P1 482261-1 P1 482261	ZYMV-FL	Provvidenti, 1991
Citrullus lanatus var. *citroides* (Egun)	-	Watermelon mosaic virus-2 (now called WMV)	Webb, R.E, 1977, 2003
	P1 244017 P1 244018 P1 244019 P1 482342 P1 482318 P1 482379 P1 485583	PRSV-W	Strange *et al.*, 2002
	-	Resistant to root-knot nematodes	Theies and Levi, 2007

Citrullus Spp.	Resistance Sources	Pathogen	References
Citrullus lanatus var. *lanatus*	PI 296341-FR	Fusarium oxysporum f. sp. niveum	Martyn and Netzer, 1991
		Race-0	
		Race-1	
		Race-2	
Citrullus colosynthis	PI 386016	WMV	Gillaspie and Wright, 1993
	PI 386024	WMV	Gillaspie and Wright, 1993
	PI 386025	ZYMV	Guner and Wehner, 2008
		WMV	Gillaspie and Wright, 1993
	PI 386026	ZYMV	Guner and Wehner, 2008
		WMV	Gillaspie and Wright, 1993
	PI 388770	WMV	Gillaspie and Wright, 1993
		Spotted spider mites	Cantu,Hector **Jr,** 2014
	PI 525080	Spotted spider mites	Cantu,Hector **Jr,** 2014
	PI 537300	Spotted spider mites	Cantu,Hector **Jr,** 2014
	PI 494528	ZYMV-FL	Provvidenti, 1991
		WMV	Lecoq *et al.,* 1998
	PI 494532	ZYMV	Lecoq *et al.,* 1998
		WMV	Provvidenti, 1989, 1991
	-	Resistant to spider mites	Lopez *et al.,* 2005
		ZYMV	Boyhan *et al.,* 1992

In *Citrullus lanatus* var. *lanatus* the accession P1 595203 has multiple resistance to ZYMV, PRSV and WMV. *Citrullus lanatus* var. *citroides* (PI 296341-FR) shows resistance to *Fusarium oxysporum f.* sp. *niveum* race-0, race-1 and race-2.

In *Citrullus colosynthis* there are several PI lines showing cobined resistance to ZYMV and WMV. Efforts are on to transfer diseases resistance of these species into cultivated watermelon.

Wild relatives of *Abelmoschus* species

The genus *Abelmoschus* includes important species such as *A. manihot, A. moschatus, A. angulosus, A. crinitus, A. ficulneus, A. tetraphyllus, A. esculentus, A. tuberculatus* and *A. caillei.* Sources of resistance to biotic and abiotic stresses have been located in different abelmoschus species which have been summarised in Table.12

Table 12 Sources of resistance to biotic and abiotic stresses in *Abelmoschus* species

Species	Ploidy level	Resistance
Abelmoschus caillei (cultivated spp.)	Hexaploid	YVMV
Abelmoschus manihot (cultivated spp)	Diploid	YVMV, PM, ELCV, Aphids
Abelmoschus crinitus (wild spp.)	Diploid	YVMV, PM, Cercospora, Alternaria blight, ELCV, Drought, Jassids and Mites
Abelmoschus tetraphyllus var. *tetraphyllus* (wild spp.)	Tetraploid	PM, YVMV
Abelmoschus tetraphyllus var. *pungens* (wild spp.)	Tetraploid	PM, YVMV, Low temperature
Abelmoschus angulosus (wild spp) var. *grandiflorus*	Diploid	ELCV, YVMV, PM, Cercospora and Alternaria blight, low temperature, Jassisds and Mites
Abelmoschus moschatus (cultivated spp.)	Diploid	PM, Jassids, Mites and Fruit borer
Abelmoschus tuberculatus (wild spp.)	Diploid	Fruit borer

YVMV = Yellow vein mosaic virus, PM = Powdery mildew, ELCV = Enation leaf curl virus

Ref. : Umesh Srivastawa, 2009, Nerkar *et al.,* 1981, Nerkar, 1990, Gautam *et al.,* 2000, Sheshadri and Srivastawa, 2002, Dhanker *et al.,*2005, R. S. Rana and T. A. Thomas, 1991, Prabhu *et al.,*1971, Singh *et al.,* 2007.

Abelmoschus caillei, A. manihot and *A. tetraphyllus* var. *tetraphyllus* have been used in developing okra varieties resistant to YVMV. In India, okra variety Punjab-7 resistant to YVMV was developed by crossing *A. esculentus* (Pusa Sawani) with *A. caillei* followed by backcrossing with Pusa Sawani (Thakur and Arora, 1987). Another variety Punjab Padmini was developed by crossing *A. esculentus* (Reshmi) with *A. caillei* followed by selection in advanced generation (Sharma, 1982),

Successful crosses have been obtained by crossing cultivated species *A. esculentus* with (a) *A. tuburculatus* (Pal *et al.,* 1982, Joshi and Hardas, 1956, Joshi *et al.,* 1974). The F_1s were either sterile or with low fertility or were obtained with embryo rescue technique (b) with A. manihot (Pal *et al.,* 1952, Jambhale and Nerkar, 1986) (c) with *A. moschatus* (Hamon and yapo, 1986, Gadwall *et al.,* 1968*, Joshi *et al.,* 1974*) (d) with *A. caillei* (Thakur and Arora, 1986) (e) with *A. tetraphyllus* var. *tetraphyllus* (Datta, 1990) f) with *A. ficulneous* and *A. angulosus* (Prabhu and warada *et al.,* 2013).

Okra variety Prabhani Kranti resistant to YVMV was developed through back cross breeding programme, using *A. manihot* as source of resistance (Jambhale and Nerker, 1986) Fig 4.

Fig. 4: Okra variety Prabhani Kranti resistant to YVMV.

A. esculentus (Pusa Sawani) ṅ *A. manihot*

F_1

BC_1

$BC_2 \longrightarrow BC_2F_8 \longrightarrow$ **Parbhani Kranti**

Okra varieties Arka Anamika and Arka Abhay resisatnt to YVMV were developed using A. tetraphyllus as sources of resisatnce (Dutta 1990) Suresh Babu & Dutta (1990) Fig 5.

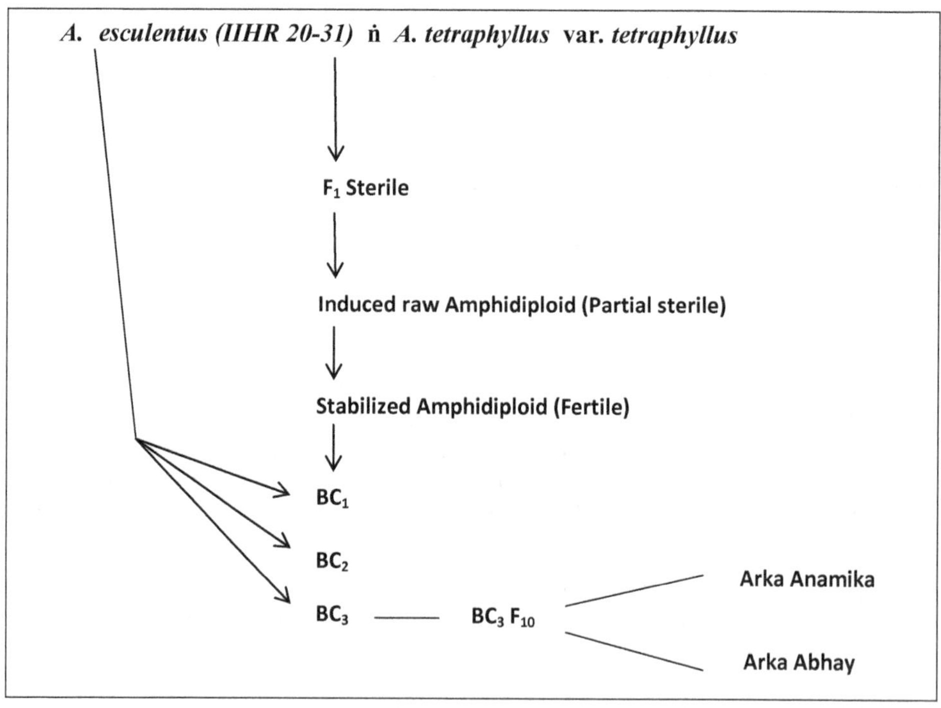

Fig 5. Arka Anamika and Arka Abhay development scheme.

Wild relatives of Cucurbits and Solanaceous vegetables used as rootstock in grafting

There is an inverse correlation between the degree of domestication of plants and damage by pests and diseases (Coley et al., 1985, Rosenthal & Dirzo, 1997). The resistance to pests and diseases in wild relatives is thus higher than in native varieties. Grafting is the fastest way of utilizing the stress tolerance of wild relatives. It is now common in several parts of the world not only to manage soil borne diseases but also to increase fruit quality, yield and resistannce to a variety at abiotic stresses including moisture stress, thermal stress and organic pollutants. It is being done on commercial scale in cucurbits such as melons, watermelon and cucumber as well as in *Solanaceous* crops such as tomato, eggplant and pepper. Rootstock breeders need to evaluate not only the genotype environment interaction (G E) of these new genotypes but also the reaction of these rootstocks along with the grafted scion across different environments (G G E) (King *et al.*, 2010). Wild relatives of cucumber and solanaceous crops used as rootstock and their resistance have been compiled in table 13.

Table: 13 Root stock tolerance to biotic and abiotic stresses

Root stock	Resistance / Tolerance	Grafting	Reference
Cucurbits			
Cucurbita maxima × *Cucurbita moschata*	*Fusarium oxysporum* f.sp. *niveum, F. oxysporum* f.sp. *melonis,* cold, heat and drought stress	Watermelon, Muskmelon, Cucumber	Davis *et al.*, 2008, King *et al.*, 2010, Lee 1994, Rouphael *et al.*, 2008
Cucurbita moschata	Bloomless cucumber	Cucumber	Sakata *et al.*, 2008
Cucurbita ficifolia (Fig leaf gourd)	Fusarium wilt, cold stress	Cucumber	Lee 1994, Tachibana 1982, Marukawa *et al.*, 1969
Sicos angulatus (Bur cucumber)	Cold stress	Cucumber	Venema *et la.*, 2008
Lagenaria siceraria (Bottle gourd)	Fusarium wilt	Watermelon, Muskmelon	Lee 1994
Cucumis metuliferus (African horned cucumber)	Root knot nematode (*Meloidogyne incognita*)	Cucumber, melon	Marukawa *et al.*, 1969
Citrullus lanatus var. *citroides*	*Fusarium oxysporum* f.sp. *niveum, F. oxysporum* f.sp. *melonis, F. oxysporum* f.sp. *cucumerinum,* root knot nematode, gummy stem blight	Watermelon	Davis *et al.*, 2008, Theis and Levi 2003, 2004, Thies *et al.*,2015

Contd...

Root stock	Resistance / Tolerance	Grafting	Reference
Solanum species			
Solanum torvum	Bacterial wilt, fusarium wilt, verticillium wilt, root knot nematode, Cold stress	Egg palnt	Lee 1994, Yamakawa 1982, King *et al.*, 2010, Okimura *et al.*, 1986
Solanum macsocarpon	Bacterial wilt, induces high phenolic content of grafted scion	Egg plant	Carmino Gilbert *et al.*, 2011, Capelli *et al.*, 1995
Solanum gilo	fusarium wilt,	Egg plant	Yamakawa 1982
Solanum sisymbrifolium	Bacterial wilt, fusarium wilt, verticillium wilt, root knot nematode	Egg plant	Yoshida *et al.*, 2004, Morra, 2004.
Solanum incanum × *S. melogena*	root knot nematode	Egg plant	Lester and Hasan 1991, Gilbert *et al.*, 2011
Solanum melongena × *S. ethiopicum*	root knot nematode	Egg plant	Behera and Singh 2002
Solanum integrifolium	Bacterial wilt, fusarium wilt, heat and humidity	Egg plant	Tachibana 1994, Yamakawa 1982
Solanum integrifolium × *S. melongena*	Low temperature, Bacterial wilt, fusarium wilt, verticillium wilt	Egg plant	Dounay *et al.*, 2001, Lee 1994, Yamakawa 1982
Solanum pimpinellifolium	Nematode, salt	Tomato	Lee 1994
Solanum habrochaites	Nematode, salt, cold, brown root rot	Tomato	Venema *et al.*, 2000, 2008, Yamakawa 1982
Solanum cheasmanii	Salt	Tomato	Munns and Tester, 2008
Solanum chilens	Fusarium oxysporum f.sp. lycopersici race 1, 2, 3, Drought	Tomato	Rai *et al.*, 2011, Reis alton *et al.*, 2004, O'connell *et al.*, 2007
Solanum lycopersicum × *Solanum habrochaites*	Low temperature (10 – 13°C), Bacterial wilt	Tomato	Okimura *et al.*, 1986, King *et al.*, 2010
Solanum lycopersicum var. *ceresiforme*	B.cockerelli Aphids, whitefly	Tomato	Hipolit cortez-Madrigal 2012

ROOT STOCK GRAFTING IN SOLANACEOUS CROPS

Tomato

Bacterial wilt, brown rot and fusarium wilt are the main limiting factors for the cultivation of tomatoes in the open field. Suitable root stocks include *Solanum chilens* resistant to *Fusarium oxysporum* f. sp. *lycopersici* race 1, 2, and 3, inter specific hybrid

Solanum lycopersicum Solanum habrochaites, resistant to bacterial wilt and *Solanum habrochaites* resistant to brown rot. (Rai *et al.*, 2011, Reis *et al.*, 2004, Okimura *et al.*, 1986, King *et al.*, 2010, and Yamakawa, 1982) AVRDC has recommended eggplant cultivar 'Surya' as a rootstock for tomato cultivation in soils infested with bacterial wilt during hot – wet season. (Black *et al.*, 2003).

Green house cultivation of tomato in the temperate climatic zone is limited when tomatoes are exposed to low temperature. Shoots of tomato cultivars start showing wilting symptoms if their roots are exposed to temperatures around 5^0C. High altitude accessions of *S. habrochaites* can tolerate low temperature and can be used as a rootstock in tomato (Venema et al., 1999 a, b). It was further observed that the stomata of *S. habrochaites* remain closed during chilling temperature, whereas, stomata of cultivated tomato susceptible to low temperature remained open (Bloom *et al.*, 2004). Tolerance to low temperature (0 - 13^0C) was also provided by the rootstock hybrid *S. lycopersicum S. habrochaites* to the susceptible scion (Okumura *et al.*, 1986). In Netherland, approximately 75% of the commercially grown greenhouse tomatoes are grafted on rootstocks that are selected for higher vigour and nutrient uptake, which in turn increases scion growth and rate of development, earliness, extended harvesting period and yield (Veneme *et al.*, 2008, Ruiz *et al.*, 1997, Nijs 1980, and King *et al.*, 2010).

Drought tests have shown that *S. chilens* is five times more tolerant to wilting than the cultivated tomato (O'connell *et al.*, 2007)

Solanum pimpinellifolium and *S. cheesmanii* showed good tolerance to salt (Lee, 1994, Munns and Tester, 2008) whereas; nematode resistance was observed in *S. habrochaites* and *S. pimpinellifolium* (Lee, 1994).

Egg Plant

The main purpose of grafting egg plant is to counter the adverse effects of soil borne diseases such as bacterial wilt, verticillium wilt, fusarium wilt, root knot nematode and low temperature (Goth *et al.*, 1991, Yamakawa, 1982). *Solanum torvum*, resistant to bacterial wilt, fusarium wilt, verticillium wilt, root knot nematode and cold stress; and *Solanum integrifolium*, resistant to bacterial wilt, fusarium wilt, high humidity and heat are the most popular and promising rootstocks for egg plant in Japan (Lee 1994, Yamakawa 1982, King *et al.*, 2010, Okimura *et al.*, 1986, Tachibana 1994). These rootstocks provide high vigour and prolonged harvest of grafted egg plant. However, in *Solanum torvum*, homogenous seed germination is a major problem. *Solanum aethiopicum* and *S. macrocorpon* on the other side provide uniform germination along with tolerance to fusarium wilt and bacterial wilt (Cappelli *et al.*, 1995, Daunay *et al.*, 1991). The rootstock *S. macrocorpon* also includes high phenolic content a powerful antioxidant in fruits of grafted scion (Cappelli *et al.*, 1995). Interspecific hybrids such as *S. incanum* and *S. melongena* x *S. melongena S. aethiopicum*, both confer resistance to root knot nematode and provide uniform seed germination (Lester and Hason, 1991, Gisbert et al., 2011, Behera and Singh, 2002). Interspecific hybrid *S. integrifolium S. melongena* provides better tolerance to low temperature (18-25^0C), bacterial wilt, verticillium wilt and fusarium wilt than the non grafted egg plant (Okimura *et al.*, 1986, Daunay *et al.*, 2001, Lee, 1994,

Yamakawa 1982). *Solanum sisymbrifolium* root stock resistant to bacterial wilt, fusarium wilt, verticillium wilt and root knot nematode gave promising results in Italy when diverse group of egg plant cultivars were grafted on it (Morro, 2004).

Sweet Pepper

Sweet pepper is not grafted commercially, except in winter greenhouse production in Korea to confer resistance to *Phytopthera capsici* (Kwon *et al.*, 2006). However there is a report from Taiwan indicating that sweet pepper when grafted on *Capsicum baccatum, C. frutesens* and *C. chacoense* can tolerate flood damage during hot-wet and hot-dry season (Palada and Wu, 2008).

Cucurbits

Grafting is commercially done in watermelon, melon and cucumber in Japan, Korea, Turkey and China (Lee 1994).

Watermelon

Most common rootstocks for watermelon are bottle gourd, interspecific hybrid *C. maxima   C. moschates* and wild watermelon *Citrullus lanatus* var. *citroides. Citrullus lanatus* var. *citroides* has multiple resistance to soil borne diseases such as *Fusarium oxysporum* f.sp. *nivea, f. oxysporum* f.sp. *melonis, F. oxysporum* f.sp. *cucumerinum,* gummy stem blight and root knot nematode (Davis *et al.*, 2008, Thies *et al.*, 2007, 2003). Bottle gourd (*Lageneria siceraria*) provides excellent resistance to fusarium wilt. However, with prolonged use of bottle gourd as rootstock for watermelon, Fusarium wilt caused by *Fusarium oxysporum* f. sp. *lageneriae* starts appearing in the field (Sakata *et al.*, 2007). Further, bottle gourd has extensive but shallow root system not conducive to use plastic mulch in watermelon cultivation due to high soil temperature during summer season. Interspecific hybrids – *Cucurbita maxima   C. moschata,* having multiple resistance to fusarium wilt as well as tolerance to cold and drought with deep root system, commonly used to advance the planting dates during cool period, has become a much preferred root stock for watermelon (Davis et al., 2008, King et al., 2010, lee, 1994, Rouphael et al., 2008). However, in this root stock, there is a loss of flesh quality of grafted watermelon as they acquire flesh colour at about the same time as non grafted plant, but sugar accumulation takes place at a later stage. Harvesting of watermelon fruit thus requires careful observations. In general, flesh firmness, shelf life and field holding capacity is better in grafted watermelon. Mini watermelon when grafted on *C. maxima   C. moschata* root stock, gave more than 60 % higher marketable yield when grown under water deficit conditions, when compared to ungrafted watermelon. The higher marketable yield recorded with grafted watermelon was mainly due to an improvement in water and nutrient uptake as indicated by higher N, K, Mg concentration in the leaf and higher CO_2 assimilation (Rouphael *et al.*, 2008).

Melon

Major soil borne diseases associated with melon cultivation includes fusarium wilt (*F. oxysporum f.sp. melonis*), gummy stem blight (*Didymella bryoniae*) and charcoal rot (*Macrofamina phaseolina*). Most common root stock for melon is *C. maxima   C.*

moschata hybrid, which provides resistance to fusarium wilt, cold and drought stresses (Davis *et al.*, 2008, King *et al.*, 2010, Lee 1994). Bottle gourd root stock provides complete resistance to fusarium wilt but affects fruit quality of grafted melon negatively. Grafted melons were more resistant to salinity and showed higher yield than the ungrafted ones (Romero *et al.*, 1997)

Cucumber

Fig leaf gourd (*Cucurbita ficifolia*) is the common root stock for cucumber as it provides superior cold tolerance and fusarium wilt resistant and has high affinity for cucumber, (Marakawa and Takatsu, 1969). In summer grown cucumber, *Cucurbita maxima   C. moschata* hybrid is quite popular as it provides resistance to fusarium wilt as well as better heat tolerance (King *et al.*, 2010). Bur cucumber (*Sicos angulatus*) is another root stock for cucumber as it provides resistance to low temperature. Several studies demonstrated that fig leaf and Bur cucumber root stocks improve vegetative growth and early yield at suboptimal temperature (den Nijs, 1980, 1984, Tachibana, 1982, Bulder *et al.*, 1990). The physiological basis of the low temperature tolerance of fig leaf gourd root stock is supposed to be associated with the maintenance of higher absorption rate of water and nutrients (Tachibana, 1987) and higher cytokinin synthesis which stimulates root meristem activities and translocation of photosynthates to the root (Tachibana, 1988). Studies also showed that the cucumber aquaporins were more sensitive than that of fig leaf gourd under low root zone temperature (Lee *et al.*, 2005c).

REFERENCES

Agrama, H. A and Scott, J. W. (2006). Quantitative trait loci for tomato yellow leaf curl virus and tomato mottle virus resistance in tomato. *J. Amer. Soc. Hort. Sci.* **131**:267-272.

Ammiraju, J., Veremis, J., Huang, X., Roberts, P. and Kaleshian, I. (2003). The heat-stable root knot nematode resistance gene M1-9 from *Lycopersicon peruvianum* is localized on the short arm of chromosome 6. *Theoritical and Applied Genetics,* **109**, 478-484.

Anbindes, L., Reuveni, M., Azari, R., Paran, L., Nahon, S., Schlomo, H., Chen, L., Labidot, M., Levin I (2009). Molecular dissection of Tomato leaf curl virus resistance in tomato line TY172 derived from *Solanum pervianum. Theor. Appl. Genet.* **119**(3):519-30.

Ano, G., Hebert, Y., Prior Philippe, Missiaen, C. M. (1991). A new source of resistance to bacterial wilt of eggplant obtained from a cross *S. aethiopicum* x *S. melongena* L. *Agromonoie.* **11**(7):550-560.

Ares, J.L.A., Martinz, A.R. and Paz, J. F. (2005). Resistance of pepper germplasm to *Phytophthora capsici* isolates collected in northwest Spain. *Spn. J. Agric. Res.* **3**:429-436.

Asins, M. J., Breto, M. P., Cambro, M. and Carbonell, E A. (1993). Salt tolerance in lycopersicon species. I. Character definition and changes in gene expression. *Theor. Appl. Genet.* **86**:737-743.

Attavian, B. N., Jelenkovick, G. and Pollack, B. L. (1983). Crossability of selected non tuberous *Solanum* species. *J. Am. Soc. Hort. Sci.* **108**(1):15-20.

Bai, Y. (2004). The genetics and mechanism of resistance to tomato powdery mildew (*Oidium neolycopersici*) in *Lycopersicon* specis. *Thesis*, Wageningen University, The Netherlands.

Bai, Y., Anang, C. C., Van der Hulst, R., Meijer_Dekens, F., Bonnema, G. and Lindhout, P. (2003). QTLs for tomato powdery mildew resistance (*Oidium lycopersici*) in *Lycopersicon parviflorum* G1.1601 co-localize with two qualitative powdery mildew resistance genes. *Molecular plant Microbe Interactions,* **16**:169-176.

Beharav, A. and Cohn, Y. (1994). Callus formation from cotyledon and hypocotyls of *Cucumis melo* (L) and *Cucumis metuliferus. Cucurbit Genet. Crop Rep.* **17**:97-100.

Behera, T. K. and Singh Narendra (2002a). Interspecific cross between eggplants (*S. melongena* L) with *Solanum* species. *Scientia Horticulturae.* **95**:165-172.

Bemis, W. P. (1973). Interspecific aneuploidy in Cucurbita. *Genet. Res.* **21**:221-228.

Ben Chaim, A., Grube, R., Lapidot, M., Jahn, M., Paran, L. (2000a). Identification of quantitative trait loci associated with resistance to cucumber mosaic virus in *Capsicum annuum. Theor. Appl. Genet.* **102**:1213-1220.

Berlinger, M. M. Mordechi, S. and Pilowsky, M. (1991). Tomato resistance to tobacco whitefly. *Gan Ssadeh VaMeshek,* **9**:56-59.

Black, L. L., Wu, D. L., Wang, J. F., Kalb, T., Abbass, D., Chen, J. H. (2003). Grafting tomatoes for production in the hot-wet season. In : *International co-operators' guide.* Asian Vegetable Research and Development Centre.

Blacker, N. S. and Hewitt, J. D. (1987). A comparison of resistance to *Phytophthora parasitica* in tomato. *Phytopathology,* **77**:1113-1116.

Bloom, A. J., Zwieniecki, M. A., Passioura, J. B., Randall, L. B., Holbrook, N. M., St. Clair, D. A. (2004). Water relations under root chilling in a sensitive and tolerant tomato species. *Plant cell Environ.* **27**:971-979.

Boyaci, F., Unlu, A., Abak, K. (2010). Screening for resistance to *Fusarium* wilt of some cultivated egg plant and *Solanum* accession in: ISHS. *Acta Horticulture,* 935. XXVIII International Horticultural Congress on Science and Horticulture for people (IHC 2010) International Symposium on new development in Plant Genetics and Breeding.

Boyhan, G., Norton, J. D., Jacobson, B. J., Abrahems, B. R. (1992). Evaluation of watermelon and related germplasm for resistance to Zucchini yellow mosaic virus. *Plant Dis.* **76**:251-252.

Brower, D. J., Jones, E. S. and St.Clair, D. A. (2004). QTL analysis of quantitative resistance to *phtophthora infestans* (late blight) in tomato and comparison with potato. *Genome,* **47**(3):475-492.

Bulder, H.A.M., Van Hassalt, P. R., Kuiper, P.J.C., Speek, E. J., den Nijs, A.P.M. (1990). The effect of low root temperature in growth and lipid composition of low temperature tolerant rootstock genotypes for cucumber. *Journal of Plant Physiology,* **138**:661-666.

Cantu, Hector Jr.., (2014). An evaluation of watermelon (*Citrullus* spp.) germplasm for additional sources of resistance to the two spotted spider mite (*Tetranychus urticae* Koch) Thesis, dissertation and student research in Agronomy and Horticulture, Paper 80.

Cappelli, C., Stravato, V. M., Rotino, GL., Bounauria, R. (1995). Sources of resistance among *Solanum* spp. to an Italian isolate of *Fusarium oxysporum* f. sp. *melongenae*. Eucarpia Meeting on Genetics and Breeding of Capsicum in Eggplant. **1X**:221-224.

Caranta, C., Palloix, A., GebreSelassie, K., Lefebvre, V., Moury, B. and Daubeze, A. M. (1996). A complementation of two genes originating from susceptible Capsicum *annuum* lines confers a new and complete resistance to pepper venal mottle virus. *Phytopathology*, **86**:739-743.

Chaine –Dogimont, C., Palloix, A., Daubze, A. M., Merchoux, G., Selassie, K. G. and Pochard, E. (1996). Genetic analysis of broad spectrum resistance to potyviruses using double haploid lines of pepper (*Capsicum annuum* L.). *Euphytica*, **88**:231-239.

Chakrabarti, A. K. and Choudhary, B. (1975). Breeding brinjal resistant to little leaf disease. *Proc. Ind. Nat. sci. Acad. B.* **41**:379-385.

Chatterjee, M. and More, T. A. (1991). Interspecific hybridization in *Cucumis* spp. *Cucrbit Genet. Co-op. Rep.* **14**:69.

Checma, D. S. Singh, D. P., Rawat, R. D. and Deshpande, A. A. (1984). Inheritance of resistance to anthracnose disease in chillies, capsicum, eggplant news, 3:44.

Chen, J. F. and Jeffrey, A. (2000). Interspecific hybridization in Cucumis- progress, problems and perspective. *Horti. Sci.* **35**(1) 2000.

Chen, J. F., Adelberg, J. W., Staub, J. E., Skorupska, H. T. and Rhodes, B. B. (1988). A New Synthetic amphidiploids in *Cucumis* from a *C. sativus* x *C. hystrix*. (Chark) F_1 interspecific hybrid. P:336-339. In: J. McCreight (ed) Cucurbitaceae. 98-Evolution and enhancement of cucurbit germplasm ASHS Press, Alexandria, UK.

Chen, J. F., Isshiki, S., Tashiro, Y. and Miyazaki, S. (1997a). Successful interspecific hybridization between *Cucumis sativus* L. and *Cucumis hystrix*. *Euphytica*, **96**:413-419.

Chen, J. F., Isshiki, S., Tashiro, Y. and Miyazaki, S. (1997b). Biochemical affinities between *Cucumis hystrix* chakr. and two cultivated *Cucumis* species. *Euphytica*, **97**:139-141.

Chen, J. F., Luo, X. D., Staub, J. E., Qian, C. T., Zhuang, F. Y. and Ren, G. (2003a). An allotriploid derived from an amphidiploid x diploid mating in *Cucumis* sp.: Production, micro propagation and verification. *Euphytica,* **131**:235-241.

Cheng, Y. G., Zhang, B. K., Zhang, E. H. and Zhao, Z. L. (2002). Germplasm innovation by interspecific crosses in pumpkin. *Cucurbit Genetics Coop. Rep.* **25**:56-57.

Chunying, Sun., Sheng Li, Mao, Zheng Hai Zheng, Alain Palloix Li Haowang, Bao Xi Zhang (2015). Resistance to anthracnose (*Colletotrichum acutatum*) of capsicum mature green and ripe fruit are controlled by a major dominant cluster of QTLs on chromosome P_5. *Scientia Hort.* **181**:81-88.

Coley, P. D., Bryant, J. P. and Chopin, F. S. (1985). Resource availability and plant antiherbivore defence. *Science*, **230**, 4728, 895-899.

Custers, J. B. M. and den Nijs, A. P. M. (1986). Effect of aminoethoxyvinylglycine (AVG), environment and genotype in overcoming hybridization barriers between *Cucumis* species, *Euphytica*, **35**:639-647.

Dane, F., Denna, D. W., Tsuchiya, T. (1980). Evolutionary studies of wild species in the genus *Cucumis* . *Zptanzenzucht.* **85**:89-109.

Datar, V. V., and Ashtaputre, J. U. (1988). Studies on resistance to *phomopsis* fruit rot in eggplant. *Indian Phytopathology*, **41**:637-638.

Daunay, M. C., Lestar, R. N., Gebhardt, C. H., Hemart, J.W., John, M., Fray, A. and Doganlar, S., (2001) Genetic resources of eggplant (*Solanum melongena* L.) and allied species: A New Challenge for Molecular Genetics and Eggplant Breeders (Solanaceae V (ed.)) Van Den Berg, R. G., Baredse, G. W., and Mariani, Nijmegen University Press, Nijmegen, The Netherlands. Pp. 251-274.

Daunay, M. C., Lestar, R. N., Laterrot, H. (1991). The use of wild species for the genetic improvement of brinjal, eggplant (*Solanum melongena* L.) and Tomato (*Lycopersicon esculentum*). In: Hawkes, J. C., Lester, R. N., Nee, M. and Estrada, N. (eds) Solanaceae III: Taxonomy, Chemistry, Evolution. PP. 389-413. Royal Botanical Gardens, Kew and Linnean society, London.

Davis, A. R., Perkins-veazie, P., Dakata, Y., Lopez-Galarza, S., Cohen, R., Lee, J. M. (2008). Cucurbit grafting. *Crit. Rev. Plant Sci.* **27**, 50-74.

Deakin, J. R., Bohn, G. W., Witakar, T. W. (1971). interspecific hybridization in Cucumis. *Econ. Bot.* **25**:195-211.

De lapena, R. C., Ebert, A. W., Gniffke, P., Hanson, P. and Symoneb, R. C. (2011, chapter 18). Genetic adjustment to changing climate: Vegetables. In: Yadav, S. S., Redden, R. J., Hatfield, J. L., Lotze-campen, H. and Hall, A. E. (eds) Crop adoption to climate change, First edn. John wiley and Sons Ltd., Chichester, West Sussex, UK, pp. 396-410.

den Nijs, A. P. M. (1980). The effect of grafting on growth and early production of cucumber at low temperature. *Acta Hortic.* **118**, 57-63.

den Nijs, A. P. M. (1984). Rootstock-Scion interactions in the cucumber: Implications for cultivation and breeding. *Acta Hortic.* **156**:53-60.

den Nijs, A.P.M. and Custers, J.B.M.,(1990). Introducing resistance into cucumber by interspecific hybridization, In: Bates, D. *et. al.* (eds) Biology and utilization of Cucurbitaceae. Cornell University Press, Ithaca, Newyork, 382-395.

Den Nijis, A. P. N. and Visser, D. L. (1985). Relationship between African species of the genus Cucumis estimated by the production vigour and fertility of F_1 hybrids. *Euphytica*, **34**:279-290.

Deshpande, A. A., Anand, N., Pathak, C. S., Sridhar, T. S. (1985). New sources of powdery mildew resistance in capsicum species. *Capsicum* Newsletter, No.4:75-76.

Dhankar, B. S., Mishra, J. P. and Bisht, I. S. (2005). Okra. In plant genetic resources: Horticultural crops, B. S. Dhillon, P. K. Tyagi, S. Saxena, and G. J. Randhwa (eds). Narosa Publishing House Pvt. Ltd. New Delhi. Pp. 59-74.

Dianese, E. C., Foennseca, M. E. N., Inoue-Nagota, A. K., Resendo, R. O. and Boiteux, L. S. (2011) Search in Solanum (Section *Lycopersicon*) germplasm for sources of broad spectrum resistance to four Tospo virus species. *Euphytica*, **180**:307-319.

Diwan, N., Fluhr, R., Eshed, Y. and Zamir, D. (1999). Maping of Ve in tomato: a gene cofering resistance to the broad spectrum pathogen *Verticillium dahlia* race 1. *Theor. Appl. Genet.* **98**:315-319.

Djian-Caporalino, C., Lefebre, V., Sage-Daubeze, A. M., Palloix, A. (2007). Capsicum In: Singh, R. J., (ed) Genetic resources, chromosome engineering and crop improvement: Vegetable crops, Vol .3. CRC, Boca Raton, FL, USA, pp. 185-243.

Dutta, O. P., (1991). Okra germplasm utilization at IIHR, Bangalore In: Report of an International workshop on okra genetic resources held at NBPGR, New Delhi, India, 8-12 Oct. 1990.IBPGR, Rome, pp 114-116.

Egashira, H., Kawashima, A. and Imanishi, S. (2000). Screening of wild accessions resistant to grey mould (*Botrytis cinera* pers) in *Lycopersicon*. *Report of the Tomato Genetics Cooperative*, **50**:14-15.

Fargette, D., Leslie, M., Harrison, B. D. (1996). Serological studies on the accumulation and localisation of three leaf curl Gemini viruses in resistant and susceptible *Lycopersicon* species and tomato cultivars. *Annals of Applied Biology*, **128**:317-328.

Fassuliotis, S. G., (1977). Self fertilization of *Cucumis metulliferus* (Neud) and its cross compatibility with *C. melo. J. Amer. Soc. Hort. Sci.* **102**:336-339.

Fernandoz-Munos, R., Dominguez, E. and Curatero, J. (2000). A novel source of resistance to the two spotted spider mite in *Lycopersicon pimpinellifolium* Mill: its genetics as affected by interplot interference. *Euphytica*, **111**:169-173.

Finkers, R., van der Berg, P., van Berloo, R. et. al. (2007). Three QTLs for Botrytis Cinerea resistance in tomato. *Theoretical and applied Genetics*, **114**:585-593.

Firdaus, S., van Heusden, A. W., Hidayati, N. et. al. (2013). Identification and QTL mapping of whitefly resistance components in *Solanum galapagense. Theoretical and Applied Genetics.* Doi: 10, 1007 IS00122-013-2067-Z.

Firdaus, S., van Heusden, A. W., Hidayati, N., Subena, E. D. J., Visser, R. G. F. and Vosman, B. (2012). Resistance to *Bemicia tabacci* in tomato wild relatives, *Euphytica*, **187**: 31-45.

Foolad , M. R., Merk, H. L. and Ashrafi, H. (2008). Genetics, genomics and breeding of late and early blight resistance in tomato. *Crit. Rev. Plant Sci.,* **27**:75-107.

Foolad, M. R. (2004). Recent advances in genetics of salt tolerance in tomato. *Plant Cell Tissue Culture*, **76**:101-109.

Foolad, M. R. (2007). Tolerance to abiotic stresses In: Razdon, M. K. and Mattoo, A. K. (eds) Genetic improvement of solanaceous crops: tomato, Vol. 2. Science Publishers, Enfield, New Hampshire, USA, pp521-590.

Foolad, M. R. and Lin, G. Y. (1997). Absence of a genetic relationship between salt tolerance during seed germination and vegetative growth in tomato. *Plant Breeding*, **116**(4):363-367.

Foolad, M. R., Ntahimpera, N., Christ, B. J. and Lin, G. Y. (2000). Comparison between field, greenhouse and detached leaflet evaluation of tomato germplasm for early blight resistance. *Plant Dis.* **84**:917-972.

Franken, J., Custors, J. B. M. and Bino, R. J. (1988). Effects of temperature on pollen tube growth and fruit set in reciprocal cross between *Cucumis sativus* and *C. metulliferous*. *Plant Breeding*. **100**:150-153.

Garwal, V. R., Joshi, A. B. and Iyer, R. D. (1968). interspecific hybrids in *Abelmoschus* through ovule and embryo culture. *Indian J. Genet. Pl. Breed.* **28**:269-274.

Gautam, P. L., Sharma, G. D., Srinivastava, U., Singh, B. M., Kumar, A., Saxena, R. K. and Srinivasan, K. (2000). 20 glorious years of NBPGR (1976-1996) National Bureau of Plant Genetic Resources, New Delhi. pp. 29-31,107, 161

Gilbert, J. C. (1958). Some linkage studies with the Mi gene for resistance to root knot nematodes. *Tomato Genet. Coop. Rep.* **8**:15-17.

Gillaspie, A. G., Wright, J. M. (1993). Evaluation of *Citrullus* species germplasm for resistance to watermelon mosaic virus 2. *Plant Dis.* **77**:52-354.

Gisbert, C., Prohens, J., Raigon, M. D., Stommel, J., Nuez, F. (2011). Eggplant relatives as sources of variation for developing new rootstocks: Effect of grafting on yield of fruit apparent quality and composition. *Scientia Horticulturae,* **128**:14-22.

Gorcia-Neria, M. A. and Rivera-Bustamante, R. F. (2011). Characterization of gemini virus resistance in an accession of *Capsicum chinense jacq. Mol. Plant Microbe Interact.* **24**(2):172-82.

Gordille, L. F., Stevenes, M. R., Millard, M. A. and Geary, B. (2008). Screening two *Lycopersicon peruvianum* collections for resistance to tomato spotted wilt virus. *Plant Dis.,* **92**:694-704.

Goth, R. W., Haynes, K. G., Barksdale, T. H. (1991). Improvement of levels of bacterial wilt resistance in eggplant through breeding. *Plant Dis.* **75**:398-401.

Gousset, C., Collonnier, C., Mulya, K., Kariska, L., Rotino, G. L., Besse, P., Servaes, A., Sihochaks, D. (2005). *Solanum torvum* as a useful source of resistance against bacterial and fungal diseases for improvement of eggplant (*S. melongena* L.). *Plant sci.,* **168**:319-32.

Gratiige, R. and O'Brien, R. G. (1982). Occurrence of a third race of Fusarium wilt of tomato in Queensland. *Plant Dis.,* **66**:165-166.

Guner, N. and Wehner, T. C. (2008). Overview of poty virus resistance in watermelon In: Proceedings of the IXth EUCARPIA meeting on genetics and breeding of Cucurbitaceae. Pitrat, M, (eds) INRA, Avignon (France), May 21-24th, 2008.

Hall, T. j., (1980). Resistance at the Tm-2 locus in the tomato to tomato mosaic virus. Euphytica, **29**:189-197.

Hamon, S. and Yapo, A. (1986). Perturbation induced within the genus *Abelmoschus* by the discovery of a second edible okra species in West Africa. *Acta Hort.* **182**:133-144.

Hanson, P. M., Bernacchi, D. Green, S., Tanksley, S, D., Muniyappa, V., Padmaja, A. S., Chen, H., Kuo, G., Fang, D. and Chen, J., (2000). Mapping a wild tomato introgression associated with tomato yellow leaf curl virus resistance in a cultivated line. *J. Amer. Soc. Hort. Sci.*, **15**:15-20.

Hanson, P. M., Licardo, O., Hanudin, Wang, J. F. and Chen, J. T. (1998). Diallel analysis of bacterial wilt resistance in tomato derived from different sources. *Plant disease*, **82**:74-78.

Hanson, P., Green, S. and Kuo, G. (2006). Ty-2. A gene on chromosome 11 conditioning Gemini virus resistance in tomato. *Rep. Tomato Genet. Coop.* **56**:17-18.

Hanson, P., Kodirvel, P., Schafleitner, R., delapenna, R., Geethanjali, S., Kenyou, L., Stsai, W., Wang, J. F., S-Muffuang F-LHO and Tan, C. W. (2013). Recent progress in mapping begomovirus resistance and marker assisted selection for bacterial wilt resistance in tomato at AVRDC in: Tomato breeding roundtable Chiang Mai, Thailand, 6-8 Feb. 2013.

Harrington, M. E., Prytz, S. and Brown, P. (1991). Resistance to papaya ring spot virus-w, Zuechini yellow mosaic virus and watermelon mosaic virus in *C. maxima. Cucurbita Genet. Coop. Rep.* **14**-123.

Hassan, A. A., Mazyad, H., Moustate, S., Nassar,S., Makha, M. and Sims, W. (1984). Inheritance of resistance to tomato yellow leaf curl virus derived from *Lycopersicon cheesmanii* and *Lycopersicon hirsutum. Hort. Science,* **19**:574-575.

Heber, Y. (1985). Comparative resistance of 9 *Solanum* species to bacterial wilt (*Pseudomonas solancearum*) and to the nematode *Meloidogyne incognita.* Importance for breeding aubergine (*Solanum melongena* L) in a humid tropical zone. *Agronomie.* 5(1):27-32.

Heinz, K. M. and Zalon, F. G. (1995). Variation in trachome based resistance to *Bemicia argentifolii* (*Homoptera aleyrodidae*) Oviposition on tomato. *J. Econ. Entomol.* **88**:1494-1502.

Heiser, C. B., Pickersgill, B. (1969). Names of the cultivated *Capsicum* species (Solanaceae). *Taxon.* **18**:852-859.

Heroshi, H. (1963). Cucurbit crosses: XVI Flower ages and reciprocal cross compatibility in *C. pepo, C. moschata* and *C. maxima. Japanese J. of Breeding,* **13**:159-167.

Hipolito Cortez-Madrigal (2012). Grafts of crops on wild relatives as base of an integrated pest management. The tomato *Solanum lycopersicun* as example, integrated pest management and pest control – current and future Tactics, Dr. Sonia Soloneski (Ed), In Tech, DOI:10 5772/30181.

Huang, C. C. and Lindhout, P. (1997). Screening for resistance in wild Lycopersicon species to *Fusarium oxysporum* f.sp. *Lycopersici* race 1 and race 2. *Euphytica*, **93**:145-193.

Huang, C. C., Van de Putta, P. M., Haanstra-van der Meer, J. G., Meijer-Dekens, F. and Lindhout, P. (2000). Characterization and mapping of resistance to *Oidium lycopersicum* in two *Lycopersicon hirsutum* accessions. Evidence for close linkage of two Ol-genes on chromosome 6. *Heredity*, **85**:511-520.

Hurd Jr, P. D., Linsley, E. G. and Whitekar, T. W. (1971). Squash and gourd bees (*Peponapis, Xenoglossa*) and the origin of the Cultivated Cucurbita. *Evolution*, **25**:218-234.

Hutton, S. F. and Scot, J. W. (2015). Recent progress in TYLCV resistance breeding and implications for tomato varieties of the future. Presentation. www.1mok. ufl.edu/media../doesl.../tomato.../t1 2015/ti2015-hutton pdf.

Hutton, S. F., Scot, J. W. and Sehuster, D. J. (2012). Recessive resistance to tomato yellow leaf curl virus from the tomato cultivar Tyking is located in the same region asTy-5 on chromosome 4. *Hort. Science*, **47**(3):324-327.

Hutton, S. F., Scot, J. W., Shekasteband, R., Lavin, L. and Lapidot, M. (2015). Combination of Ty resistance genes generally provide more effective control against begomoviruses then do single gene. ISH *Acta Hort.* 1069:IV International Symposium on Tomato diseases.

Jahn, M., Paran, L., Hoffmann, K., Radwanski, E. R., Livingstone, K. D., Grube, R. C., Afetrgoot, E., Lapidot, M. and Moyer, J. W. (2000). Genetic mapping of the Tsw locus for resistance to the tospovirus tomato spotted wilt virus in *Capsicum* spp. and its relationship to the Sw-5 gene for resistance to the same pathogen in tomato. *Mol. Plant Microb. Interactions*, **13**:673-682.

Jambhale, N. D. and Nerkar, Y. S. (1981). Inheritance of resistance for okra yellow vein mosaic disease in interspecific crosses of *Abelmoschus. Theor. Appl. Genet.* **60**:313-316.

Jambhale, N. D. and Nerkar, Y. S. (1986). "Parbhani Kranti" a yellow vein mosaic resistant okra. *Hort. Science.* **21**:1470-1471.

Janzac, B., Faber, M. F., Palloix, A. and Moury, B. (2009). Phenotype and spectrum of action of the Pvr y resistance in pepper against *potyvirus* and selection for virulent variants. *Plant* Pathology, **58**(3):443-449.

Jeong, H. J., Kwon, J. K., Pandeya, D., Hwong, J., Hoong, N., Bae, J. H. et al. (2012). A survey of natural and ethyl methane sulphate induced variation of elF 4E using high resolution melting analysis in Capsicum. *Molecular Breeding*, **29**(2):349-360.

Ji, Y., Schuster, D. J. and Scot, J. W. (2007a). Ty-3, a *begomovirus* resistance locus near the tomato yellow leaf curl virus resistance locus Ty-1 on chromosome 6 of Tomato. *Mol. Breed.*, **20**:271-284.

Ji, Y., Scot, J. W., Hanson, P., Graham, E. and Maxwell, D. P. (2007b). In tomato yellow leaf curl virus diseases: Management, Molecular biology breeding for resistance, Sources of resistance, Inheritance and location of genetic loci conferring resistance to members of the tomato Infecting *begomoviruses*, ed Czosnek, H. (kluwer, Dordrecht, The Netherlands). pp 343-362.

Ji, Y., Scot, J. W., Maxwell, D. P. and Schuster, D. J. (2008). Ty-4, a tomato yellow leaf curl virus resistance gene on chromosome 3 of tomato. *Rep. Tomato Genet. Coop.*, **58**:29-31.

Ji, Y., Scot, J. W., Schuster, D. J. and Maxwell, D. P. (2009). Molecular mapping of Ty-4, a new tomato yellow leaf curl virus resistance locus on Chromosome 3 of tomato. *J. amer. Soc. Hort. Sci.*, **134**:281-288.

Joshi, A. B. and Hardes, M. W. (1956). Alloploid nature of okra. *Abelmoschus esculentus* (L.) Moench. *Nature*, **178**:1190.

Joshi, A. B., Godwal, V. R. and Hardas, M. W. (1974). Okra. In: Evolutionary studies in world crops. Diversity and change in the Indian subcontinent. J. B. Hutchinson (ed) Cambridge University press. Cambridge. Pp 99-105.

Julian, O., Hessair, J., Corella, S., Lolli, I. d., Soler, S., Diez, M. S. and Perez-de castro, A. (2013). Initial development of a set of introgression lines from *Solanum peruvianum* P1 126944 into tomato exploitation of resistance to viruses. *Euphytica*, **193**:183-196.

Kadirvel, R., delaPena, R., Schafleitner, R. et al (2013). Mapping of QTLs in tomato line FLA456 associated with resistance to a virus causing tomato yellow leaf curl virus disease. *Euphytica*, **190**:297-308.

Kang, W. H., Hoong, N. H. Yang, H. B., Kwon, J. K., Jo, S. H., Seo, J. K., Choi, D. and Kang, B. C. (2010). Molecular mapping and characterization of a single dominant gene controlling CMV resistance in pepper (*Capsicum annuum* L.). *Theor. Appl. Genet.* **120**:1587-1596.

Kennedy, G. G. (2007). Resistance in tomato and other lycopersicon species to insect and mite pests. In: Razdox, M. K., Mattoo, A. K. (eds) Genetic improvement of Solanaceous crop, vol. 2 *Tomato Science Publ.*, Exfield, NH, USA, pp. 488-519.

Kenyon, Lowrence, Kumar, Sanjeet, Tsai, Wenshi, Hughes, Jacueline, dA. (2014). Virus diseases of peppers (*Capsicum spp.*) and their control. *Adv. Virus Res.* **90**:297-354. doi:10-1066/B978-0-861246.00006-8.

Kho, Y. D., den Nijs, A. P. M. and Franken, J. (1980). Interspecific hybridization in Cucumis (L) 2. The crossabbility of species on investigation of in vivo pollen tube growth and seed set. *Euphytica*, **29**:661-672.

Kim, S. H., Yoon, J. B. and Park, H. G. (2008b). Inheritance of anthracnose resistance in a new genetic resource, *Capsicum baccatum* P1594137. *J. Crop. Sci. Biotechnol.* **11**:13-16.

Kim, S. H., Yoon, J. B., Do, J. W. and Park, H. G. (2008a). A major recessive gene associated anthracnose resistance to *Colletotrichum capsici* in chilli pepper (*Capsicum annum*). *Breed. Sci.* **58**:137-141.

Kim, S., Kim, K. T., Kim, D. H., Yang, E. Y., Chao, M. C., Jamal, A., Cahe, y., Pae, D. H., Oh, D. G. and Hwang, J. K. (2010). Identification of QTL associated with anthracnose resistance in chilli pepper (*Capsicum* spp.). *Korean J. Hort. Sci. Technol.*, **28**:1014-1024.

King, S. R., Davis ,A. R., Zhang, X. and Croshy, K. (2010). Genetics, breeding and selection of rootstocks for Solanaceae and Cucurbitaceae. *Sci. Hort.* **127**:106-111.

Kishi, Y. and Fujishta, N. (1969). Studies on interspecific hybridization in the genus *Cucumis*. I. pollen germination and pollen tube growth in selfing and incompatibility crossing. *J. Jpn. Soc. Hort. Sci.* **28**:329-334.

Kishi, Y. and Fujishta, N. (1970). Studies on interspecific hybridization in the genus *Cucumis*. II. Pollen tube growth, fertilization and embryogenesis of post fertilization stage in incompatible crossing. *J. Jpn. Soc. Hort. Sci.* **39**:51-57.

Kroon, G. H., Custers, J. B. M., Kho, Y. O., den Nijs, A. P. M. and Varekamp, H. Q. 1979, Interspecific hybridization in *Cucumis sativus* (L). Need for genetic variation, biosystematic relations and possibilities to overcome crossability barriers. *Euphytica*, **28**:723-728.

Kumar, N. K. K., Sadashiva, A. T. (1996). *Solanum macrocarpon* a wild species of brinjal resistant to brinjal shoot and fruit borer, *Leucinodes orbonail* (Guen.). *Insect Envron.* **2**(2):41-42.

Kumar, N. K. K., Ullman, D. E. and Cho, J. J. (1995a). Resistance among *lycopersicon* species to *Frankliniella occidentalis* (Thysanoptera: Thripidae). *J. Econ. Entom.*, **88**:1057-1065.

Kumchai, J., Wei, Y. G., Lee, C. Y., Chen, F. C. and Chin, S. W. (2013). Production of intraspecific hybrids between commercial cultivar of the eggplant (*Solanum melongena* L.) and its wild relatives *S. torvum. Genet. Mol. Res.* **12**(1):755-764.

Kwon, T. R., Pae, D. H., Shin, Y. A., Oh, D. G. and Lee, J. M. (2006). Capsicum pepper: vital crop of Korea. *Chron. Horticul.* **46**:16-19.

Labeda, A. (1996). Resistance in wild *Cucumis* species to two spotted spider mites (*Tetranychus urticae*). *Acta Phytopathol. Entom. Hung.* **31**, 247-252.

Labeda, A., Kubalakova, M., Kriskova, E., Navratilova, B., Dolazel, K., Dolazel, J. and Lysak, M. (1999). Morphological and physiological characteristics of plant issued from interspecific hybridization of *Cucumis sativus X C. melo. Acta Hort.* **492**, 149-155.

Labeda, A., Windrlechner, M. P., Staub, J., Ezura, H., Zalapa, J. and Kriskova, E. (2007). Cucurbits (*Cucurbitaceae, Cucumis* spp, *Cucurbita* spp) Chapter 8, 271-376 In: Singh, R. (ed) Genetic resources, chromosome engineering and crop improvement. Series, Vol. 3- Vegetable crops. CRC press, Boca Raton, FL, USA.

Laterrot, H. (1978). Resistance gux maladies. B. *pyrenochaeta lycopersici*. Report d' Activite INRA station d'Ameloration desplantes Maraicheres, 1977-1978:102-103.

Laterrot, H. and Pecaut, P. (1969). Gene Tm2 new source. *Tomato Genet. Coop. Rep.* **19**:13-14.

Laura Toppino, Giampiero, Vale, Giuseppe, Leanordo, Rotino. (2008). Inheritance of *fusarium* wilt resistance introgressed from *Solanum aeothiopicum* Gilo and Aculeostum groups into cultivated eggplant (*S. melongena*) and development of associated PCR- based markers. Mol Breeding, **22**:237-250.

Lecoq, H., Wisler, G., Pitrat, M. (1998). Cucurbit viruses: the classics and the emerging. In: McCreight, J. D. (Ed) Cucrbitaceae. Evaluation and enhancement of cucurbits germplasm, ASHS press, Alexandria, pp. 126-142.

Lee, J. M. (1994). Cultivation of grafted vegetables. I. Current status, grafting methods and benefits. *Hort. Science,* **29**:235-239.

Lee, J. M. (2003). Advances in vegetable grafting. *Chronica Horticulturae,* **43**(2):13-19.

Lee, J., Hong, H., Do, J., Yoon, J., Lee, J. D., Hong, J. H., Do, J. W. and Yoon, J. B. (2010). Identification of QTLs for resistance to anthracnose to two *Colletotrichum* species in pepper. *J. Cro Sci. Biotechnolo.,* **13**:227-233.

Lee, S. H., Chung, G. C. and Steudle, E. (2005c). Low temperature and mechanical stress differently gate aquaporins of root cortical cells of chilling sensitive cucumber and resistant fig leaf gourd. *Plant Cell Environ.* **28**:1191-1202.

Lester, R. N., Hasan (1991). Origin and domestication of the brinjal eggplant, *Solanum melongena* from *S. incanum* in Africa and Asia : In: Hawkes, JG., Lester, R. N., Nees, M., Estrade, N. (eds) *Solanaceae.* III. Taxonomy, chemistry, evolution. Royal Botanical Garden, Kew, pp. 369-387.

Li, J., Liu, L., Bai, Y., et al (2011). Identification of mapping of quantitative resistance to late blight (*Phytophthora infestans*) in *Solanum habrochaites* LA 1777. Euphytica, **179**(3):427-438.

Lin, Q., Kanchana-udomkam, C., Jaunet, T., Mongkolporn, O., Lin, Q., Chutchamas, K. U. and Orarat, M. (2002). Genetic analysis of resistance caused by *Colletotrichum capsici. Thai. J. Agric. Sci.,* **35**(2002):259-264.

Lin, S. W., Gniffke, P. A. and Wang, T. C. (2007). Inheritance of resistance for pepper anthracnose caused by *Colletotrichum acutatum. Acta Hortic.* **760**:329-334.

Lopez, R., Amnon, L., Shepard, B. M., Simmons, A. M. and Jackson, D. M. (2005). Sources of resistance to two spotted spider mite (*Acarii tetranychidae*) in *Citrullus* spp. *Hort. Science,* **40**(6):1661-1663.

Madhure, S., Chowdappa, P. and Pavani, K. V. (2015). Genetic structure, aggressiveness and fungicidal sensitivity of *phytophthora* associated with foliar blight of hot peppers. *Third International Symposium on phytophthora*, Bangalore, 9-12 Sept. 2015.

Mahasuk, P., Khumpeng, N., Waree, S., Taylor, P. W. J. and Mongkolporn, O. (2009a). Inheritance of resistance to *anthracnose* at seedling and green and red chilli fruit stages. *Plant Breed.* **128**:701-706.

Mahasuk, P., Taylor, P. W. J. and Mongkolporn, O. (2009b). Identification of two new genes conferring resistance to *Colletotrichum acutatum* in *Capsicum baccatum. Phytopathology,* **99**:1100-1104.

Malathi, V. G. (2013). Biology and Pathogenesis of whitefly transmitted Begomoviruses causing tomato leaf curl disease- Indian and global scenario. Presentation in: *Tomato Breeders Round table* 6-8 Feb, 2013, Chiang Mai, Thailand.

Maliepaard, C., Bas, N., Van Heusden, S., Kos, J., Pet, G., VerKer K, . R., Vrielink, R., Zabel, P. and Lindhout, P. (1995). Mapping of QTLs for grandular trichome densities and *Trialeurodes vaporariorum* (green house white fly) resistance in an F_2 from *Lycopersicon esculentum x L. hirsutum F glabratum. Heredity,* **75**:425-433.

Martyn, R. D. and Netzer D. (1991). Resistance to races 0, 1 and 2 of fusarium wilt of watermelon in *Citrullus* sp P1296341-FR. *Hort. Sci.* **26**:429-432.

Marukawa, S. and takatsu, l. (1969). Studies on the selection of *Cucurbita* species as cucumber root stock. *Bul. Ibaraki Hort. Expt. Sta.* **3**:11-18.

McCandless, L. (1998). Geneva releases 'Whitekar' summer squash at the NYS vegetable conference. 2011.

McCarthy, W. H., Wehner, T. C., Xie, J. and Daub, M. E. (2001). Isolation and callus production from cotyledon protoplasts of *Cucumis metulliferous. Cucurbit. Gent. Coop.* **24**:102-106.

Merk, H. L., Ashrat, H. and Foolad, M. R. (2012). Selective genotyping to identify late blight resistance genes in an accession of tomato wild species *Solanum pimpinellifolium, Euphytica,* **187**:63-75.

Mohammed Ali and Kunimitsu Fujieda (1989-1990). Cross compatibility between eggplant (*Solanum melongena* L.) and wild relatives. *J. of the Jap. Soc. For Hort. Sci.* **58** (4):977-984.

Momotaz, A., Scott, J. W. and Schuster, D. J. (2010). Identification of QTL conferring resistance to *Bemisia tabaci* in an F_2 population of *Solanum lycopersicum* x *S. habrochaites* accessions LA 1777. *J. Amer. Soc. Hort. Sci.* **135**:134-142.

Mongkolporn, O.and Taylor, P. W. J. (2011). Capsicum. In: Kole, C. (ed) Wild Crop Relatives. Genomic and Breeding Resources, vol. 5, Vegetables, Springer, Berlin, pp. 43-57.

Morra, L. (2004). Grafting in vegetable crops. In: Tognoni, F., Pardassi, A., Mensuali-sodi, A., Dimauro, B. (Eds0, The production in the greenhouse after the era of the methyl bromide, Comiso, Italy, pp. 147-154.

Moury, B. and Verdin, E. (2012). Viruses of pepper in the Mediterranean basins: a remarkable stasis. *Adv. Virus Res.* **89**:127-162.

Munger, H. M. (1976). *Cucurbita martinezii* as a source of disease resistance. *Veg. Imp. Newsletter,* **18**:4.

Muniyappa, Jalikop, V., Saika, S. H., Chennarayappa, A. K., Shivashankar, G., Ishwara, Bhat, A. and Ramappa, H. K. (1991). Reaction of *Lycopersicun* cultivars and wild accessions of tomato leaf curl virus. *Euphytica,* **56**:37-41.

Munns, R. and Tester, M. (2008). Mechanism of salinity tolerance. *Annual Review of Plant Biology,* **59**:651-681.

Murphy, J. F., Blauth, J. R., Livingstone, K. D., Lackney, V. K. and John, M. K. (1994) .Genetic mapping of *pvr1* locus in *Capsicum* spp. and evidence that distinct *potyvirus* resistance loci control response that differ at the whole plant land cellular levels. *Mol. Plant Microbe Interact.* **11**:943-951.

Narasegowda Maruthi, M., Czosnek, H., Vidovski, F., Tarba, S. Y., Milo, J., et al., (2003). Comparison of resistance to tomato leaf curl virus (India) and tomato yellow leaf curl virus (Israel) among lycopersicon wild species, *Breeding Lines and Hybrids,* **109**:1-11.

Nerkar, Y. S. (1991). The use of related species in transferring disease and pest resistance genes to okra. In: Report of an *International workshop on okra genetic resources* held at NBPGR, New Delhi, India, 8-12 Oct. 1990. IBPGR. pp. 110-113.

Nicot, P. C., Moretti, A., Romiti, C., Bardin, M., Caranta, C. and Ferriere, H. (2002). Differences in susceptibility of pruning wounds and leaves to infection by *Botrytis cinerea* among wild tomato accession. *Tomato Genet. Coop. Rep.* **52**:24-26.

Niemirowicz-Szczytt, K. and Kubicki, B. (1979). Cross fertilization between cultivated species of genera *Cucumis* (L) and *Cucurbita* (L). *Genetica polonica,* **20**:117-125.

Nishio, T., Mochizuki, H. and Yamakova, K. (1984). Interspecific crosses between eggplant and related species, *Bull, Veg. Ornamental crops, Res. Stn.* A. No.**12**, 59-64.

Niss den, APN and Visser, D. L. (1985). Relationship between African species of the genus *Cucumis* L. estimated by the production, vigour and fertility of F_1 hybrids. *Euphytica,* **34**:279-290.

Nombela, G., Beitia, F. and Muniz, M. (2000). Variation in tomato response to *Bemisia tabaci* (Homoptera: Aleyrodidae) in relation to acylsugae content and presence of the nematode and potato aphid resistance gene Mi. *Bulletin of Entomological Res.,* **90**:161-167.

Nombela, G., Williamson, V. M. and Muniz, M. (2003). The root knot nematode resistance gene Mi-1.2 of tomato is responsible for resistance against the white fly *Bemisia tabaci. Molecular Plant Microbe Interactions,* **16**:645-649.

Norton, J. D. and Granberry, D. M. (1980). Characteristics of progeny from an interspecific cross of *Cucumis melo* with *Cucumis metuliferus. J. Amer. Soc. Hort. Sci.* **105**:174-180.

O'connell, M. A., Medina, A. L. and Sanchez Pena Pand Trevino, M. B. (2007). Molecular genetics of drought resistance in tomato and related species. In: Razdan, M. K. and Mattoo, A. K. (eds). Genetic improvement of Solanaceous crops, vol.2, Tomato: Science publishers, Enfield, USA, 261-283.

Okimura, M., Matso, S., Arai, K. and Okitsu, S., (1986). Influence of soil temperature on the growth of fruit vegetable grafted on different rootstocks. *Bull. Veg. Ornam. Crops. Res. Stn. Japan.,* C**9**:43-58.

Olimpia Gomez Consuegra, Mayte Pinon Gomez, Yamila Martinez Zubiaur (2015). Pyramiding TYLCV and TSWV resistance genes in tomato genotypes. *Rev. Protection Veg.* Vol.30, No. 2, La Habana mayo-ago.

Pakdeevaraporn, P., Wasee, S., Taylor, P. W. J. and Mongkolporn, D. (2005). Inheritance of resistance to anthracnose caused by *Colletotrichum capsici* in Capsicum. *Plant Breed.* **124**:206-208.

Pal, B.P., Singh, H.B., and Swarup, V., (1952). Taxonomic relationships and breeding possibilities of species *Abelmoschus* related to okra (A. *esculentus*). *Bot. Gaz.* **113**:455-464.

Palada, M.C. and Wu, D.L.(2008). Evaluation of chilli rootstocks for grafted sweet pepper production during the hot wet and hot dry season in Taiwan. *Acta Horticulturae,* **767**:167-174.

Park, H. K., Kim, B. S. and Lee, W. S. (1990a). Inheritance of resistance to anthracnose (*Colletotrichum* spp) in pepper (*Capsicum annuum* L). I. Genetic analysis of anthracnose resistance by diallel crosses. *J. Kor. Soc. Hortic. Sci.* **31**:91-105.

Park, H. K., Kim, B. S. and Lee, W. S. (1990b). Inheritance of resistance to anthracnose (*Colletotrichum* spp) in pepper (*Capsicumannuum* L). II. Genetic analysis of resistance to *Colletotrichum dematium. J. Kor. Soc. Hortic. Sci.* **31**:207-212.

Park, K. S. and Kin, C. H. (1992). Identification, distribution and etiological characteristics of anthracnose fungi of red pepper in Korea. *Korean J. Plant Pathol.***8**:61-69.

Parlevliet, J. E. (2002). Durability of resistance against fungal, bacterial and viral pathogen; present situation. *Euphytica,* **124**:147-156.

Parrella, G., Ruffel, S., Moretti, A., Morel. C., Palloix, A. and Caranta, C. (2002). Recessive resistance genes against *potyviruses* are localized in collinear genomic regions of the tomato (*Lycopersicon* spp) and pepper (*Capsicum* spp) genomes. *Theor. Appl. Genet.* **105**:855-86.

Paterson, R. G. Scott, S. G., Gergerich, R. C. (1989). Resistance in two *Lycopercicon* species to an Arkanses isolate of tomato spotted wilt virus. *Euphytica,* **43**:173-178.

Peral-Treves, R., and Galun, E. (1985). The cucumis plastome: physical map, intrageneric variation and phylogenetic relationships. *Theor. Appl. Genet.* **71**:417-429.

Peral-Treves, R., Zamir , D.,Navot, N., and Galun, E. (1985). Phylogeny of cucumis based on isozyme variability and its comparison with plastome phylogeny. *Theor. Appl. Genet.* **71**: 430-436.

Pereira-Carvalho, R.C.,Boiteux, L.S.,Fonseca, M. E. N., et al., (2010). Multiple resistance to *Meloidogyne* spp. In wild solanum (*Lycopersicon*) accessions. *Plant Disease,* **94**(2):179-185.

Perez de Castro A, Julian, O.,Diez, M.(2007). Genetic control and mapping of *Solanum chilense* LA 1960 and LA 1971-derived resistance to Tomato yellow leaf curl disease-*Euphytica* :1-12.

Pernezny, K., Robert, P. D., Murphy, J. F., Goldberg, N. P. (2003). Compendium of pepper diseases. *The American Phytopathological Society Minnesto.* pp. 24-25.

Persley, D.M.,McGroth, D.,Sharman,M.and Walker I.O.(2010). Breeding for tospo vorus resistance to capsicum & tomato in Australia. IXth International symposiums on Thysanoptera & Tospo viruses,31st August- 4th Sept.2009.,*Journal of Insect Science,* **10**:166, Available online; Insectscience.org/10.166.

Pico, B., Diez, M.J.&Nuez, F.(1996). Viral diseases causing the greatest economic losses to the tomato crop.II the tomato yellow leaf curl virus-a review. *Scientia Horticulturae,* **67**:151-196.

Pilowsky, M. &Cohen, S.(1990). Tolerance to tomato yellow leaf curl virus derived from *Lycopersicon peruvianum. Plant Disease,* **74**:248-250.

Pitblado,R.E.,Kerr,E.A.(1980). Resistance to bacterial speck (*Pseudomonas* tomato) in tomato. *Acta Horticulturae,* **100**:379-382.

Prabhu, A.S., Pathak, K.D., Singh, R.P., (1971). Powdery mildew of 'bhindi' (*Abelmoschus esculentus* (L.) Moench) in Delhi state. *Indian Journal of Horticulture,* **28:** 310-312.

Prabhu, T., and Warde, S.D. (2013). Crossability studies in genus *Abelmoschus. Vegetable science,* **40**(1):11-16.

Prasanna, H.C., Sinha, D.P.,G.K.,Krishna, Ram,Kashyap;S.P.Singh, Nikhil,Singh, M and Malathi,V.G.(2014). Pyramiding of tomato yellow leaf curl virus resistance gene derived from *Solanum habrochaites* and *Solanum chilens* by marker assisted selection. Indian Society of Vegetable science: National symposium on abiotic and biotic stress management in Vegetable crops.

Prohens, J., Whitekar, B. D., Plazzas, M., Vilonova, S., Hurta do, M., Blasco, M., Gramazio, P., Stommel, J. R. (2013). Genetic diversity in morphological characters and phenolic acid content resulting from an interspecific cross between eggplant, *Solanum melongena* and its wild ancestors (*S. incanum*). *Ann. Appl. Biol.* **162**:242-257.

Provvidenti, R. (1980). Viral diseases and genetic sources of resistance in cucurbits. Program and Abstracts: Conference on the biology and chemistry of *Cucurbitaceae.* August: 3-6, 1980, CornellUniversity, Ithaca, Newyork.

Provvidenti, R. (1989). Sources of resistance to viruses incucumber, melon, squash and watermelon. Pages 29-36. In: Proceedings of *Cucurbitaceae* spp.Evaluation and enhancement of cucurbit germplasm : 29-36.us Dept.Agric.Res. Serv., U.S.Vegetable laboratory, Charleston, SC.

Provvidenti, R. (1990).Viral diseases and genetic sources of resistance in Cucurbita species. Pg. 427-435. Biology and utilization of the Cucurbitaceae (D.M.Bates, R.W.Robinson and Ch.Jeffery, eds) Comstock publ. Assoc. Cornell university press, Ithaca&London.

Provvidenti, R. (1991). Inheritance of resistance to the florida strain of zucchini yellow mosaic virus in Watermelon. *Hort. Science,* **26**:407-408.

Puangmalai, P., Potapohn, N., Akarapisarn, A., Pascho, H. J. (2013). Inheritance of tomato necrotic ring virus resistance in *Capsicum annuum. J. of Agricultural Sci.* 5(2):129-133.

Puangmalai, P., Potapohn, N., Akarapisaru, A., Cheewa chaiwit, S. and Insuan, N. (2013). Screening capsicum accession for tomato necrotic ringspot virus resistance **Cmv**. *J. Nat. Sci.***12**(1)35.

Q.I, Zhang. Enda, Yu., and Andy Media (2012). Development of advanced interspecific bridge lines among *Cucurbitapepo, C. maxima, and C. moschata. Hort. Science,* **47** (4):452-458

Rai, N., Tiwari, S. K., Kumar, R., Singh, M. and Bhardwaj, D. R. (2011). Genetic resources of *Solanaceous* Vegetables in India. *National symposium on vegetable biodiversity,* Jawaharlal Nehru Krishi Vishwa Vidyalaya, Jabalpur, MP, April, 4-5, 2011, pp. 91-303.

Rai,V.P., Kumar,R.,Singh, S.P., Kumar, Kumar,S., Singh,M and Rai, M. (2014). Monogenic recessive resistance to pepper leaf curl virus in an interspecific cross of capsicum. *Scientia Horeticulrae,* **172**: 34-38.

Rajsekharan, S. (1969). Cytogenetic studies on the interrelationships of some common solanum species occurring in South India. *Annamalai Uni. Agri. Res. Annual.* **1**:49-60.

Rana, R. S., Thomas, T. A. and Arora, R. K. (1991). Plant genetic resources activities in okra- an Indian perspective. In: Report of an *International workshop on okra genetic resources,* held at NBPGR, New Delhi, India, 8-12 Oct, 1990, IBPGR, Rome, pp. 38-47.

Rao,E.S. Kadirvel, P., Symonds, R.C. & Ebert, A.W. (2013). Relationship between survival and yield related traits in *Solanaum pimpinellifolium* under salt stress . *Euphytica,* **190** (2):215-228.

Reddy, K. M. and Reddy, M. K. (2010). Breeding for virus resistance, in: Kumar, R., Rai, A. B., Rai, M. and Singh, H. P. eds. Advances in chilli research, Stadium Press vpt. New Delhi, India, pp. 119-132.

Reddy, M. K., Srivastava, A., Kumar, S., Kumar, R., Chawda, N., Ebert, A. W. and Vishwakarma, M. (2014). Chilli (*Capsicum annuum* L.) breeding in India: An overview. SABRAO Journal of Breeding and Genetics, **46** (2): 160-173.

Reifschneider, F. J. B., Boiteux, L. S., Veechia, P. T. P., Poulos, J. M. and Kuroda, N. (1992). Inheritance of adult plant resistance to *Phytophthora capsici* in Pepper. *Euphytica,* **62**:45-49.

Reis, A., Girdano, L. B., Lopes, Carlos. and Boiteux, L. (2004). Novel sources of multiple resistances to three races of *Fusarium oxysporum* f.sp. *lycopersici* in Lycopersicon germplasm. *Crop Breeding and Applied Biotechnology,* **4**:495-502.

Rhodes, A. M. (1959). Species hybridization and interspecific gene transfer in the genus Cucurbita. *Proc. Amer. Soc. Hort. Sci.* **74**:546-552.

Ribeiro, C.S.C, Lopes, C.A. Carvalho, S .I.C., Henz, G.P., Reifschneider, F.J.B. (eds) Pimentas Capsicum (2008). 200P, *Embrapa Hortalices.*

Rick, C.M. (1982). The potential of exotic germplasm for tomato improvement . In: Vasil, I.K. , Scowcroft, W.R. & Frey, K.J. (eds, Plant improvement and somatic cell genetics. Academic Press, New York, USA, pp. 1-28

Rizza, F., Mennella, G., Collonnier, C., Chiachakar, D., Kashyap, V., Rajan, M. V., Presterc, M. and Rofino, G. L. (2002). Androgenic dihaploid from somatic hybrids between *S. melongena* and *S. aethiopicum* group gilo as a source of resistance to *Fusarium oxysporum* f. sp. *melongena*. *Plant Cell Rep.* **20**:1022-1032.

Robertson,L.D & Labate, J.A. (2007). Genetic Resources of tomato *Lycopersicon esculentum* Mill) and wild relatives. In: Razdan, M.K. and Mattoo, A.K. (eds), Genetic improvement of solanaceous crops. Tomato. Vol.2. Science Publishers Inc. Enfield, New Hampshire, USA, pp: 25-75.

Robinson, R. W. and Decker-walter, D. S. (1999). Cucurbits. CAB International, New York, NY.

Romero, L., Belaklier, A., Ragala, L. and Ruiz, M. J. (1997). Response of plant yield and leaf symptoms to saline conditions: Effectiveness of different rootstocks in melon plants (*Cucumismelo* L.). *Soil Science Plant Nutrition*, **43**:855-862.

Rosello, S., Diez, M.J Nuez, F., (1998). Genetics of tomato spotted wilt virus resistance coming from *Lycopercon peruvianum*, European Journal of *Pant Pathology* , **104**:499-509

Rosenthal, J. P. and Dirzo, R. (1997). Effects of life history, domestication and agronomic selection of plant defence against insects: Evidence from maizes and wild relatives. *Evolutionary Ecology*, **11**, 337-355.

Rouphael, Y. M., Carderelli, M., Colla, G. (2008). Yield, mineral composition water relations and water use efficiency of grafted mini watermelon under deficit irrigation. *Horti.Sci.***43**:730-736.

Rowe, R.C., Farley,J.D (1981). Strategies for controlling *Fusaruim* Crown and root rot in green house tomatoes. *Plant Dis.*, **65**: 107-112.

Rubio, Manuel., Nicolai, Maryse., Caranta, Carole., and Palloix Alain (2009). Allele mining in the pepper gene pool provided ne complementation effects between pvr2-el F4E and pvr6-el F (ISO) 4 E alleles for resistance to pepper venal mottle virus. *Journal of general virology*, **90**: 2808-2814.

Ruiz, J. M. Belakhir, L., Rangala, J. M. and Romer, L. (1997). Response of plant yield and leaf pigments to saline conditions: Effectiveness of different root stocks in melon plants (*Cucumismelo* L). *Soil Science Plant Nutrition*, **43**:855-862.

Sadashiva, A. T. (2014) Breeding tomato for resistance to early blight in combination with bacterial wilt and tomato leaf curl virus presentation:ASRT-2014-Hotel Lalit Ashok, Bangalore, India.

Sakata, Y., Ohara, T. and Sugiyama, M. (2007). The history and present state of grafting of cucurbitaceous vegetables in Japan. In: McConchie, G. R. (ed), *Proceedings of the third international Symposium on Cucurbits. Acta. Hortic.* 159-170.

Sandlerink, J.M., Van Ooijen , J.W., Purimahua, C. C. *et al.*, (1995). Localization of genes for bacterial canker for resistance to *Lycopersicom peruvianum* using RFLPs. *Theoretical and Applied Genetics,* **90**: 444-450

Sarath Babu, B., Pandravada, S., Prasad Rao, R., Antha, K., Chakrabarthy, S. and Varaprasad, K. (2011). Global sources of pepper genetic resources against arthropods, nematodes and pathogens. *Crop Port.*, **30**: 389-400.

Scholth of,K.B.G., Adkins, S., Czosnk,H., Palukaitis, P., Jaquot,E., (2011).Top 10 plant viruses in molecular plant pathology. *Molecular plant Pathology***12**: 938-954.

Scott, J. W., and Jones, J.P. (1990). Soil borne fungal resistance to *Lycopersicon pennellii* accessions . *Hort sci.*, **25**:1068.

Scott, J. W., Gardener R.G. (2007). Breeding for resistance to fungal pathogen. In : Razdan, M.K. and Mattoo, A.K. (eds) Genetic improvement of solanacieous crops. Vol.2: Tomato. Science Publisher, Enfield, NH, USA. pp. 421- 425.

Scott, J.W. (1993). Breeding tomatoes for resistance to high temperature, biotic and abiotic diseases. *Horticulture Brazileria*, **11**: 167- 170.

Scott, J.W. 2007. Breeding for resistance to viral pathogens,. In. M.K. Razden & A.K. Mattoo (eds) Genetic improvement of *Solanaceous* Crops.: Tomato. Science Publisher, Enfield, NH.P. **2**: 447-474.

Seshadri, V. S. and Srivastava, U. (2002). Evaluation of vegetable genetic resources with special reference to value addition. In: *Proceedings of international conference on vegetables: Vegetable for sustainable food and nutrition security in the new millennium,* Organized by *Prem Nath Agricultual Science foundation, Bangalore,* 11-14 Nov, 2002, CDROM publication paper, No 1-2-L. pp. 41-62.

Sharma, B. R. (1982). Punjab Padmini- a new variety of okra. *Prog. Farm.* **82**:15-16.

Shekasteband ,R., Hotton, .S.F., Levin, I., Lapidot, M. and Scott, J.W. (2014). Begomovirus resistance and yield of tomato lines with various combinations of Ty-3 , Ty-4 and Ty-5 genes and update of the fine mapping of the Ty-4 genes . Proc.Tomato breeders Round table (TBRT) , Mountain Horticultural Crops Research & Extension etc., MillsRiver , NK:7.

Shifriss., O. (1987). Interspecific hybridization- Notes on squash breeding. *Cucurbit Genetic Coop. Rep.* **10**:93.

Singh, A. K. and Yadav, K. S. (1984b). Cytogenetics of Cucumis: IV Comparative study of natural and induced polyploids. *Cytologia.* **49**:183-192.

Singh, B., Rai, Muthura, Kalloo, G., Satpathy, S. and Pandey, K.K., (2007). Wild texa of okra (*Abelmoschus* species) reservoir of genes for resistance to biotic stresse: *Acta. Horti.* **752**:323-328.

Smith, P. G. (1944). Embryo culture of tomato species hybrid. *Proc. American Soc. Hort. Sci.* **44**:413-416.

Smith, P. G., Kimble, K. A., Grogan, R. G., and Millet, A. H. (1967). Inheritance of resistance in peppers to phytophthora root rot. *Phytopathology*, **57**:377-379.

Snyder, J. C., Simmons, A. M. and Thacker, R. R. (1998). Atrractancy and ovipositional response of adult *Bemisiaargentifolii* (Homoptera Aleyrodidae) to type IV trichom density on leaves of *Lycopersicon hirsutum* grown in three day length regimes. *Journal of Entomological Science.* **33**:270-281.

Soria, C., Gomez-Guillamen, M. L., Esteva, J. and Nuez, F. (1990). Ten interspecific crosses in the genus Cucumis:A preparatory study to seek crosses resistant to melon yellow disease. *Cucurbit Gent. Coop. Rep.* **13**:31-33.

Srivastava, Umesh (2009). Genetic resources In: Okra Handbook: *Global production, processing and crop improvement* (eds) B. S. Dhanakar and Ram Singh. HNB, New York.

Stebbins, G. L. (1956). Artificial polyploidy as a tool in plant breeding and genetics. In: *Plant breeding Book haven symposium in Biology*. pp. 37-52.

Stemova, L.,(2005). Resistance to verticillium dahliae race-2 and its introgression in to processing tomato cultivar. *Acta Horticulturae*, **695**: 257- 262.

Stevens, M.R., Lamb, E.M., and Rhoads, D.D.(1995). Mapping the Sw-5 locus for tomato spotted wilt virus resistance in tomato using RAPD and RFLP analysis. *Theoretical and Applied Genetics,* **90**:451-456.

Stevens, M.R., Scott S. J., Gergerich. R.C. (1992). Inheritance of a gene for resistance to Tomato Spotted Wilt Virus (TSWV) from *Lycopersicon pervianum* Mill. *Euphytica,* **59**:9-17.

Stevens, R., Price, D. L., Memmott, F.D., Scott, J. W. and Obson, S. M. (2007). Identification of markers linked to SW 7, a new tomato spotted wilt virus resistance gene, derived from *S. chilense. Tomato Breeders Round table*, The Pennsylvania State University, Penn-Sylvania.

Strange, E. B., Guner, N., Pesic-VanEsbroek, Z., Wehner, J. C. (2002). Screening the watermelon germplasm collection for resistance to papaya ringspot virus type-w. *Crop Sci.,* **42**:1324-1330.

Stravato, V. M., Cappelli, C., Polverari, A. (1993). *Fusarium oxysporum* f. sp. *melongenae,* against agent of wilting of Aubergine. *Inf. Fitopatol.,* **43**:51-54.

Sureshbabu, K. V. and Dutta, O. P. (1990). Pollen fertility studies in *Abelmoschus* spp. *South Indian Hort.* **38**:109.

Sureshbabu, K. V. and Dutta, O. P.(1990). Cytogenetic studies of the F_1 hybrids (*Abelmoschusesculentus* L. moench) x *Abelmoschus tetraphyllus* and its amphidiploid. *Agric. Res. J. Kerala.,* **28**:22-25.

Sy, O., Bosland, P., Steiner, R. (2005). Inheritance of resistance to *Capsicum annuum. J. Amer. Soc. Hort. Sci.,* **130**:75-78.

Tachibana, S. (1994). Eggplant. In: Koshini, K., Iwahori, S., Kitagawa, H., Yakuwa, T. (eds.), Horticulture in Japan. AsaKura Puplishing, Tokyo, Pp. 63- 66.

Tachibana, S., (1988). Cytokinin concentration in roots and root xylem exudates of cucumber and fig leaf gourd as a factor by root temperature. *J. Japan. Soc. Hortic. Sci.* **56**:417-425.

Tachibana, S., (1982). Effect of root temperature on the rate of water and nutrient absorption in cucumber cultivars and fig leaf gourd. *J. Japan. Soc. Hortic. Sci.* **55**: 461-467.

Tal, M., Katz, A., Heikin, H. and Dehan, K. (1979). Salt tolerance in the wild relatives of the cultivated tomatoes : Proline accumulation in *Lycopersicon esculantum* Mill, *Lycopersicom peruvianum* Mill and *Soalnum perniellii* Cor. Treated with Nacl and polyethylene glycol. *New Phytologist,* **82**:349- 355.

Taylor, P.W.J., Mongkolporn, O., Than, P.P., Montri, P., Ranathunga, N., kanchana-udonkan, C., Ford, R., Pongsupasamit, S., Hyda, K.D.(2007). Pathtypes of Colletotrichum spp. Infecting chilli pepper and mechanism of resistance, In: Oh, D.G. and Kim, K.T.(eds) First international symposium on chilli anthracnose, Abstracts….. Seoul, *South Korea.National Horticultural Research Institute.* P.29.

Thachibana, S., (1987). Comparision of effects of root temperature on the growth and mineral nutrient absorption in cucumber cultivars and fig leaf gourd. *J. Japan. Soc. Hortic. Sci.* **51**:299-308.

Thakur , M.R. and Arora, S.K. (1986). Okra, In : vegetable crops in India. (eds). Bose, T.K. and Som, M.G.. Naya Prakash, Calcutta. India. pp. 606 -622.

Thakur , M.R. and Arora, S.K. (1988). Punjab-7, a virus resistant variety of okra. *Prog. Farming.* **24**: 13.

Thies, J.A. and Levi, A. (2003). Resistance of watermelon germplasm to the peanut root knot nematode. *Hort. Sci.* **38**: 1417- 1421.

Thies, J.A. and Levi, A. (2007). Characterization of watermelon (*Citrullus lanatus* var. citroides) germplasm for resistance to root knot nematode. *Hort. Sci.* **42** : 1530- 1533.

This J.A., Levi Amon, Jennifer J. Aries., and Richard L. Hassell. (2015). RKVL- 318, a root nematode resistant watermelon line as rootstock for grafted watermelon. *Hort. Sci.* **50** (1): 141- 142.

Van Heuseden , A.W., Koornneef,M., Voorrips, R.E., et al., (1999). Three QTLs from *Lycopersicon peruvianum* confer a high level of resistance to *Clavibacter michiganesis* spp. *Michganensis. Theoretical and Applied Genetics,* **99**: 1068-1074.

Venema, J. H., Boukelien, E. D., Bax, J. E. M., Hasselt, P. R. V. and Elzenga, J. T. M. (2008). Grafting tomato (*Solanum lycopersicum*) onto the rootstock of a high altitude accession of Solanum habrochaites improves subtropical temperature tolerance. *Environmental and Experimental Botany,* **63**:359-367.

Venema., J.H., Posthumus, F., de Vries, M., Van Hasselt P.R. (1999b). Differential response of domestic and wild *Lycopersicon species* to chilling under low light : growth of carbohydrate content and xanthophyll cycle . *Physiol. Plant.,* **105**: 81-88.

Venema., J.H., Posthumus, F., Van . Hasselt, P.R., (1999a). Impact of suboptimal temperature on growth, photosynthesis, leaf pigments and carbohydrates of domestic and high altitude wild *Lycopersicon species. J. plant physiol.,* **155**: 711-718.

Verlaan, M.GHutton,S.F., Ibrahem,R.M., Koremelink, R., Visser, R.G.F., Scott.J.W *et al.,* (2013). The tomato yellow leaf Curl Virus Resistant Genes Ty-1 and Ty-3 are allec and Code for DFDGD-class RNA-dependent RNA polymerases. PLOS Genet 9(3): e 1003399.doi:101377/ journal .pgen .1003399.

Vidavsky,F., and Czosnek , H. (1998). Tomato breeding lines resistant and tolerant to tomato Yellow leaf curl virus issued from *Lycopersicon hirsutum. Phytopathology* . **88**: 910-914.

Voorrips,R.E., Finkers, R., Sanjaya, L.., Groenwald, R. (2004). QTL mapping of anthracnose (*Colletotrichum spp*) resistance in a cross between *Capsicum annuum* and *C. chinese* . *Theoretical and Applied Genetics*, **109**:1275-1282.

Walker, S., Bosland, P., (1999). Inheritance of phytopthora root rot and foliage blight resistance in peppers. *J. Amer.Soc.Hort.Sci.*, **1240**:14-18.

Wang J.F., Ho, F.I., Troung, H.T.H., *et al.*, (2013a). Identification of major QTLs associated with stable resistance to tomato cultivar "Hawaii7996" to *Ralstonia solanacearum* . *Euphytica*, **190**(2): 241-252.

Wang, J.F, Ho, F.L., Truong H.T.H., Huang S.M. *et al.*, Botatero C.H., Dittapongpitch, V.,*et al.*, (2013). Identification of marker QTLs associated with stable resistance of tomato cultivar "Hawaii 7996" to *Ralstonia solanacearum*. *Euphytica,* **190**: 241-252.

Wang, Y., Yang, W., Zhang, W. Han, Q., Feng, M, and Shen, H. (1913b). Mapping of a heat stable gene for resistance to Southern root knot nematode in *Solanumlycopersicum.Plant molecular Biology Reporter*, **31**(2): 352-362.

Webb,RE. (1997). Resistance to watermelon mosaic virus in *Citrullus lanatus* (Absts). *Proc. Am. Phytopatol. Soc.* **4**: 220.

Wessel_Beaver, L., Luevas, H. E., Andres, T. C. and Piperno, D. R. (2004). Genetic compatibility between *Cucurbita moschata* and *C. argyrosperma*. Progress in cucurbit genetics and breeding research, *Proc. of Cucurbitaceae 2000, The 8ᵗʰ EUCARPIA meeting on cucurbit genetics and breeding*, Palceky University, Olomone Czech Republic, pp.393-400.

Whener, T. C., Cade, R. M. and Locy, R. D. (1990). Cell, tissue and organ culture techniques for genetic improvement of cucurbits. In: Bates, D.,(eds) Biology and utilization of the *Cucurbitaceae,* Cornell University Press, Ithaca, New york, pp.367-381.

Whiteker, T.W. and G,N, Davis (1962). Cucurbits, Botany, Cultivation and utilization , . Interscience Publisher, New York.

Whiteker, T.W., Bemis, W.P. (1964). Evolution in the genus Cucurbitae. *Evolution,* **18**: 553- 559.

Williamson, V.M., (1998). Root knot nematode resistance genes in tomato and this potential for future use. *Annual Review of phytopathology*, **36**: 227-293.

Xu,Y.Kang, D., Shi,Z., Shen,H. Wehner, T.C.(2004). Inheritance of resistance to Zucchini yellow mosaic virus and watermelon mosaic virus in watermelon. *J. Hered.* **96**:498-502.

Yamakawa, K. (1982). Use of rootstocks in solanaceous fruit – vegetable production in Japan. *Japan. Agric. Res. Q.* **15**:175-179.

Yamakawa, K. and Mochizuki, H. (1979). Nature and inheritance of fusarium wilt resistance in eggplant cultivars and related wild Solanum species. *Bull. Veg. Orn. Crops. Res. Sta.*, **6**:9-27.

Yamakawa, K., and Nagata, N., (1975). Three tomato lines obtained by use of Chronic gamma radiation with combined resistance to TMV and *Fusarium* race 1-3. *Tech News Inst. Radiat. Breed.*16:2.

Yang, W. and Francis, D.M. 2007. Genetics and breeding for resistance to bacterial wilt diseases in tomato: prospects for marker assisted selection. In: Razdan, M.K. & Mattor, A.K. (eds), Genetic Improvement of Solanaceous Crops Vol.2. Tomato. Science Publishers, Enfield, pp. 379-419.

Yao, Minghua., Li, Ning., Wang, Fei., Ye, Zhibiao (2013). Genetic analysis and identification of QTLs for resistance to Cucumber Mosaic Virus in Chilli pepper (*Capsicum annuum* L.) *Euphytica,* **193**: 135-145.

Yashido, T., Monma, S., Matsunaga, H., Sakata, Y., Sato, T., Saito, T., Saito, A. and Yamada, T. (2004). New root stock 'Eggplant Ano2' with high resistance to bacterial wilt and fusarium wilt. *Proceedings of the 12th Eucarpia meeting on Genetics and Breeding of Capsicum and Eggplant,* May 17-19, 2014, Noordwigkerhout, The Netherlands, p88. (Abstract).

Yoon, J. B. and Park, H.G. (2005). Trispecies bridge crosses (*Capsicum annuum* X *C. chinese*) X *C. baccatum* as an alternative for introgression of anthracnose resistance from *C. baccatum* in to *C. annuum. J Kore.Sco.Hortic. Sci.,* **46** :5-9.

Yu, Z. H. Wang, J. F., Stall, R.E. and Vallejor, C.E.(1995). Genomic localization of tomato genes that control a hypersensitive reaction to *Xanthomonas campestris* pv. *Vesicatoria* (Doidge) Dye. *Genetics,* **141**: 675-682.

Zamir, D., Ekstein Michelson, I., Zakay, Y., Navot, N., Zeidan, M., Sarfetti, M., Eshed , Y., Harel, E., Pelban, T., Anoss, H. V., Kedar, N., Ralbin Owiteh, HD. and Czonsnek, H. (1994). Mapping & introgression of a tomato yellow leaf Curl Virus tolerance gene , Ty-1 . *Theor. Appl. Genet.,* **88**: 141-146.

11

Wide Hybridisation in Eggplant (*Solanum* spp.)

Padmini, K, Sadashiva, A.T, Singh, T.H and Dhananjay. V. Naik

Division of vegetable crops,
ICAR-Indian Institute of Horticultural Research,Bangalore-560 089.

INTRODUCTION

Eggplant (*Solanum melongena* L.) is one of the important vegetable crops all over the world especially in Asia and Africa. The basic chromosome number of brinjal is n = 12 and 2n =24 in many of the domesticated and a few wild species (Oyelana and Ogunwenmo, 2005). Cultivated brinjal is susceptible to a number of pests and diseases. Wide hybridization refers to crossing individuals belonging to two different species or genera. Wild relatives of brinjal have desirable traits such as disease resistance, which could be transferred into the eggplant genome by introgression (Singh *et al.*, 2006). The ultimate goal of wide hybridization is to transfer useful genes from wild species into cultivated species of same *Solanum* genera or from other genera to *Solanum* to tackle problems caused mainly due to biotic and abiotic stresses.

ORIGIN

Eggplant first originated in India over 4000 years ago, where it continues to grow wild. Brinjal is believed to have primarily originated in the Indian subcontinent, in the Assam region in India and Burma (Choudhary, 1976).

Cultivated egg plant and their wild relatives belong to The Genus *Solanum* , a large genus with 1500 species.The Genus *Solanum* includes both tuberous and non tuberous group of species. sub genus *Leptostemonum* (dunal) bitter, which includes more than 450 species distributed among 22 sections (Kantharajah and Golegaonkar, 2004;Van eck and Snyder ,2006). The sections and species of subgenus *Leptostemonum* present in India has been presented in Table.1 1.

Solanum species found in Asia and India: There are 38 asian species of *Solanum* out of which 22 belong to India and are listed below.

1 *Solanum aethiopicum* L.

2. *Solanum americanum* Mill.

3. *Solanum anguivi* Lam.

4. *Solanum blumei* Nees.

5. *Solanum dulcamara* L.

6. *Solanum ferox* L.

7. *Solanum incanum* L.

8. *Solanum indicum* L.

9. *Solanum lasiocarpum* Dunal

10. *Solanum macrocarpon* L.

11. *Solanum mammosum* L.

12. *Solanum marginatum* L

13. *Solanum melongena* L.

14. *Solanum nigrum* L.

15. *Solanum sisymbriifolium* Lam.

16. *Solanum stramonifolium* L.

17. *Solanum torvum* Swartz

18. *Solanum trilobatum* L.

19. *Solanum undatum* Jacq. non Lam.

20. *Solanum viarum* L

21. *Solanum violaceum* Ortega

22. *Solanum xanthocarpum* Schrad. Wendl.

(Source :Choudhary, 1976)

Table 1. Sections and species of subgenus *Leptostemonum* present in india

Section *Acanthophora* Dunal	Section *Anisantherum* Bitter	Section *Cryptocarpum* Dunal	Section *Lasiocarpa* (Dunal) D'Arcy	Section *Melongena* (Miller) Dunal	Section *Monodolichopus* Bitter	Section *Oliganthes* (Dunal) Bitter	Section *Torva* Nees
S. aculeatissimum Jacq. *S. viarum* Dunal	*S. pubescens* Willd.	*S. sisymbriifolium* Lam.	*S. lasiocarpum* Dunal	*S. cumingii* Dunal* *S. heteracanthum* Dunal *S. incanum* L.* *S. insanum* L.* *S. macrocarpon* L.* *S. marginatum* L.f.* *S. ovigerum* Dunal* *S. virginianum* L.*	*S. coagulans* Forsskal	*S. aethiopicum* L.* *S. cordatum* Forsskal *S. forskalii* Dunal *S. hovei* Dunal *S. kurzii* Prain *S. trilobatum* L. *S. violaceum* Ortega*	*S. torvum* Sw.* *S. barbisetum* Nees *S. sarmentosum* Nees *S. verbascifolium* L.

(Source : Choudhary, 1976)

WIDE HYBRIDIZATION AND ALIEN GENE INTROGRESSION IN EGGPLANT (*SOLANUM* spp.)

i) Intergeneric hybridization in *Solanum* (eggplant)

Intergeneric hybridization refers to the transfer of desirable genes between the genus Solanum and related genera either within the family Solanaceae or any other genera belonging to a different familiy.Intergeneric crosses in *Solanum* are desirable only in situations where the desirable genes are not present in the same genus, but present in allied genera of the same family *Solanaceae*. However, there are no reports on success in intergeneric hybridisation in *Solanum* by conventional hybridization due to the existence of very strong crossability barriers. F_1 hybrids of this type of crosses are always sterile. However, intergeneric hybridization has been successful only in the development of the hybrids involving *Solanum melongena* and *Nicotiana tabaccum* and *Solanum melongena* crossed with an F1 hybrid involving *Lycopersicon esculentum* and *Lycopersicon pennelli* by somatic hybridisation(Liu *et al.*, 2015) .

ii) Interspecific hybridization in *Solanum* (eggplant)

Eggplant wild relatives have much higher genetic diversity than the cultivated species (Mutegi et al., 2015) and they   represent sources of variation for resistance to traits of interest for eggplant breeding (Daunay and Hazra, 2012). Interspecific hybridization refers to the transfer of desirable genes within the genus *Solanum* from one species into the genome of another species.

PRESENT STATUS OF WILD GENE POOLS

Depending on phylogenetic relationships and crossability with eggplant, wild relatives are classified as primary, secondary and tertiary genepools (Table 2).

Primary Gene Pool(GP-1)

The species in primary gene pool are closely related and   crosses by artificial pollination between the species are successful .The hybrids are fertile with complete chromosome pairing .The primary genepool is constituted by only two species,*viz.*, *S. incanum* and *S. insanum*, which provide fertile hybrids with *Solanum melongena*. *Solanum incanum* grows in desertic environments in the Middle East and North Africa and is tolerant to drought, while *S. insanum* is considered as the ancestor of eggplant. Both species are phylogenetically closest to the cultivated eggplant and are part of the "eggplant clade".

Secondary Gene Pool (GP-2)

The members of secondary gene pool are probably normally classified as different species than the crop species under consideration (the primary gene pool). The F1 hybrids have partial F1 sterility due to incomplete chromosome pairing . Crosses by conventional crossing are difficult between members of the primary and secondary gene pools due to the existence of reproductive barriers and may require embryo culture. Around twenty species are present in secondary gene pool. These secondary genepool species are of interest in breeding as this group is genetically very diverse and within it there is a wide diversity in tolerance to abiotic stresses, resistance to pests and diseases, fruit size, fruit shape and composition.

Tertiary Gene Pool(GP-3)

The tertiary genepool is an admixture of species from subgenus *Leptostemonum* , including species from the Old World as well as from the New World, with which occasionally it may be possible to obtain highly sterile or low fertility hybrids after embryo rescue or colchicine treatment and somatic hybridization (Daunay and Hazra, 2012; Rotino *et al.*, 2014). Finally, among these tertiary genepool species, *S. torvum* and *S. sisymbriifolium* are of great interest in breeding owing to its resistance to multiple diseases (Blestsos *et al.*, 2003; Gousset *et al.*, 2005). *Solanum macrocarpon* also belongs to the tertiary genepool (Grubben, 2004) which has been reported to have resistance to shoot and fruit borer (Kumar and Sadashiva ,1996) , mites, nematodes ,wilts and drought resistance (Gowda *et al.*, 1990).Also, *S. elaeagnifolium*, an invasive weed with high tolerance to drought (Christodoulakis *et al.*, 2009) may represent a genetic resource of interest in eggplant breeding.

Table 2 . Wild Gene pool classification in *Solanum*

Primary gene pool	Secondary gene pool	Tertiary gene pool
Solanum incanum	*Solanum anguivi*	*Solanum sisymbriifolium Solanum torvum*
Solanum insanum	*Solanum dasyphyllum*	*Solanum macrocarpon*
	Solanum lichtensteinii	*Solanum elaeagnifolium*
	Solanum linnaeanum	
	Solanum pyracanthum	
	Solanum aethiopicum	
	Solanum tomentosum	
	Solanum violaceum	
	Solanum campylacanthum	
	Solanum vespertlio	
	Solanum violaceum	

(**Source** –EGGNET database)

ROLE OF WILD SPECIES IN SOLANUM

The wild relatives of eggplant possess a number of unexploited resistant genes for biotic and abiotic stresses. Because of their tolerance to abiotic and biotic stresses, eggplant wild relatives have been used for eggplant grafting (Gisbert *et al.*, 2011). Since eggplant wild relatives are found in a much wider range of environmental conditions than those of cultivated eggplant, these wild relatives could play a major role in breeding for adaptation to climatic change, which is a major problem in many developing countries. Also, wild relatives have higher contents of phenolic acids which are of interest for developing new eggplant varieties with improved bioactive properties and insect resistance (Prohens *et al.*, 2013).

However, contrary to other important vegetable crops, like tomato (*Solanum lycopersicum*), the use of wild relatives in eggplant breeding has been very limited, and no commercial varieties have neen developed out of gene introgressions from wild related species. Several wild *Solanum* species have been used as germplasm in breeding programs to improve cultivated brinjal which are presented in Table 3.

Table 3. Genetic resources of desirable genes in wild species (donors) for introgression in cultivated egg plant

Disease / Pathogens/pests	Resistant wild species	References
Fungal wilts Fusarium wilt	*S. indicum, S. integrifolium, S. incanum*	Yamakawa and Mochizuki (1979)
Verticillium wilt	*S. caripense, S. periscum, S. scabrum, S. sisymbrifolium, S. torvum*	Fassuliotis and Dukes (1972), Petrov *et al.* (1989), Sakata *et al.* (1989)
Bacterial wilts	*S. integrifolium, S. torvum*	Yamakawa (1982), Sheela *et al.* (1984)
Fruit rot	*S. gilo, S. integrifolium*	Ahmad (1987)
Phomopsis vexans	*S xanthocarpum ,S.indicum, S. gilo, S.nigrum, S.khasiannum,S. sisymbrifolium*	Kalda *et al.*(1966)
Cercospora solani	*S. macrocarpon*	Madalageri *et al.* (1988)
Fruit and shoot borers (*L. orbonalis*)	*S. xanthocarpum, S. khasianum,* *S. integriifolium,* *S. sisymbriifolium* *S.macrocarpon* *S.gilo, S.indicum, S.incanum* *S. indicum* *S.gilo*	Khan *et al.* (1978), Sharma *et al.* (1980), Chelliah and Srinivasan (1985), Schaff *et al* (1982), Gowda *et al* (1990), Kumar and Sadashiva (1996) Bletsos *et al.*, (2003), Padmini *et al.*, (2015) Behera *et al.*(2002) Kalda *et al* .(1966)
Root-knot nematodes *Meloidogyne spp.*	*S. sisymbriifolium, S. torvum,* *S. aethiopicum,* *S. warscewiczii*	Fassuliotis and Dukes (1972), Daunay and Dalmasso (1985), He´bert (1985) Ahuja *et al.* (1987), Di Vito *et al.* (1992)

Disease / Pathogens/pests	Resistant wild species	References
Spider mite	*S. macrocarpon, S. integriifolium, S. mammosum, S. pseudocapsium, S. sisymbriifolium*	Schalk *et al.* (1975), Schaff *et al.* (1982)
Other pests/ pathogens	*S. sisymbrifolium, S. mammosum*	Sambandam and Chelliah (1983)
Aphis gossypii		
Tetranychuns urticae	*S. macrocarpon*	Schaff *et al.* (1982)
Egg plant mosaic virus (Tymovirus)	*S. hispidum*	Rao (1980)
Mycoplasma	*S. hispidum, S. integrifolium*	Khan *et al.* (1978) ,Rao (1980)

GENETIC DIVERSITY STUDIES IN WILD SPECIES OF EGGPLANT

The genetic diversity in wild species is large compared to *Solanum melongena*. Lower number of polymorphisms using SSR markers was observed in primary genepool species *S. insanum* (0 to 1), while the largest number was found in tertiary genepool species *S. sisymbrifolium* (10). The secondary genepool *S. lichtensteinii* and *S. linnaeanum* recorded lower number of polymorphic loci with *S. melongena* than the single accession of the primary genepool species *S. incanum*. Also, *S. torvum* ,the tertiary genepool species recorded the lowest number of polymorphic loci with *S. melongena*. Genetic relatedness *viz.*, the percentage of similar genes shared between *S. dasyphyllum* and *S. macrocarpon* and between *S. aethiopicum* and *S. macrocarpon* are important in breeding. Close genetic relationships associated with high similarity (> 0.80) among S. *aethiopicum*, S. *anguivi* and S. *integrifolium* provides scope for introgression of high yielding and resistant genes into commercial varieties

NATURAL INTERSPECIFIC HYBRIDIZATION IN SOLANUM spp (EGGPLANT)

Although basically brinjal was classified earlier in self pollinated crops, since natural cross pollination to the extent of 48% (Agrawal, 1980) exists ,of late it is classified under often cross pollinated crop (Peter,2009). Studies on gene transfer between cultivated and wild brinjal accessions suggest that natural interspecific hybridization is a possibility. Occurrence of natural hybridization between *S. incanum* and *S. melongena* has also been reported (Viswanathan, 1975). Davidar *et al* .(2015) also reported that natural hybridization is possible between cultivated and wild brinjal in southern India which was found as a part of the risk assessment process in Bt crop of brinjal, and there is all possibility that transgenes from the crop could spread to wild brinjal populations that occur nearby.

CROSSABILITY IN INTERSPECIFIC HYBRIDISATION OF EGGPLANT

Crossability among some of the important species has been reviewed by Bhaduri(1951), Rao(1968) and Rao (1979). Crossability among ten species was studied by Rao (1979) and was reported that *S. melongena* cultivar as female parent hybridized with *S. melongena* variety *S. insanum, S. incanum, S. integrifolium* and *S. gilo* which resulted in viable seeds. However, only shrunken seeds were produced with *S. khasianum . S.melongena* was not crossable with *S. indicum ,S. zuccagnianum* and *S. sisymbrifolium. S. melongena* as male parent was crossable with *S. indicum* which produced a number of hybrids. Rajasekharan (1969) studied the interrelationship of some common *Solanum* species occurring in South India. Eight *Solanum* spp. were crossed in all possible combinations out of which the crosses   *S. indicum* x *S. melongena; S. xanthocarpum* x *S. trilobatum* crosses were successful. Schaff *et al.* (1980) produced interspecific hybrids in reciprocal crosses by using 11 *S. melongena* types and an African accession of *S. macrocarpon.* Nishio *et al* (1984) classified 11 *Solanum* species into three groups on the basis of their interspecific compatibility (1) *S. melongena, S. incanum* and *S. macrocarpon;* (2) *S. integrifolium, S. gilo* and *S. nodiflorum;* and (3) *S. indicum, S. mammosum, S. toryum, S. sisymbrifollium,* and *S. toxicarium.* Crosses were compatible within and between the first and second groups, but were otherwise incompatible. Based on the various reports ,it is understood that *S. melongena* is more closely related to *S. incanum* than to any other species.A wild type of brinjal with deeply variegated leaves *S. sisymbrifolium* Lam. is considered the ancestral species of *S. melongena* L. Out of 14 different combinations attempted, only the crosses *Solanum incanum* x *Solanum melongena* and *Solanum incanum* x *Solanum indicum* were successful resulting in production of viable seedlings.

CROSSABILITY BARRIERS IN WIDE HYBRIDISATION OF SOLANUM

Several wild species are not crossable with the commercial cultivars due to various isolation barriers. The isolation barrier may be pre-zygotic that prevents fertilization and zygote formation or postzygotic in which fertilization takes place, hybrid zygotes are formed but they are inviable or give rise to weak or sterile hybrids.

Pre- zygotic barriers

1. Failure of pollen germination

2. Slow growth of the pollen tube

3. Inability of the pollen tube to reach the ovary

4. Arrest of pollen tube in the style, ovary and ovule.

These are due to genic differences or differences in ploidy between species.

Post-zygotic barriers

1. Hybrid inviability and weakness leading to chromosome elimination, lethality and embryo abortion.

2. Hybrid sterility

3. Hybrid breakdown with weak or sterile individuals in F_2 owing to recombination of the gene complements of the parental species.

Table 4. Crossability barriers in interspecific hybridisation

PREZYGOTIC BARRIERS	Reference
F_1 cannot be obtained	
Solanum melongena x Solanum mammosum	Sequeria (2014)
S. melongena x Solanum pyracanthos	Plazas *et al.* (2014)
S. melongena x Solanum sisymbriifolium	Plazas *et al.* (2014)
S. melongena x Solanum elaeagnifolium	Plazas *et al.* (2014)
S.torvum x S. melongena	Plazas *et al.* (2014)
POST ZYGOTIC BARRIERS	
F_1 obtained	
Solanum melongena x S. insanum	Ali and Fujieda (1990)
S. melongena x S. gilo	Ali and Fujieda(1990)
S. integrifolium x S. melongena	Ali and Fujieda (1990)
S. sodomeum x S. melongena	Turdor and Tomescu(1995)
S.incanum x S .indicum	Behera and Singh (2002)
S.melongena x S.indicum	Behera and Singh (2002)
F_1 obtained by embryo culture	
S. melongena x S. khasianurn	Sharma *et al* (1980)
S.aethiopicum x S. melongena	Ano *et al.*(1991)
S.melongena x S.torvum	Kumchai *et al.*(2013)
F_1 obtained by somatic hybridisation	
S. melongena x S. torvum	Guri (1988)
S. melongena x Solanum aethiopicum	Collonnier (2001)
F1 hybrid seedling mortality	
S.indicum X S.melongena	Behera and Singh(2002)
F_1 sterile parthenocarpy	
S.melongenax S. macrocarpon	Rajasekaran (1961)
	Schaff *et al* (1982)
S. gilo x S. melongena	Omidiji (1981), Rao and
	Bakshak(1 981)
S. melongena x S. hispidum	Rao (1980)
S. melongena x S. torvum	McCammona nd Honma
	(1983)
S. melongena x S. gilo	Nasrallah and Hopp (1963)
S. integifolium x S. melongena	Kirti and Rao (1982)

PREZYGOTIC BARRIERS	Reference
S. anomalumx S.melongena	Behera and Singh(2002)
Functional seeds	
S. melongena x S. insanum / S. íncanum / S. integrifolium / S. gilo	Rao(1979)
F_1 Partially sterile	
S. incanum,	Rao and Rao (1987)
S. pubescence, S.indicum.	
F_1 **–highly sterile –no seed set**	
S.melongena x *S.torvum*	Kumchai *et al.*(2013)
F1-highly sterile but set very few viable seeds	
S.macrocarpon x *S.melongena*	Padmini *et al* .(2015)
F_2 **obtained**	
S. melongena x *S. khasianurn*	Sharma *et al.*(1980)
F_4 **progenies obtained**	
S. melongena x *S. indicum*	Rao and Kumar (1980)
F_5 **progenies obtained**	
S.macrocarpon X *S.melongena*	Padmini *et al* .(2015)

FACTORS INFLUENCING CROSSABILITY IN INTERSPECIFIC HYBRIDIZATION OF EGGPLANT (SOLANUM spp)

1.Gene Pool

The success of interspecific hybridization, seed production and germination is dependant on the gene pool in which the wild species of *Solanum* is present.

Other researchers have also reported the use of   *S. linnaeanum* as maternal parent to obtain interspecific hybrid plants with *S. melongena* (Liu *et al.*, 2015). No germination was obtained in the hybrids with S. *linnaeanum* when *S. melongena* was used as maternal parent, while good germination was obtained in the reciprocal cross. Hence, S. *linnaeanum* should be used as female parent in interspecific hybridization of eggplant.

Solanum pyracanthos is the most phylogenetically distant species of all the secondary genepool species tested and till now no interspecific hybrid plantlets had been obtained with this species. Artificial sexual hybrids of eggplant with some phylogenetically distant Old World species of *Solanum* section *Leptostemonum* can be obtained without needing embryo rescue and expands the range of species for which interspecific hybrids with eggplant can be obtained (Plazas *et al.*2016).

Kumchai *et al.*(2013)   obtained interspecific hybrids between *S. melongena* and the New World tertiary genepool species and *S. torvum* using the former as maternal parent, although a low fruit set percentage was obtained followed by embryo rescue .However, these interspecific hybrids are highly sterile (Kumchai *et al.*, 2013), which may be difficult to backcross with *S. melongena.*

Table .5 Crossability between *Solanum* wild species (direct and reciprocal crosses)

Male/Female	*S. melongena*	*S. melongena var. insanum*	*S. incanum*	*S. integrifolium*	*S. gilo*	*S. indicum*	*S. zuccagnianum*	*S. xanthocarpum*	*S. khasianum*	*S. sisymbriifolium*
S. melongena	N	G	G	G	G	U	U	U	R	U
S. melongena var. insanum	G	N	G	G	G	U	U	U	U	U
S. incanum	G	G	N	G	G	G	U	U	U	U
S. integrifolium	G	G	G	N	S	S	U	U	U	U
S. gilo	S	S	S	S	N	U	G	U	S	S
S. indicum	G	G	D	G	G	N	D	S	U	U
S. zuccagnianum	U	G	U	U	U	U	N	S	U	U
S. xanthocarpum	S	G	S	U	U	G	U	N	U	U
S. khasianum	U	U	O	U	U	U	U	U	N	U
S. sisymbriifolium	U	U	O	U	O	U	U	U	O	N

Source : (Rao,1979)
N-Normal seeds set on selfing,
G-Normal seeds which grew well.
D-Normal seeds but seedlings dead,
S-Normal seeds but did not grow,
R- Fruits set but seeds are shrunken ,
O-Fruits set but parthenocarpic,U-Cross-unsuccessful

2. Direction of the cross

An important issue in interspecific hybridization is the direction of the cross, as this affects the success of the cross, the number of seeds produced, as well as influences the dormancy of the seeds due to maternal effects. Hence, using *S. melongena* as female parent is usually preferred, as it allows the recovery of the first generation of the *S. melongena* cytoplasm. This avoids potential sterility problems in backcross generations and produces more seeds than small fruited wild species without dormancy and seed germination issues (Gisbert *et al.*, 2011).

The crossability in direct and reciprocal crosses using different wild species of *Solanum* has been presented below in Table 5 .

3. Differences in ploidy level

Differences in the hybridization success may also be due to the differences in the ploidy level between the diploid *S. melongena* and some of the wild species.. On the other side, no hybrids were obtained with *S. sisymbrifolium and S. elaeagnifolium*, confirming that both species are very distant from eggplant .

Table 6. Ploidyof Solanum species

Species	Chromosome number	Reference
S. incanum	2n= 24	Omidiji (1975)
S. aethiopicum	2n=24	Okoli, (1988)
Solanum macrocarpon	2n=24	Rao *et al.*, (1976)
Solanum elaeagnifolium	2n = 24	Heap and Carter (1999)
Solanum sisymbriifolium	2n=24	*Ugur Bal and Kazim Abak* (2003)
Solanum torvum	2n=24	Vasudevan, (1975)
S. erianthum	2n= 24, 66	Vasudevan, (1975)
S. gilo	2n=24	Rao *et al.*, (1976)
S. indicum	2n=24	Krishnappa and Chennavaeraiah, (1975)
S. nigrum	2n=48,54	Labadie (1976)

4. Differences among genotypes of a species

Differences among genotypes of a given species are important in obtaining interspecific hybrids of eggplant (Behera and Singh, 2002; Bletsos *et al.*, 1998; Devi *et al.*, 2015; Kumchai *et al.*, 2013). Some accessions which are better than others for obtaining interspecific hybrids could be used as a recurrent parent for introgression breeding in eggplant or for acting as a bridge for introgression in other *S. melongena* accessions.

Table 7. *Solanum* species used in eggplant breeding for introgression/rootstock purpose

Species	Introgression for desirable traits/ rootstock purpose	Reference
S. anguivi Lam	Backcross to obtain cytoplasm substitution male-sterile lines	Khan and Isshiki (2011)
S. incanum L.	Interspecific hybrids as rootstock Backcross introgression of high content in bioactive phenolic acids	Gisbert *et al.*, (2011) Prohens *et al.*,(2013)
	Resistance to *Fusarium* wilts	
	Resistance to shoot and fruit borer	Rao and Kumar(1980) Behera and Singh (2002)
S. linnaeanum Hepper &	Backcross introgression of resistance to *Verticillium dahlia*	Sunseri *et al.*,(2003) , Liu *et al.*(2015)
P.- M.L.Jaeger	Development of introgression lines	Mennella *et al.*,(2010)
S. sisymbriifolium Lam.	Sexual and somatic hybridization for resistance to *Ralstonia solanacearum* and *Verticillium ahlia* .	Bletsos *et al.*(1998) ; Collonnier *et al.*,(2003)
	No backcrosses or selfings of hybrids obtained Resistance to *Ralstonia solanacearum* and *Verticillium ahlia* wilts	
	Resistance to fruit and shoot borers (*Leucinodes orbonalis*)	Collonnier *et al.*,(2003)
	Resistance to root-knot nematodes (*Meloidogyne* spp.)	Chelliah and Srinivasan, (1985)
	Resistance to spider mite	Ahuja *et al.*,(1987) Schalk *et al.*, (1975)
S. violaceum Ortega	Selection of selfings of interspecific hybrids resistant to *Fusarium* wilt	Rao and Kumar, (1980)
	Backcross to obtain cytoplasm substitution male-sterile lines	Khan and Isshiki,(2009)
S. torvum Sw.	Sexual and somatic hybridization for resistance to *Ralstonia solanacearum*,	Bletsos *et al.*, (2003) ; Jarl *et al.*, (1999) ;
	Verticillium ahlia and root-knot-nematodes. No backcrosses or selfings of hybrids obtained.	Kumchai *et al.*, (2013).
	Resistance to soil pathogens *R. solanacearum* and *Verticillium* and as graftstock.	Jarl *et al.*, (1999) ; Collonnier *et al.*, (2003) ;
	Resistance to root-knot nematodes	Gisbert *et al.*, (2011)
	(*Meloidogyne* spp.)	Ahuja *et al.*, (1987)

Species	Introgression for desirable traits/ rootstock purpose	Reference
S.macrocarpon	Successful record of fertile F2 and fertile BC1 progenies in reciprocal cross (*Solanum macrocarpon* x *Solanum melongena*) which is highly sterile hybrid for use in breeding for shoot and fruit borer resistance	Padmini *et al* .,(2015)
S.viarum	Successful progenies obtained in direct cross using *S.melongena* as female parent	Pugalendhi *et al.*, (2010)
Hybrids Between S. incanum and S. melongena	Highly vigorous rootstock	Gisbert et al (2011)

BREEDING FOR RESISTANCE TO BIOTIC STRESSES

1. Bacterial Wilt, *Ralstonia solanacearum* Smith (*Pseudomonas solanacearum*)

Bacterial wilt is a widespread soil borne tropical disease which invades the plant vessels and provokes the complete wilting of the plant followed by its death. Various degrees of varietal resistance have been reported (Yamakawa, 1979).

Eggplant wild species has been only partially but already successfully investigated and sometimes also used as rootstocks. The highest level of resistance was found in *S. torvum* and *S. toxicarium* i.e. *S. stramonifolium*(Yamakawa, 1979). He reported that the resistance of *S. torvum* (few or no wilting of plants) was found in several accessions, but that the bacteria was present (serological test) in the roots of symptomless plants; this suggests that the mechanism of resistance of this species is tolerance. Date *et al.* (1994) pointed out that the resistance of *S. torvum* is broken down by a combination of some bacterial strains (belonging to group IV) and high ambient temperatures.

The resistance of *S. integrifolium* (i.e. *S. aethiopicum* Aculeatum Group) to this disease and to *Fusarium* wilt is known for a long time in Japan and French West Indies conditions (He bert, 1985).

Collonnier *et al.* (2001) transferred the resistance of two accessions of *S. aethiopicum* into *S. melongena* by somatic hybridization . Various levels of resistance were found in a wild species *S. sisymbrifolium* (Yamakawa, 1979), *S. viarum* (Hebert, 1985), *S. xanthocarpum* from India , *S. viarum* (Yamakawa, 1979), *S. warzcewiczii* , *S. hispidum)* (Hébert, 1985), *S. mammosum* (Mondal *et al.,* 1991) and *S. surratense* (*S. virginianum*) (Mondal *et al.*, 1991). Sakata *et al.* (1989) identified resistance in *Solanum violaceum,* as well as in distant species (*S. carolinense, S. pseudolulo, S. tequilense).* Resistance to *Ralstonia solanacearum* has been reported in *S. anguivi* (Schippers, 2000). Resistance to bacterial wilt was also found in *S. nigrum* and *S. maroniense* (Mochizuki and Yamakawa, 1979), but these species are so distantly related to eggplant that they cannot be used in breeding programs however, could be attempted for use as rootstocks in eggplant.

2. Multiple disease resistance

Among the related wild species, *S.sisymbriifolium* and *S.torvum* are particularly interesting on account of their resistance to three of the most serious eggplant diseases (bacterial wilt, verticillium wilt and nematodes). Unfortunately, these two species do not give fertile progenies when crossed with *S. melongena* by conventional hybridisation. Although somatic crosses were realized for both species, owing to the sterility of the somatic hybrids ,no progenies could be obtained for breeding .Hence, strategies to overcome crossability barriers via ploidy manipulations and biotechnological toolsneed to be explored .

3. Shoot and fruit borer resistance

Eggplant (*Solanum melongena* L.) is susceptible to the most destructing pest eggplant fruit and shoot borer (EFSB). *i.e. Leucinodes orbonalis* being the most destructive pest in brinjal crop There is a difficulty in developing resistant cultivars by traditional plant breeding methods within *Solanum melongena* varieties. No commercial cultivars with appreciable levels of resistance have been developed from wild species due to inadequate screening and crossability and reproduction barriers. The resistance in traditional Indian brinjal landraces, which are genetically close to *S. incanum* L. (the putative progenitor of brinjal) may prove fruitful, as this wild species itself displays EFSB (Samuels ,2014).

To date, around ten EFSB resistant wild relatives have been identified in brinjal. (Samuels ,2014) .The solanum wild species, *Solanum incanum* (Behera *et al* .,2002) , *Solanum viarum* (Praneetha *et al.*,2001) and *Solanum macrocarpon* (Padmini *et al* .2015) have been used in interspecific hybridization for transferring genes of EFSB resistance to cultivated eggplant.

Identification of resistance is possible through quantifying the biochemical components present in the genotype. The biochemical constituents like glycoalkaloid (solasodine), phenols, phenolic oxidase enzymes namely polyphenol oxidase and peroxidase are available in brinjal and these biochemical constituents possess insect resistance properties (Kalloo, 1988).

Rao (1981) produced interspecific crosses by using *S. incanum* and advanced until F4 generation. Since the breeding features of subsequent progenies were not satisfactory they could not be used in egg plant cultivation although they were resistant under field fruit rot and fruit borer .

Open field screening followed by artificial challenging confirmed that *Solanum macrocarpon* is resistant to brinjal shoot and fruit borer, supported by biochemical studies (Padmini *et al.*,2015). Introgression of genes from *Solanum macrocarpon* to *Solanum melongena* through interspecific hybridization for host plant resistance is bottlenecked due to F1 hybrid sterility barrier resulting in difficulty in obtaining F2 seeds or fertile progenies which is a challenge for crop improvement in brinjal.

Padmini *et al.*, (2015) identified that prefertilisation barriers existed in the interspecific F1 hybrid *Solanum macrocarpon* and *Solanum melongena* which was overcome by conventional breeding methods. Fertility was restored in sterile F1 hybrid resulting in fertile progenies. Fertile backcrosses (BC1) were also evolved which were capable of producing BC2 seeds. Globally, this is the first successful record in evolving fertile F2 and fertile BC1 progenies in reciprocal cross (*Solanum macrocarpon* X *Solanum melongena*) when *Solanum macrocarpon* is used as female parent with out biotechnological interventions. Further, the fertile progenies are being evaluated for resistance to shoot and fruit borer, yield and quality(Padmini *et al* .,2015).

An understanding of the precise mechanisms involved in host plant EFSB resistance or tolerance and more focused search for resistant lines by open field screening followed by insect challenge inoculation technique would make it possible to develop resistant brinjal varieties by traditional plant breeding methods. Several of these studies are on- going (Padmini *et al.*, 2015) and continue to assess the augmentation of fertility and yield in these hybrids, whereby they may attain full commercial potential (Samuels ,2014)

4. Root Knot Nematodes, *Meloidogyne* spp.

Eggplant is infested by various *Meloidogyne* species causing heavy damage to the crop. Upadhyay et al. (1977) screened numerous cultivars for resistance to *M. javanica* and found that no variety was immune, but that there were strong varietal differences for susceptibility (from 25% galling to 100%). Daunay and Dalmasso (1985) identified high level of resistance in *S. warzcewiczii* (*S. hispidum*). Boiteux and Charchar (1996) reported *S. torvum* to be resistant to *M. javanica*. Immunity or high resistance to *M. incognita* of *S. khasianum* (i.e. probably *S. viarum*), *S. toxicarium* - *i.e. S. stramonifolium* and again *S. torvum* has been reported by Ali *et al.* (1992).

BREEDING FOR RESISTANCE TO ABIOTIC STRESSES

Eggplant wild relatives grow in a wide range of conditions, including extreme conditions, like desert areas, environments with wide ranges of temperatures including night temperatures below 0 °C, waterlogged and swampy areas, *etc* (Lester *et al.*, 2011).Wild species possessing drought tolerance has been observed in *S. lichtensteinii* and tolerance to salinity in *S. linnaeanum*. The wild species suitable for tolerating adverse climatic conditions need to be screened and identified for incorporating abiotic stress tolerance in cultivated *Solanum* (eggplant).

STRATEGIES TO OVERCOME CROSSABILITY BARRIERS IN SOLANUM

1.	Selection of compatible parents

2.	Reciprocal crosses

3.	Backcrosses

4.	Manipulation of ploidy

5.	Restoration of hybrid fertility in sterile crosses

6.	Bridge crosses

7.	Use of growth regulators

8.	Protoplast fusion

9. Embryo rescue and

10. Somatic hybridisation

LIMITATIONS OF WIDE HYBRIDIZATION OF SOLANUM

1. Utilizing the wild gene pool in breeding programmes may also be constrained by collection gaps in wild species, with no information on genome relationships, limited screening of wild species, linkage drag and genetic complexity of the traits.

2. Interspecific incompatibility is one the major limiting factor when wild relatives are used as a donor sources in wide hybridisation. However, using embryo rescue and other techniques to overcome interspecific crossing barriers, it could be possible to make new hybrid combinations involving different species and to transfer many new traits.

3. Gene transfer through wide crosses is a long and tedious process.

FUTURE PROSPECTS AND SCOPE OF WIDE HYBRIDIZATION

In summary ,. cultivated eggplant is amenable to interspecific hybridization (Daunay and Hazra, 2012; Rotino *et al.*, 2014), as hybrids could be obtained between *S. melongena* and all the wild species from the primary and secondary genepools as well as with *S. torvum* , which is a New World species from the tertiary genepool (Daunay and Hazra, 2012; Whalen, 1984). The genetic diversity present in these wild species is large compared to the cultivated *Solanum melongena* and there is ample scope for broadening the narrow genetic base of eggplant (Munoz-Falcon *et al.*, 2009) and for obtaining interspecific hybrids for use as rootstocks (Gisbert *et al.*, 2011).

Cultivated brinjal can be hybridized with many wild relatives, including the phylogenetically distant *S. torvum*.Wild species would play a major role in adapting to climate change in eggplant breeding. With the advent of molecular biology tools, reliable molecular markers, the hybridity of the plants could be efficiently confirmed at the seedling stage by DNA analysis in addition to the observation of morphological characters. Wide hybridization would be of interest for eggplant breeding aimed at introgression breeding as well as on the feasibility of using eggplant wild relatives for developing interspecific hybrid rootstocks.

In future, collection of wild species which are under the threat of wild crops genetic erosion has to be made for use in wide hybridisation. Since ploidy level differences exist either between or within species, the chromosome counts and/or flow cytometry DNA content measures are needed for all eggplant wild relatives. Also, molecular work using highly repeatable co-dominant markers like SSRs and SNPs can complement morphological and chromosome cytology studies in order to understand the relationships between species as well as the genetic structure of populations within species. The reliance on a limited number of accessions, some of which lack exact provenance data, for much of the work done on eggplant origins and evolution means that new field collections with accurate provenance and good field observations at both population and individual level are a priority for improved eggplant breeding in the future.

There are not much interspecific hybridization studies involving several *S. melongena* accessions and a large number of wild accessions from different genepools including comprehensive quantitative data of fruit set, seed yield, and germination, as well as hybridity confirmation. At present, no modern commercial cultivars of eggplant with traits introgressed from crop wild relatives have been released. (Syfert *et al.,* 2016).

Since the origin of Brinjal is India and wild relatives are in abundance, prebreeding research efforts should be strengthened to introgress the derisable genes in to the cultivated brinjal to tackle the problems in cultivation of brinjal *viz.,* biotic and abiotic stresses, bacterial wilt *etc.* Ploidy level differences have not been investigated in the eggplant wild relatives which would be a priority for germplasm enhancement and wide hybridization.The use of improved interspecific hybridization techniques could lead to an increase in use of secondary and tertiary gene pools in wide hybridization of crops. However, biological constraints still prevent successful use of wild relatives in *Solanum* spp, where blocks to hybridization and hybrid sterility have not yet been overcome. Regardless of inter-specific crossability, hybrid inviability, hybrid sterility and retention of undesirable agronomic traits which are limitations in wide hybridization, strategies such as ploidy manipulations,production of amphidiploids, bridge crosses, reciprocal crosses , growth regulators and biotechnological interventions such as embryo rescue, protoplast fusion by somatic hybridization,double haploids may be attempted to tackle the cross incompatibility barriers in wide hybridization of *Solanum* for introgression of desirable genes.

REFERENCES

Agrawal, R.L .1980. Seed Technology,Oxford and IBH publishing Co.,Newdelhi., pp.198-201.

Ahmad, Q. 1987. Sources of resistance in brinjal to phomopsis fruit rot. *Ind. Phytopathol.,* **40**: 98.

Ahuja, S., Mukhopadhya, M.C., Singh, A and Ahuja, S.P. 1987. Effects of infestation of eggplant (*Solanum melongena*) with root knot nematode (*Meloidogyne incognita*) on the oxidative enzymes and cell wall constituents in their roots. *Cap. Newslett.* , **6**: 98–99.

Ali, M and Fujieda, K. 1990. Cross compatibility between egg-plant (*Solanum melongena* L.) and wild relatives. *J. Japan.Soc. Hort. Sci.,* **58**(4):977-984.

Ali, M., Matsuzoe, N., Okubo, H and Fujieda, K. 1992. Resistance of non-tuberous *Solanum* to root knot nematode. *J. Japan.Soc. Hort. Sci.,***60**(4) : 921-926.

Ano,G., Hebert,Y., Prior.P., and Messiaen,C.M.1991. A new source of resistance to Bacterial wilt of eggplants obtained from a cross: *Solanum aethiopicum* L. x *Solanum melongena* L. *Agronomie* :555-560.

Behera,T. K and Singh, N. K. 2002. Inter-specific crosses between eggplant (Solanum melongena L.) with related Solanum specie. Scientia Horticulturae .,95(1):165-172 .

Bhaduri, P.N. 1951. Inter-relationship of non-tuberiferous species of *Solanum* with some consideration on the origin of brinjal. *Indian J. Genet.*, **2**: 75-86.

Blestsos,F., D. Roupakias., M.L. Tsaktsira., A. B. Scaltsoyjannes and Thanassoulopoulos, C. C. 2003. Interspecific hybrids between three eggplant (*Solanum melongena* L.) cultivars and two wild species (*Solanum torvum* Sw. and *Solanum sisymbrifolium* Lam.). *Plant Breeding* .,**117** : 159 – 164 .

Boiteux, L.S and J.M. Charchar. 1996. Genetic resistance to root-knot nematode (*Meloidogyne javanica*) in eggplant (*Solanum melongena*). *Plant Breed.*, **115**:198–200.

Chelliah , S and Srinivasan, K. 1985. Resistance in bhendi, brinjal and tomato to major insect and mites pests. *In* Proceedings of National Seminar in Breeding Crop Plants for Resistance to Pests and Diseases (May 25-27, 1983). School of Genetics, Tamil Nadu Agricultural University , Coimbatore, Tamil Nadu, India :pp. **43** - 44.

Choudhary, B. 1976. Evolution of Crop Plants, Ed. N.W. Simmonds, Longman Inc., London and New York: 278-279.

Christodoulakis, N. S. , P.N. Lampri. and C. Fasseas. 2009. Structural and cytochemical investigation of the leaf of silverleaf nightshade (*Solanum elaeagnifolium*), a drought-resistant alien weed of the Greek flora. *Australian Journal of Botany.*, **57** : 432 – 438.

Collonnier, C., Mulya, K., Fock, I., Mariska, I., Servaes, A., Vedel, F., Siljak-Yakovlev, Souvannavong, V., Ducreux, G and Sihachakr, D. 2001. Source of resistance against *Ralstonia solanaceraum* in fertile somatic hybrids of eggplant (*Solanum melongena* L.) with *Solanum aethiopicum* L. *Plant Sci* .,**160**(2) : 301–313.

Date ,H., Nasu,H and Hatamoto,M.1994. Breakdown of resistance of eggplant rootstock (*Solanum torvum* Swartz) to bacterial wilt by high ambient temperature .*Ann.Phypathol.Soc.Japan.*, **60**(4):483-486.

Daunay, M. C and Hazra, P. 2012. "Eggplant," in *Handbook of Vegetables*, eds K. V. Peter and P. Hazra (Houston, TX: Studium Press): 257–322.

Daunay, M.C and Dalmasso, A.1985. Multiplication of *Meloidogyne javanica, M. incognita* and *M. arenaria* on several *Solanum* species. *Rev. Nematol.*, **8** (1): 31–34 (in French).

Davidar, P., Snow AA, Rajkumar M, Pasquet R, Daunay, M.C and Mutegi, E.2015. The potential for crop to wild hybridization in eggplant (*Solanum melongena;* Solanaceae) in southern India. *Amer.J.Bot* .,**102**(1):128-139.

Devi, C. P., Munshi A. D., Behera T. K., Choudhary H and Vinod Gurung, B. 2015. Cross compatibility in interspecific hybridization of eggplant, *Solanum melongena*, with its wild relatives. *Sci. Hort.*, **193:** 353–358.

Di Vito, M., Zaccheo, G and Catalano, F. 1992. Source for resistance to root knot nematodes *Meloidogyne* spp. In eggplant. *In: Proceedings of the Eighth Meeting on Genetics and Breeding of Capsicum and Eggplant,* Rome, Italy: pp 301–303.

Eggplant database. www.bgard.science.ru.nl/**eggnet/eggnet**01.html

Fassuliotis,G and Dukes, P.D. 1972. Disease reaction of *Solanum melongena* and S. *sisymbrifolium* to *Meloidogyne incognita* and *Verticillium alboatrum* . *J. Nematol.*, **4**(4): 222–223.

Gisbert , C. , J. Prohens , M. D. Raigón , J. R. Stommel and F. Nuez . 2011. Eggplant relatives as sources of variation for developing new rootstocks: Effects of grafting on eggplant yield and fruit apparent quality and composition. *Scientia Horticulturae* .,**128** : 14 – 22.

Gousset, C., Collonnier, C., Mulya, K., Mariska, I., Rotino, G.L., Besse, P., Servaes, A., Sihachakr, D. 2005.*Solanum torvum*, as a useful source of resistance against the bacterial and fungal diseases for improvement of eggplant (*S. melongena* L.) .*Plant Sci.*,**168** :319–327.

Gowda, P.H.R., Shivashankar, K.T and Joshi.S. 1990. Interspecific hybridization between *Solanum melongena* and *Solanum macrocarpon* study of the F1 hybrid plants. *Euphytica.*, **48**:59–61.

Grubben ., G. J. H. 2004.Vegetables PROTA: 672.

Guri,A., Sink, K. C.1988. Interspecific somatic hybrid plants between eggplant (*Solanum melongena*) and *Solanum torvum*. *Theor Appl Genet.* 1988 Oct., **76**(4):490-496.

He bert, Y. 1985. Comparative resistance of mine species of the genes Solanum to bacterial wilt (*Psudomonas solanacearum*) and the nematode *Meloidogyne incognita*. Implications for the breeding of aubergine (*S. Melongena*) in the humid tropical zone. *Agronomie .*, **5**: 27.

Heap, J.W and Carter, R.J. 1999. The biology of Australian weeds. 35. *Solanum elaeagnifolium*

Jarl , C. I. , Rietveld , E. M and de Haas, J. M. 1999. Transfer of fungal tolerance through interspecific somatic hybridisation between *Solanum melongena* and S. *torvum. Plant Cell Reports.*, **18**: 791 – 796 .

Kalloo, G.1988. Biochemical basis of insect resistance in vegetables. Vegetable Breeding Volume II, CRC Press Inc Boca Raton, Florida:125.

Kantharajah, A.S and Golegaonkar, P.G. 2004. Somatic embryogenesis in eggplant. *Scientia Horticulture.*, **99**: 107-117.

Khan , M. M. R and Isshiki , S. 2011. Functional male-sterility expressed in eggplant (*Solanum melongena* L.) containing the cytoplasm of *S. kurzii* Brace & Prain. *Journal of Horticultural Science & Biotechnology.*, **84** : 92 – 96 .

Khan, R., Rao and G.R., Baksh, S. 1978. Cytogenetics of *Solanum integrifolium* and its possible use in eggplant breeding. *Indian J. Genet. Plant Breeding.*, **38**: 343–347.

Khan,R. 1978. *Solanum melongena* and its ancestral forms. In: Hawkes JC, Lester JG & Skelding AD (eds) The biology and taxonomy of the *Solanaceae*, Linean Soc., Academic Press, London : 629–638.

Kirti,P.B., Rao and B.G.S. 1982. Cytological studies on F1 hybrid *Solanum integrifolium* with *S. melongena* and *S. melongena* var. *insanum*. Genetica., **59**: 127-131.

Krishnappa, D. G. and Chennaveeraiah, M. S. 1975. Cytotaxonomy of *Solanum indicum* Complex. *Cytologia.*, **40**: 322-331.

Kumar, N.K.K., Sadashiva, A.T.1996. *Solanum macrocarpon*: a wild species of brinjal resistant to brinjal shoot and fruit borer, *Leucinodes orbonalis* (Guen.). *Insect Environment.*, **2** (2): 41-42.

Kumchai, J., Wei, Y.C., Lee, C.Y., Chen, F.C and S. W. Chin .2013. Production of interspecific hybrids between commercial cultivars of eggplant (*Solanum melongena* L.) and its wild relative *S. torvum. Genetics and Molecular Research.*, **12** : 755 – 764 .

Labadie, J. 1976. In: 10PB Chromosome number reports LIV *Taxon.*, **25**: 631-649.

Lester, R.N and Hasan, S.M.Z. 2011. Origin and domestication of the brinjal-eggplant, *Solanum melongena*, from *S. incanum* in Africa and Asia. In: J.G. Hawkes, R.N. Lester, M. Nee and N. Estrada-R (eds.), *Solanaceae III: Taxonomy, Chemistry, Evolution.* Royal Botanic Gardens, Kew, UK: 369-387.

Liu, J., Z. Zheng., Zhou, X., Feng, C and Zhuang . Y. 2015. Improving the resistance of eggplant (*Solanum melongena)* to Verticillium wilt using wild species *Solanum linnaeanum. Euphytica.*, 201 : 463 – 469 .

MacCammon, K.R and Honma, S .1982. Morphological and cytogenetic study of an interspecific hybrid eggplant *Solanum melongena* L. *Solanum torvum* Sw. In: Proc XXIst *Int. Hortic. Congr* (Abstract I 1458). Hamburg, Germany.

Madalageri, B.B., Dharmatti, P.R., Madalagri, M.B and Padaganur, G.M.1988. Reaction of eggplant genotypes to *Cercospora solani* and *Leucinodes orbonalis. Plant Pathol. Newslett.*, **6**: 26.

Magoon, M.L., Ramanujah,S and Cooper,D.C.1962. Cytogenetical studies in relation to the origin and differenciation of species in the genus *Solanum* L. *Caryologia* .,**15**: 151–252.

McCammon,K.R and Honma,S.1983. Morphological and cytogenetic analyses of an interspecific hybrid eggplant, *Solanum melongena x Solanum torvum. – Hort Science.*, **1** (8): 894-895.

Mennella, G. , Rotino, G. L.,M. Fibiani, Alessandro, A. D., Francese, G., Toppino, L., Cavallanti.F. 2010.Characterization of health-related compounds in eggplant (*Solanum melongena* L.) lines derived from introgression of allied species. *Journal of Agricultural and Food Chemistry.*, **58** : 7597 – 7603 .

Mochizuki,H and Yamakawa,K.1979.Resistance of selected eggplants Cultivars and related *Solanum* species to bacterial wilt (*Pseudomonas solanacearum*).*Bull. veg.Ornamental Crop Res.Stn.A.,***6**:10.

Mondal,S.N.,Khan,M.A.,Rahman,M.T.,Rashid,.M.A and Nahar,S.1991. Reaction of eggplant cultivars /lines and wild species of *Solanum* to bacterial wilt (*Pseudomonas solanacearum* smith).*Ann.bangladesh agric.*,**1**(2) :65-68.

Munoz-Falcon, J.E., Prohens, J., Vilanova, S., Ribas, F., Castro, A., Nuez, F.2009. Distinguishing a Protected Geographical Indication vegetable (Almagro eggplant) from closely related varieties with selected morphological traits and molecular markers. *J. Sci. Food Agr.*, **89**:320–328.

Mutegi, E., A.A. Snow, M. Rajkumar, R. Pasquet, H. Ponniah, M.C. Daunay and P. Davidar. 2015. Genetic diversity and population structure of wild/weedy eggplant (*Solanum insanum Solanaceae*) in southern India: Implications for conservation. *Amer. J. Bot.*, **102**:140-148.

Nasrallah,M and Hopp,R.J.1963.Interspecific crosses between *Solanum melongena* L. (eggplant) and related *Solanum* species. - *Proc. Am. Soc. Hortic. Sci.*, **83**: 571-574.

Nishio,T.,Mochizuki,H and Yamakawa, K .1984. Interspecific cross of eggplant and related species. Bull. Veg. Ornamental Crop Res. Stn., **12**: 57–64.

Okoli, B. E. 1988. Cytotaxonomic studies of five West African Species of *Solanum* L. (Solanaceae). *Feddes Repertorium.*, **99**(5-6): 183-187.

Omidiji, M. O. 1975. Yield and nutritive quality of leafy vegetables from interspecific Solanuin L. crosses. *Nigerian J. of Sci.*, **2**(1/2): 197-208.

Omidiji, M.O.1981. Cytogenetic studies on the Fl hybrid between the African eggplant, *Solanum gilo* Raddi, and *Solanum melongena* Z. - *Hortic. Res.*, **21**:75-82.

Oyelana, O.A and Ogunwenmo, K.O.2005. Comparative assessment of induced mutants from *Solanum macrocarpon* L. (Solanaceae). *Acta Satech.*, **2**: 50-56.

Padmini,K., Rajasekharan,P.E., GangaVishalakshi,P.N., Singh,T.H and Upreti,K.K. 2015.Unlocking the potential of *Solanum macrocarpon* L.for genetic enhancement in interspecific hybridization of brinjal (*Solanum melongena* L) Challenges and approaches. In Souveiner of lead papers and abstracts: National meet on distant hybridization in horticultural crop improvement at ICAR- IIHR Bengaluru on 22nd and 23rd Janurary 2015: pp. 251.

Pearce, K.G. 1975. *Solanum melongena* L. and related species. In: Ph. D. Thesis. Univ. Birmingham, UK: 251.

Peter,K.V.Ed.   2009.Vegetable Science .In Basics of horticulture. New India publishing .pp: 762

Petrov, C., Nakov, B and Krasteva, L. 1989. Studies on the resistance of eggplant introductions to Verticilium wilt. In: Proceedings of the Seventh Meeting on Genetics and Breeding of Capsicum and Eggplant, Yugoslavia: pp 27.

Plazas, Mariola., Santiago,Vilanova., Pietro, Gramazio., Adrián, Rodríguez.,-Burruezo., Ana Fita, Francisco J., Herraiz, Rajakapasha., Ranil, Ramya, Fonseka., Lahiru., Niran, Hemal Fonseka, Brice Kouassi., Abou, Kouassi., Auguste, Kouassi and Jaime Prohens. 2016. Interspecific Hybridization between Eggplant and Wild Relatives from Different Genepools . *J. Am. Soc. Hortic. Soc.,,.*, **1**:34-44.

Praneetha .S., Rajashree.V and Pugalendhi .L., 2011. Association of charecters on yield and shoot and fruit borer resistance in brinjal (*Solanum melongena* L.) *Electronic Journal of Plant Breeding .*, **2**(4) :574-577.

Prohens , J. , B. D. Whitaker. , Plazas, M. S., Vilanova , M., Hurtado , M., Blasco , Gramazio , P and. Stommel , J. R. 2013 . Genetic diversity in morphological characters and phenolic acids content resulting from an interspecific cross between eggplant, *Solanum melongena* , and its wild ancestor (*S. incanum*). *Annals of Applied Biology.,* 162 : 242 – 257 .

Pugalendhi , L., Veeraragavathatham, D., Natarjan ,S and Praneetha ,S. (2010) Utilizing wild relative((Solanum viarum)as resistant source to shoot and fruit borer in brinjal *(Solanum melongena* Linn.) *Electronic Journal of Plant Breeding.,* **1**(4): 643- 648.

Rajasekaran, S. 1969.Cytogenetic studies in *Solanum melongena* L., S. *melongena* var. bulsarensis Argikar and their hybrid and study of colchicine induced polyploidy in *S. melongena* Z. - M. Sc.(Ag.) dissertation, University of Poona, India.

Rajasekaran, S.1970. Cytogenetic studies of the F_1 hybrid *Solanum indicum* L. *S. melongena* L. and its amphidiploids .*Euphytica.* ,**19**(2):217-224.

Rajasekaran, 1961. Cytological Studies on the F1 hybrid (*Solanum xanthocarpum* Schrad. and Wendl. X *S. melongena* L.) and its amphidiploid, *Caryologia.,* **24**(3):261-267.

Rao ,G.R.1981. Results of interspecific cross pollination between *S. melongena* L. and S. *incanum*L. in eggplant breeding. *Proceedings of Indian National Science Academy.,* **47**: 893-898.

Rao ,S.V and Rao,B.G.S. 1984. Studies on the crossability relationships of some spinous *Solanum. Theor. Appl. Genet.,* **67**: 419–426.

Rao G.R and Bakshak,S.1981. Relationship between *Solanum melongena* L. and *Solanum integrifolium* poir. - *Ind. J. Genetics & Plant Breed.,* **41**(1): 46-53.

Rao, G. R and A. Kumar .1980 . Some observations on interspecific hybrids of *Solanum melongena* L. *Proceedings of the Indian Academy of Sciences (Plant Sciences).,* **89** : 117 – 121 .

Rao, G. R., Khan, A. H. and Khan, R. 1976. Interrelationship between tetraploid *Solanum nigrum* L. and *S. villosum* Mills. *Science & Culture .,***42**: 559-561.

Rao, G.R.1980. Cytogenic relationship and barrier to gene exchange between *Solanum melongena* L. and *Solanum hispidium* Pers. *Caryologia.,* **33**: 429–433.

Rao, N.N. 1979. The barriers to hybridisation between *Solanum melongena* and some other species of *Solanum.* In: Hawkes JG, Lester RN & Skelding AD (eds) The Biology and Taxonomy of the *Solanaceae* Acad. Press, London :605–614.

Rotino, G. L., Sala, T., Toppino, L. 2014. Eggplant, in Alien Gene Transfer in Crop Plants, Vol. 2, eds Pratap A., Kumar J., editors. NewYork, NY: *Springer ;*381–409.

Sakata, Y., Nishio, T and Mon'ma, S. 1989. Resistance of *Solanum* species to Verticillium wilt and bacterial wilt. In: Proceedings of the Eucarpia 55th meeting on Genetics and Breeding of Capsicum and Eggplant, Kragujevac, Yugoslavia:177.

Sambandam, C.N and Chelliah, S. 1983. Breeding brinjal for resistance to *Aphis gossypii* G. In: Proceedings of the National Seminar on breeding crop plants for resistance to pests and diseases. Tamil Nadu Agricultural University: pp. 15.

Samuels, J. 2014. Genetic resources for EFSB resistance in brinjal . *Current Science.,* **107** :12- 25.

Sarvayya, J. 1936. The first generation of an interspecific hybrid cross in *Solanums,* between *Solanum melongena* and *S. xanthocarpum. Madras Agric. J.,* **24**: 139-142.

Schaff, D.A., Jelenkovic, G., Boyer, C.D and Pollack, B.L. 1982. Hybridization and fertility of hybrid derivatives of *Solanum melongena* L. and *Solanum macrocarpon* L. *Theoret. Appl. Genet.,* **62**: 149–153.

Schalk, J.M., Stoner, A.K., Webb, R.E and Winters, H.F. 1975. Resistance in eggplant, *Solanum melongena* L. and nontuber-bearing Solanum species to carmine spider mite. *J. Am. Soc. Hortic. Soc.,* **100**: 479–481.

Schippers., R.R. 2000.African Indigenous Vegetables.An Overview Of The Cultivated Species.Natural Resources Institute/Acp-Eu. Technical Centre for Agricultural and rural Cooperation, Chatham, UK, **5**:214.

Sequeria, G.S.2014. Studies on floral biology pollination and seed set in *Solanum mammosum* M.Sc thesis submitted to university of horticultural sciences bagalkot: pp. 104.

Sharma, D.R, Sareen, P.K and Chowdhury, J.B.1984. Crossability and pollination in some non-tuberous *Solanum* species. *Indian J. agric. Sci.,* **54**(6): 514–517.

Sharma, D.R., Chawdhury, J.B., Ahuja, U and Dhankhar, B.S.1980. Interspecific hybridization in the genus   *Solanum.* A cross between *S. melongena* and *S. khasianum* throughout embryo culture. *Z. Pfanzenzuecht.,* **85**:248–253.

Sheela, K.B., Gopalkrishana, P.K and Peter, K.V. 1984. Resistance to bacterial wilt in a set of eggplant breeding lines. *Indian J. Agric. Sci .,***54**: 454- 457.

Singh,A.K., Singh, M., Singh, A.K and Singh, R. 2006. Genetic diversity within the genus *Solanum* (*Solanaceae*) as revealed by RAPD markers. *Curr. Sci.,* **90**: 711-716.

Sunseri , F. , A. Sciancalepore , G. Martelli , N. Acciarri , G. L. Rotino , D. Valentina , and G. Tamietti . 2003 . Development of a RAPD-AFLP map of eggplant and improvement of tolerance to *Verticillium* wilt. *Acta Horticulturae.,* **625** : 197 – 198 .

Syfert,Mindy. M., Nora P. Castaneda-Alvarez, Colin K. Khoury, Tiina Sarkinen, Chrystian C. Sosa, Harold A. Achicanoy, Vivian Bernau, Jaime Prohens, Marie-Christine Brand-Daunay, Sandra Knapp.2016. Crop wild relatives of the brinjal eggplant *Solanum melongena*) poorly represented in genebanks and many species at risk of extinction. American *Journal of Botany, Botanical Society of America.* , **103** (4), pp:635-651.

Tudor, M., and Tomescu, A. 1995. Studies on the crossing compatibility of the species *Solanum   sodomeum* and *Solanum melongena* var. "Lucia". Eucarpia, IXth Meeting on genetics and breeding on Capsicum and Eggplant, Budapest (Hungary): pp. 39-40.

Ugur, Bal and Kazim, Abak . 2003. Attempts of Haploidy Induction in Tomato (*Lycopersicon esculentum* Mill.) Via Gynogenesis I: Pollination with Solanum sisymbriifolium Lam. Pollen. *Pakistan Journal of Biological Sciences.,* **6**: 745-749.

Upadhyay,K.D.,Singh,G and Pandey,R.C. 1977.A note on reaction of brinjal varities to attack of root knot nematode ,meloidygne javanica.*Indian J.Hort.*,**34**(4):455-456.

Van Eck J, Snyder. A. 2006. Eggplant (*Solanum melongena*L.). Methods in Molecular BiologyClifton, N.J., **343**:439-447.

Vasudevan, K. N. 1975. Contribution of the cytotaxonomy and cytogeography of the flora of the Western Himalayas (with an attempt to compare it with the flora of the Aips) part II *Ber. Schweiz Bot.Ges.*, **85**: 210-252.

Vishwanathan, T.V. 1975. On the occurance of natural hybridization between *Solanum incanum* L. and *Solanum melongena* L. *Current Science.*, **44**: 134.

Wanjari, K.B. 1976. Cytogenetic studies on F1 hybrids between *Solanum melongena* L. and *S. macrocarpon* L., *Hort. Res.*, **15**:77–82.

Whalen, M.D. 1984. Conspectus of species groups in *Solanum* subgenus *Leptostemonum*. *Gentes Herbarum.*, **12**: 282.

Yamakawa, K. 1982. Use of rootstocks in *solanaceous* fruit-vegetable production in Japan. *Jpn. Agric. Res. Quart.* ,**15**: 175–179.

Yamakawa, K., and Mochizuki, H. 1979. Nature and inheritance of Fusarium wilt resistance in eggplant cultivars and related wild *solanum* species. *Bull. vegetable ornamental Crops station.*, **6**: 19.

Distant Hybridization in Horticultural Crops   *Pages* **189–199**
Editors: **M.R. Dinesh & M. Sankaran**
Published by: **ASTRAL INTERNATIONAL PVT. LTD., NEW DELHI**

12

Wide Hybridization in the Genus *Abelmoschus* Medikus

Joseph John K[1], Roy YC[1], M Latha[1] and KV Bhat[2]

ICAR - National Bureau of Plant Genetic Resources,
Regional Station, Thrissur - 680656, India
ICAR - National Bureau of Plant Genetic Resources, Pusa Campus,
New Delhi - 110012, India

INTRODUCTION

The genus *Abelmoschus* in India comprises about 12 species, 3 of them cultivated and the rest wild. Further, subspecific entities in *A. angulosus*, *A. moschatus*, *A. tetraphyllus* and *A. tuberculatus* have been described, each of them with unique distribution range, restricted adaptation and specific traits. Almost all the entities have their natural distribution in India by virtue of different microclimates and adaptive niches across various eco-geographic regions. Okra crop is facing challenges of YVMV and ELCV which require search for resistance in the wild related taxa and their incorporation in cultivated okra. Attempts to develop hybrids of okra with different wild species were struck by sterility of the F1 hybrid. Crosses of cultivated okra genotypes with *A. angulosus* var. *grandiflorus*, *A. caillei*, *A. tuberculatus*, *A. tetraphyllus* and *A. mizonagensis* sp. nov. (Mizoram), *A. moschatus* subsp. *rugosus* and Guinean okra with *A. esculentus*, *A. angulosus* var. *grandiflorus* and *A. mizonagensis* sp. nov. (Mizoram) were raised but could not be advanced as the interspecific hybrid failed to produce viable seeds upon selfing and back cross. However fertility was restored in the above crosses by colchicine treatment and diploidization. Application of 1% colchicine for 3 days was lethal with less than 1% survival, however application of 0.15% colchicine for 4 days for 4 hours from 7.00am to 11.00 am was found effective and with over 90% seedling recovery. Hybridization of okra with *A. enbepeegearensis* and *A. crinitus* were also made and the seedlings were subjected to colchiploidization and these are under evaluation. Most of the hybrid combinations were intermediate for quantitative traits with prolific bearing in almost all cases. Inheritance of prickly hairs on fruits is a deterrent to their direct use as vegetable. Hence, back crossing to okra and selection among segregants is required before their direct use can be attempted. Another set of crosses involving okra with wild species as maternal parent were attempted in order to transfer the cytoplasm of wild species and in these seeds were harvested from the crosses *A. tetraphyllus* × *A. esculentus* and *A. tuberculatus* × *A. esculentus*, and fertility was restored through colchiploidization. Based on morphological characters, 8 new synthetic species were described.

Distant hybridization or mating between individuals of different species/ genera helps to combine diverse genomes in to one individual. It is often a man-made attempt to break the species barrier to gene-flow that exists in nature, thereby effecting new genomic combinations. Many of the cultivated plants have an allo-polyploid origin resulting from inter-specific hybridization that occurred in nature. The gene-flow facilitating the incorporation of genes from one species to another is called introgression, which can be either natural or man-mediated. Introgressive hybridization, the incorporation of genetic material from one species to another through wide hybridization and repeated back-crossing, plays an important role in the evolution of plant species, genetic modification and enriching the gene-pool for breeding. Breeders are increasingly interested in introgressing genes conferring desirable traits from wild to crop species in breeding programmes.

The genus *Abelmoschus* comprises about 12 species including the cultivated species *A. esculetus* (Okra) and *A. caillei* (Guinean/ African Okra). Differential ploidy level and often genotype specific variation in chromosome number has been noticed in Okra as well as some of its wild relatives (Keisham *et al.*, 2014). Even though there is a general consensus regarding the distinctness of species, lack of consensus on infra-specific taxonomy leads to misreporting of wide crosses within species as inter-specific hybridization. In general, Okra wild species had many desirable traits like dark green colour, high mucilage, extended bearing, perennation tendency, high branching, reduced fruit length, drought resistance, high temperature tolerance and YVMV resistance to be incorporated from the wild gene-pool. Kuwada (1966) has developed an amphi-diploid of Okra with *Abelmoschus tuberculatus* and described it as *Abelmoschus x tuberculatus*. Further, F_1 hybrids of *A. tuberculatus* with *A. manihot* were also attempted (Kuwada, 1974). Pal *et al.* (1952) have attempted wide hybridization between cultivated Okra and its closest progenitor, *A. tuberculatus*. Akhond *et al.* (2000) and Samarjeeva (2003) described the success of wide hybridization attempt involving Okra with *A. moschatus* and *A. angulosus*, respectively.

Wide hybridization was attempted between various *Abelmoschus* species including Okra, primarily for establishing the genetic relationship based on F_1 fertility and secondarily to derive useful new combinations (Joseph *et al.*, 2013). The material could be of use for YVMV and enation leaf curl virus resistance besides cytoplasmic male sterility in cultivated Okra. Screening of wild *Abelmoschus* germplasm at hot-spot localities (New Delhi & Varanasi) showed field resistance of *A. enbeepeegearensis* and *A. angulosus* var. *grandiflorus* to YVMV and ELCV where as all other species and Okra genotypes were susceptible to the virus. *A. caillei* had very high mucilage in the fruits besides extended bearing trend beyond 12 months. While aspects like floral biology, pollen pistil interaction, crossability barriers, *etc.* has been covered in an earlier publication (Joseph *et al.* 2013), the scope of this paper is restricted to describing and listing the successful inter-specific cross combinations in *Abelmoschus* .

MATERIALS AND METHODS

Hybridization

Fifteen taxa of *Abelmoschus* viz., *A. esculentus*, *A. caillei*, *A. angulosus* var. *grandiflorus*, *A. tuberculatus*, *A. ficulneus*, *A. enbeepeegeearensis*, *A. crinitus*, *A. tetraphyllus* var. *tetraphyllus*, *A. tetraphyllus* var. *pungens*, *A. moschatus* subsp. *moschatus*, *A. moschatus* subsp. *tuberosus*, *A. moschatus* subsp. *rugosus*, *A. palianus*, *A. mizonagensis* (sp. nov.) and *Abelmoschus* sp. nov. (Dehra Dun) were raised in pots at NBPGR Regional Station, Thrissur during *Kharif* season of 2008-2011. Mature flower buds were emasculated by removing the perianth and shaving off the anthers in the previous evening before anthesis and covered with butter paper cover. Ready to open flower buds were also covered at the same time with butter paper cover to collect pollen. Pollen from male parent was collected and dusted on receptive stigma of prepared flowers between 9.00 and 10.00 AM and again covered with butter paper cover and tagged. Upon maturity, capsules were harvested, dried and hybrid seeds stored for 3-6 months in a cool place. Such seeds were scarified using sand-paper and soaked overnight in distilled water before sowing. Seedlings were raised in poly-bags and half the seedlings in all cross combinations were subjected to colchicine treatment. The untreated seedlings were transplanted to pots and allowed to grow under optimal conditions.

Colchicine Treatment

Emerging seedlings at the 3rd leaf stage (epicotyl leaves fully grown and third leaf emergence stage) were subjected to colchicine treatment using cotton swab method. Three doses viz., 1% colchicine for 3 days from 6.00 AM to 6.00 PM at 30 minutes intervals and 0.10% and 0.15% colchicine for 3 days from 7.00 AM to 11.00 AM at 25-30 minutes intervals were applied. After completion of application each day, cotton swab was washed with distilled water and on 4th day cotton swab was removed. Surviving seedlings were transplanted to standard pots and provided optimal growth environment following package of practices recommended for Okra.

The normal hybrids and colchicine treated hybrids were compared and characterised for morphological traits. Fruit and seed set were studied. Seeds from mature dry capsules were harvested and after a period of 4 months storage studied for germination percentage.

The inter-specific hybrids were named following International Code of Botanical Nomenclature as specified by Spencer (2007). Locally adapted Okra landraces such as IC260106 (Tamil Nadu), IC265626, IC265648, IC412987, IC260039, IC413591 and IC265147 (Kerala) and IC536721 (Karnataka) were used as parents.

After care

The treated plants were regularly sprayed with artificial growth formulations of mineral supplements for boosting growth. All treated seedlings were provided with support to help them stand erect and plants were given optimal growth environment by shifting them to well lighted mist house.

Five plants each in C_0 generation were characterised for plant height, branching pattern, stem pubescence, stem thickness, stem colour, node pigmentation, seedling vigour, first and last flowering date, stipule shape, leaf length, leaf width, leaf lobing, petiole pigmentation, petiole length, leaf vein colour, stalk length, number of epicalyx segments, epicalyx shape, epicalyx length, epicalyx width, epicalyx pigmentation, flower diameter, length and width of calyx, calyx pigmentation, flower shape, corolla length, corolla width, flower colour, length and width of staminal column, anther colour, anther abundance, stigma colour, number of stigmatic lobes, fruit colour, fruit length, fruit diameter, stalk length, single fruit weight, number of ridges per fruit, fruit ornamentation, number of pods per plant, number of seeds per pod and reaction to fruit borer, leaf-roller caterpillar, *Cercospora* and YVMV.

Fig 1: New Synthetic species developed with okra as female parent

Abelmoschus x caizoramensis (*caillei* **x** *mizonagensis nom. nudum*)	*Abelmoschus x esculophyllus* (*esculentus* **x** *tetraphyllus* **var.** *tetraphyllus*)

Abelmoschus x esanguflorus (*esculentus* **x** *angulosus* **var.** *grandiflorus*)	*Abelmoschus x escullei* (*esculentus* **x** *caillei*)

Abelmoschus x cangulosus (*caillei* x *angulosus*)	*Abelmoschus x esculatus* (*esculentus* x *tuberculatus*)

RESULTS AND DISCUSSION

The crossability relationships of Indian taxa of *Abelmoschus* with cultivated okra and between various wild species were worked out at NBPGR RS, Thrissur (Table 1 & 2). Colchiploids were developed in crosses of okra and caillei with six other taxa. Eight synthetic species were produced using various locally adapted genotypes of okra as maternal parents. The characters of the amphidiploids thus obtained are presented in Table 3. The fertile amphi-diploid needs to be further selfed for one generation to stabilize the genetic architecture. Colchiploids of direct and reciprocal crosses of Okra with *A. tuberculatus* and *A. tertraphyllus* var. *tetraphyllus* were selfed for a second generation. Inherited morphological characters of the hybrid (both colchicine treated and untreated) were used to establish hybridity. Of course, failure of F_1 hybrids to set viable seeds upon selfing, open pollination and back-crossing and subsequent recovery of fertility in colchicine treated lines of the same seed lot was the best evidence for inter-specific hybridity.

Treated seedlings started showing abnormal symptoms like swelling of collar region (from 3rd day of completion of treatment). The emerging leaves were severely crinkled, retarded and often the tip was dried up. Petiole of the treated leaves (epicotyl) was also swollen, epicotyl leaves became very thick, turgid and dark green in colour. Compared to normal plants, the petioles of treated leaves were shortened as there was no further growth. Due to the abnormal swelling of the collar region, the root system could not hold the plants in position leading to lodging, which in turn invite seedling rot due to secondary fungal infection. Some of the wild species are represented by type collection or only a few collections, which need to be increased by covering the whole distribution range of the species. For further utilization in okra breeding there is scope for study of mitotic and meiotic behaviour of amphi-diploids. Detailed cytogenetic characterisation and GISH analysis will throw further light on utilization of these distant hybrids in okra breeding programme. In the case of distant hybrids of various *Abelmoschus* species other than cultivated okra, they may serve as bridge species in transferring specific traits to okra. Normally okra is raised as a *kharif* crop and its cultivation in *rabi* –summer season is not rewarding due to poor yield. Photosensitivity leading to reduced plant growth and very low yield, high incidence of African mealy bug, fruit borer and high temperature and drought susceptibility are the major problems in okra crop beyond October.

There is an urgent need to adequately and systematically collect wild genepool of *Abelmoschus* from across the country. Voucher herbarium and photographic documentation and seed conservation of synthetic species have been done for further utilization and reference.

Our reports of successful production of fertile amphi-diploids of okra with various *Abelmoschus* species except *A. tuberculatus* are new reports. However we could also repeat *A x esculotuberculatus* reported by Kuwada *et al.* (1966) and also its reciprocal. Further okra hybrids with *A.crinitus* and *A.enbeepeegeeearensis* were also produced in bulk but F_1 is yet to be evaluated. The flowers are amenable to hand emasculation and generally there is no flower drop due to corolla and anther removal. Flower bud size in the previous evening of anthesis is indicative of floral maturity. Among the synthetic hybrids involving wild species, *A. mizonagensis* was found to have extended bearing, continued fruit availability due to extended bearing and flowering and fruiting in off/ lean season. Inter-specific hybrids of Okra with *A. angulosus* and *A. mizonagensis* were found to flower and fruit continuously beyond the normal season of Okra (August-October) extending to peak summer months (March) offering scope for perennial okra. However, susceptibility to African mealy bug was a problem in summer crop. A perusal of the growth performance of wild species and fruit set upon distant hybridization indicates that the agro-climatic conditions prevailing in Thrissur, Kerala (NBPGR) is highly conducive for wild species maintenance and performance except a few niche-specific high altitude taxa like *A. angulosus* var. *angulosus* and *A. angulosus* var. *purpureus*. Some of the inter-specific hybrids, not involving okra genome may serve as bridge species to transfer useful traits to cultivated okra as done in root-stock breeding of many horticultural crops. Additional characters like nematode resistance, increased period of flowering, extended harvest duration, drought resistance, *etc.* can be achieved in okra by incorporation of genes from wild related species. Hybrid inviability in a few crosses like *A. esculentus* x *A. ficulneus* and *A. esculentus* x *A. moschatus* needs to be addressed through embryo rescue. Further failure of *A. crinitus* x *A. esculentus* needs to be reinvestigated.

According to the biological species concept of Mayr (1991), natural species are groups of actually or potentially inter-breeding populations which are reproductively isolated from other such groups. Often our designations of species do not ensure reproductive isolation and is rather based on a set of morphologically distinct characters. If we strictly follow the biological species concept, many of the species may fall under infra-specific categories and the resultant hybrids may not qualify for grouping under inter-specific hybrids. Hence, it is very important to have a clear knowledge of the identity of the species before embarking any distant hybridization programme.

Future thrust

Evaluation of the existing holding of wild *Abelmoschus* germplasm in hot-spots and further confirmation of results in the laboratory. Resistance to YVMV and ELCV has to be confirmed through sap inoculation.

A network project on development of F_1 inter-specific hybrids and seed supply to users after stabilizing the amphi-diploids.

Characterisation of F_2 populations of inter-specific hybrids and studies on association of phenotypic and molecular data.

Identification of genomic regions associated with resistance to devastating pest and diseases, especially YVMV and ELCV.

CONCLUSION

In the genus *Abelmoschus* distant hybridization is fairly easy and pre-fertilization barriers are less except in a few combinations of *A. esculentus* x *A. ficulneus* and *A. esculentus* x *A. moschatus*. Restoration of fertility through chromosome doubling using 0.15% colchicine treatment for 3 days from 7.00 AM to 11.00 AM on emerging seedlings at 2-leaves stage gave very good results and seedling recovery. Eight such synthetic amphi-diploids with okra as maternal parent have been developed.

After stabilizing the amphi-diploids and multiplication, the inter-specific hybrids will be accessioned as germplasm and will be made available to breeder under MTA for bonafide use. There is a need to study the genetics of YVMV resistance in *A. enbeepeegeearensis*, *A. angulosus* var. *grandiflorus* and any other species/ lines found to have YVMV/ELCV resistance in hot-spots.

Table 1: Inter and intra- specific hybrids developed at NBPGR RS Thrissur

Female parent	Male parent	Fruit set percentage	Status of F_1 (un-treated 2n plants)
A. esculentus	*A. angulosus* var. *grandiflorus*	100.00	F_1 shy seed setting
A. angulosus var. *grandiflorus*	*A. esculentus*	16.43	Seeds unfilled
A. esculentus	*A. tetraphyllus* var. *tetraphyllus*	92.31	F_1 vigorous with prolific bearing. Sterile seeds with aborted embryo. Pods spiny, 12-15 cm long. Fertility restored through colchicine treatment
A. tetraphyllus var. *tetraphyllus*	*A. esculentus*	19.23	Vigorous with prolific bearing. Sterile seeds with aborted embryo, pods spiny, 12-15 cm long.
A. esculentus	*A. moschatus* subsp. *moschatus*	57.14	Few filled seeds
A. moschatus subsp. *moschatus*	*A. esculentus*	11.54	Not germinated
A. esculentus	*A. moschatus* subsp. *tuberosus*	47.06	Fruit set but no seed set
A .esculentus	*A.caillei*	38.89	Vigorous F_1, seed set but no germination
A. caillei	*A. esculentus*	25.00	Filled seeds (Vigorous F_1s but sterile)
A. esculentus	*A. ficulneus*	35.48	Unfilled seeds
A. esculentus	*A. tetraphyllus* var. *pungens*	100.00	Fruit set but no seed set.
A. esculentus	*A. tuberculatus*	30.00	F_1 vigorous with prolific bearing. Sterile seeds with aborted embryo. Pods spiny, 12-15 cm long. Fertility restored through colchicine treatment.
A. tuberculatus	*A. esculentus*	85.71	F_1 vigorous with prolific bearing. Sterile seeds with aborted embryo, pods spiny, 12-15 cm long. Fertility restored through colchicine treatment.
A. esculentus	*A. crinitus*	25	F_1 sterile
A. esculentus	*A. enbeepeegearensis*	30.0	F_1 sterile
A. enbeepeegearensis	*A. esculentus*	25.0	F_1 prolific flowering, no fruit set resembling maternal parent.

Table 2 : Crossability relationship at a glance among the different taxa of Abelmoschus

Crossability in *Abelmoschus*	*A. angulosus var. angulosus*	*A. angulosus var. grandiflorus*	*A. angulosus var. purpureus*	*A. cailliei*	*A. crinitus*	*A. esculentus*	*A. ficulneus*	*A. enbeepeegearensis.*	*A. moscahtus subsp. moschatus*	*A. moschatus subsp. tuberosus*	*A. tetraphyllus*	*A. tetraphyllus var. pungens*	*A. tuberculatus*	*A. mizonagensis sp. nov.*	*A. moschatus subsp. rugosus*
A.angulosus var. *angulosus*	5	-	-	-	3	5	-	3	-	-	-	3	-	-	-
A. angulosus var. *grandiflorus*	-	5	-	0	-	0	0	-	0	0	4	-	0	-	-
A .angulosus var. *purpureus*	-	-	5	-	-	-	-	-	-	-	-	-	-	-	-
A. cailliei	5	6/8	-	5	0	5/8	-	-	2	0	6	0	0	8	-
A. crinitus	3	-	-	-	5	-	-	3	3	0	-	-	-	-	-
A. esculentus	5	8	-	8	3	5	2/4	-	3	2	8	8	8	8	8
A. ficulneus	-	-	-	-	0	3	5	0	0	0	4	0	7	-	-
A. enbeepeegearensis	-	-	-	-	5	-	-	5	6	6	-	-	-	-	-
A. moscahtus subsp. *moschatus*	0	1	-	4	3	4	4	2	5	7	4	0	0	-	-
A. moschatus subsp.*tuberosus*	-	-	-	0	3	3	-	3	7	5	0	0	4	-	-
A. tetraphyllus var. *tetraphyllus*	7	0	-	5	-	8	0	-	-	-	5	7	3	-	-
A . tetraphyllus var. *pungens*	0	-	-	0	0	0	-	0	0	0	-	5	-	-	-
A. tuberculatus	-	-	-	3	0	8	5	3	-	-	-	3	5	-	-
A. mizonagensis sp. nov.	-	-	-	-	-	-	-	-	-	-	-	-	-	-	-
A. moschatus subsp. *rugosus* (Dehra Dun)	-	-	-	-	-	-	-	-	-	-	-	-	-	-	-

0 - No fruit set; 1 - Fruit set; 2 - Fruit set but no seed set; 3 - Fruit set & seed set; 4 - Fruit set, seed set but no germination; 5 - Fruit set, seed set & F_1; 6 - Fruit set, seed set, F_1 but no F_2/BC_1; 7 - Fruit set, seed set, F_1 & F_2/BC_1; 8 - Colchichoploids – C_1 germinated and viable seeds harvested

Table 3: Characteristics of Synthetic Amphidiploids

#	Amphidiploids	Characters
1	*Abelmoschus* x *esoramensis* (*esculentus* x *mizonagensis* sp. nov.)	• Extended bearing, beyond February • Dark green fruits, good taste, short fruits, robust plants, prolific
2	*Abelmoschus* x *caizoramensis* (*caillei* x *mizonagensis* sp. nov.)	• Attractive ovate fruits – Baby okra • Good taste, high mucilage, dark green colour
3	*Abelmoschus* x *esculophyllus* (*esculentus* x *tetraphyllus* var. *tetraphyllus*)	• Intermediate between parents • More of *A .tetraphyllus* appearance • Basal branching (6-14 branches) • Reduced fruit length • Dark green fruits
4	*Abelmoschus* x *esanguflorus* (*esculentus* x *angulosus* var. *grandiflorus*)	• Plant to plant variation in photoperiodic sensitivity • Reduced fruit length, blocky fruits • Intermediate between parents • Two maternal genotypes used • Highly branched (25-35 tertiaries)
5	*Abelmoschus* x *escullei* (*esculentus* x *caillei*)	• Oval-globular fruits – Baby okra • High mucilage, good raw taste • Non-photoperiod sensitive • More affinity to *A.caillei* • 2-4 branches per plant
6	*Abelmoschus* x *cangulosus* (*caillei* x *angulosus*)	• Intermediate between parents • Ovate fruits – baby okra • Stiff hairy • Petals spot only inside • Non-persistent epicalyx • Non photoperiod sensitivity
7	*Abelmoschus* x *esculatus* (*esculentus* x *tuberculatus*)	• Red and green fruited plants • Studded with tubercles • Prolific, above 50 fruits / plant
8.	*Abelmoschus* x *tuberculentus* (*tuberculatus* x *esculentus*)	• Fruit surface studded with tubercles • 10-12 cm long fruits • Prolific bearing • 45-60 fruits/ plant

REFERENCES

Akhond Y. A. M, M. A. H. Molla, M. O. Islam and Mohammad Ali 2000. Cross compatibility between *Abelmoschus esculentus* and *A. moschatus. Euphytica* **114 (3):** 175 – 180.

Joseph John K, V A Muhammed Nissar, M Latha, Pravin Patil, SK Malik, KS Negi, Merita Keisham, S R Rao, SR Yadav and KV Bhat. 2013. Genetic Resources and Crossability Relationship among various species of *Abelmoschus. Current Horticulture.* **1(1):** 35-46.

Keisham Merita, Joseph John K., S.R. Yadav, K.V. Bhat and S.R. Rao 2014. Comparative analysis of heterochromatin distribution in wild and cultivated *Abelmoschus* species based on fluorescent staining methods. *Protoplasma* DOI 10.1007/ s00709-014-0712-2.

Kuwada, H. 1961. Studies on the interspecific crossing between *Abelmoschus esculentus* (L.).

Moench and *A. manihot* (L.) Medikus and the various hybrids and polyploids derived from the above two species. Fac. Agric. Kagawa Univ. Hem. 8, 91 p. (in Japanese).

Kuwada, H. 1966. The new amphidiploid plant named *"Abelmoschus tubercular-esculentus"*,obtained from the progeny of the reciprocal crossing between *A. tuberculatus* and *A. esculentus. Jap. J. Breed.* 16(2):21-30.

Kuwada, H. 1974. F_1 Hybrids of *Abelmoschus tuberculatus*, and *A. manihot* with reference to the genome relationship. *Jap. J. Breed.* **24:** 207-210.

Mayr, E., 1991. Principles of Systematic Zoology. McGraw-Hill, New York.

Pal, B.P., H.B. Singh and V. Swarup 1952. Taxonomic relationships and breeding possibilities of species of Abelmoschus related to okra (*A. esculentus*). *Bot. Gaz.,* 113:455-464.

Samarajeewa PK 2003. Wild *Abelmoschus* species in the improvement of okra *In* Jayasuriya AHM and DA Vaughan (Ed.) Conservation and Use of Crop Wild Relatives. Proceedings of the Joint Department of Agriculture, Sri Lanka and National Institute of Agrobiological Sciences, Japan, workshop held on 3-02-2003 pp. 97-108.

Spencer, R., Rob Cross and Peter Lumley 2007. Plants Names – A Guide to Botanical Nomenclature. CSIRO, Australia.

Distant Hybridization in Horticultural Crops Pages **201–219**
Editors: **M.R. Dinesh & M.Sankaran**
Published by: **ASTRAL INTERNATIONAL PVT. LTD., NEW DELHI**

13

Onion and Garlic Improvement Through Species Utilization

Vijay Mahajan, Ashwini Benke,
S. Anandhan and Jai Gopal

ICAR-Directorate of Onion & Garlic Research,
Rajgurunagar, Pune-410 505, M.S.

The genus *Allium* consists about 750 species including *A. cepa, A. sativum* (Rabinowitch and Brewster 1990, Hanelt *et al.* 1992). Long back from history *Alliums* were used as vegetables, spices, condiments, medicine and ornament. Besides onion and garlic nearly 35-40 different *Allium* species were found in temperate and Alpine region of Himalaya in India. However amongst all *Allium* species. Onion (*Allium cepa*), Garlic (*A. sativum*) Leek (*A. porrum*), Shallot (*A. ascalonicum*) and Chives (*A. schoenoprasum*) have been cultivated in India in different packets. *A. clarki, A. Hookeri, A. chinens, A.macaranthun, A ochinni* were some of the minor *Allium* species used by particular region for vegetable purpose as semi-domesticated and wild species (Pandey *et al.*, 2005, Pandey *et al.*, 2008). Onion (*Allium cepa* L.) is one of the oldest cultivated species, and it has been in use as a food for over 5000 years (Jones, 1983). These crops are mostly strong flavoured due to presence of sulphur containing compounds that are responsible for distinctive aroma and pungency (Rabinowitch and Currah, 2002). Primary centre of origin for onion is in Central Asia and secondary centre in the near East (Mc Collum, 1976). It occupies a vast area in Western Asia, extending from Palestine to India. Onion as medicinal herb is known to ancient world, as it finds in the medicinal treaties like *"Charak Samhita"* of third or fourth century A. D. (Ray *et. al.,* 1980 and Ray and Gupta, 1980). Sanskrit language equivalent signifying Vedic period and Aryan usage is available for onion as *"Palandu"* (Aiyar and Vaganarayana, 1956).

Onion (*Allium cepa* L.) is a valuable vegetable crop preceded in the list of worldwide cultivated vegetables after tomato (FAOSTAT, 2011). Over all, onion breeding concentrated on improvement of existing cultivar or varieties for various biotic and abiotic stress. Simultaneously breeders are concentrating on some universally desirable traits like uniformity in shape, colour, size, date of maturity, attractive and intact skin, bolting and sprouting resistance, single centered *etc*. But the cultivated onion and garlic, no resistance sources are available and exploited till date particularly in short day onion and garlic against pest and major diseases. There is scope in wild or other Allium species where screening for diseases of our interest needs to be done. In this direction collection of wild species and systematic screening work has been taken up at DOGR which can be utilized in future breeding programme. Some of the Indian originated species identified as a disease resistance or tolerant source for diseases like purple blotch, stemphyllium blight and garlic mosaic virus by NBPGR (Singh and Rana, 1994). However resistance or tolerance source for many of diseases and pests such as neck rot, basal rot, black mould *etc*. yet to be identified. As the importance of climatic change is increasing, breeder has to expand the search to find tolerance sources to various kinds of stresses like heat, cold, frost, drought, flood *etc*. Among the species which belong to *Allium*, *A. cepa*, *A. vavilovii*, *A. galanthum*, *A. roylei*, *A. fistulosum*, *A. altaicum*, *A. pskemense* and *A. oschaninii* have been considered as the most important gene pools of onion (Shigyo and Kik, 2008). Considering all above facts, introgression of foreign gene into cultivated species may give a way to mitigate biotic and abiotic stresses.

Interspecific hybridization within the genus *Allium* has a long history, and has been an important tool for increasing genetic variation. It was used with a view to create new variability, introgression of desirable genes as well as for transferring agronomically useful traits from suitable wild relatives (Brewster, 1994; Kik, 2002; Kiełkowska & Adamus, 2006). However, techniques of sexual hybridization often have been accompanied by several impediments to overcome pre- and post-fertilization barriers, resulting in a limited numbers of obtained F_1 hybrids, and further difficulties with back-crossing. Meiotic studies of hybrids and back-cross progenies within *Allium* revealed interesting information about the nature of alien introgression and improved understanding of "foreign" chromatin transmission into cultivated species such as *A. cepa* (Kik, 2002).

The purpose of present paper is to give a concise review on interspecific hybrids made in major *Allium* species irrespective of its success.

INTROGRESSION FROM *A.ROYLEI*

Allium roylei is a wild species from the Indian subcontinent which posses disease resistance against downy mildew (*Peronospora destructor*), leaf blight (*Botrytis squamosa*) and anthracnose (*Colletotrichum gloesporioides*) (van der Meer and de Vries, 1990; de Vries *et al.*, 1992; Galvan *et al.*, 1997; Scholten *et al.*, 2007; Kohli and Gohil, 2009). The first crosses between *A. roylei* and *A. cepa* were made in 1985 to achieve introgression of downy mildew resistance into onion (van der Meer and de Vries, 1990). Subsequent studies on reciprocal crosses of *A. cepa* and *A. roylei* hybridization (Kofoet *et al*.., 1990; Kofoet and Zinkernagel, 1990) led to the conclusion that resistance to downy mildew is determined by a single dominant gene, *Pd1* which was later confirmed by AFLP marker technique combined with SCAR analysis (van Heusden, 2000a,b). The years of investigation on interspecific hybridization and downy mildew resistance led to F_1BC_5-generation plants expected to be resistant to *Peronospora destructor* (Scholten *et al.*, 2007). First use of embryo rescue technique *in vitro* to produce hybrid between *A.roylei* into *A.cepa* was reported by Chuda and Adamus (2012). Ovule culture on media MS and BDS with 20g/l sucrose resulted in 906 well acclimatized plants out of 1768 regenerated embryos. Hybridity of plants were confirmed by genomic *in situ* hybridization where hybrid had sixteen chromosomes, out of which 8 were from *A. cepa* and 8 from *A. roylei*. The hybrids were also verified by three molecular markers viz., SIR, ACS and A.

INTROGRESSION FROM *A. fistulosum*

Allium fistulosum L. (Japanese bunching onion, Welsh onion) is a diploid species (2n=2x=16). *A. fistulosum*, similar to onion, belongs to section *Cepa* of the genus *Allium* (Jones, 1990) and subgenus Rhiziridium (Hanelt, 1990). Progenies from generative introgression of *A. cepa* with *A. fistulosum* have been studied most extensively among all interspecific crosses in the genus *Allium*. Those two species can be easily hybridized and the success rate can be enhanced by the use of embryo-rescue (Dole el *et al..*, 1980). The very first hybrid between *A. cepa* and *A. fistulosum* were reported in 1935 by Emsweller and Jones. First developed hybrid was reported as a sterile where seed set was occasional. In microscopic observations it is observed that the synaptonemal complex of hybrids showed heteromorphic bivalent. Chiasmata frequency is less compared to both the parents and that are mainly interstitial and distal. The F1 hybrids between *A. cepa* and *A. fistulosum* can be easily produced, but have low pollen fertility, often not exceeding 10% (Jones & Mann, 1963; Levan, 1941; Maeda, 1937; Peffley, 1986; van der Valk *et al..*, 1991b). Attempts to backcross (BC) F1 hybrid (*A. cepa* x *A. fistulosum*) to A. *cepa* were unsuccessful (Levan, 1941). Further studies showed that difficulties in this type of cross were due to strong pre-fertilization barrier, as the growth of onion pollen tubes in the style of the hybrid was inhibited (van der valk *et al.*, 1991a). Cytological analysis of back cross showed that three paracentric inversions and one translocation are present. RFLP and isozyme study of same revealed that the homeologous recombination and distorted segregation was present. Ulloa *et al.*, 1995 concluded that failure of this cross was due to nucleo-cytoplasmic incompatibility. Maeda (1937) backcrossed F1 with *A. fistulosum instead of a cepa* and could able to obtain viable seeds successfully. Emsweller and Jones (1945) obtained several BC progenies, but only when the hybrid was used as the pollen source. Khrustaleva and Kik (1998) were introgressed genes from *A. fistulosum* to *A. cepa* using *A. roylei* as a bridge species in crossing. Multi-colour GISH was performed to analyze genome organization of the hybrid. Results showed 8 *cepa* chromosomes, 1 metacentric chromosome from *roylei*, and 7 recombinant chromosomes, further recombination took place only in interstitial and distal portion of chromosome of interspecific hybrid between *A.fistulosum* to *A. roylei*. Pollen fertility of the second generation bridgecross hybrid was variable, which might be caused by the use of CMS onions as the female parent and/or the lack of *roylei* chromosome segments containing male sexual reproduction genes (Khrustaleva & Kik, 2000; Ulloa *et al.*, 1995. Monosomic addition lines of *fistulosum*

– *cepa* (Shigyo *et al.*, 1998) and *cepa* – *fistulosum* (Peffley *et al.*, 1985) were reported. These lines are valuable to study genome organization in *A. cepa* and *A. fistulosum*. Unfortunately, subsequent backcrosses of the obtained F_1 *A. cepa* *A. fistulosum* hybrids to *A. cepa* have proved problematic. The difficulties observed in genome transmission in advanced backcross generations have been explained by the sterility of F_1. *A. cepa* *A. fistulosum* hybrids probably resulting from one of three possible mechanisms: stylar incongruity (van der Valk *et al.*, 1991a), nuclear cytoplasmic incompatibility (Ulloa *et al.*, 1995) or central cell nuclear cytoplasmic incongruity (Mangum and Peffley, 2005). Due to problems in introgression of Japanese bunching onion traits into onion, so far only four new varieties of *A. fistulosum* have been developed: Beltsville Bunching, Delta Giant, Top Onion and Wakegi Onion (Kik, 2002).

Beltsville Bunching is a tetraploid hybrid where two set of chromosome comes from *A.cepa* and two were from *A.fistulosum*, which is cultivated in USA. Interspecific hybrid Delta Giant is a triploid in nature and grown in USA as a minor crop (Kik 2002). Top onion, a home garden crop of Europe and north-eastern Asia has been termed as *Allium proliferum* which is cross between *A.cepa* x *A. fistulosum* (Fritsch and Friesen, 2002) and has been confirmed by genomic studies. Top onion is commonly known as tree onion, Egyptian onion or walking onion as it produces bulblets or topsets where normal onion posses flowers. Introgression between *A. fistulosum* x shallot type *A.cepa* is a Wakegi onion which is diploid hybrid. Hybridity of these species have been confirmed by molecular and chromosomal studies. In Japan and China Wakegi onion is used for salad purpose.

'Pran' is also interspecific hybrid species (*Allium* x *cornutum*) known in Kashmir region of India, however in evolution studies seed parent of this interspecific cross is still unclear while female parent is *A.cepa* (Havey 1991).It produces very slender and attractive dark green color leaves and grown as a kitchen garden crop at Jammu ,Europe , Canada etc (Fritsch and Friesen, 2002).

INTROGRESSION FROM *A. AMPLELOPRASUM*

Species *A. ampleloprasum* comprises five different alliums such as leek, Kurrat, Taree, the great-headed garlic and pearl onion group (Brewster, 1994) which are interfertile in nature. Leek , Kurrat, Taree are allotetraploid allium species (2n=4x=32) whereas, great-headed garlic is hexaploid in nature with 48 chromosomes.

Interspecific hybrids between *Allium cepa* and *A. ampeloprasum* were generated with the aim of introduc S-cytoplasm from onion into leek. Few attempts of sexual hybridization of onion and leek failed to produce F_1 progeny (Dole el *et al.*, 1980; Currah, 1986). But use of *in vitro* culture to rescue the F_1 embryos at the early stages of development (7-14 days after pollination) was successful (Peterka *et al.*, 1997). Seven hybrid plants obtained in the study were triploids with 24 chromosomes. When the hybrid plants were used for transferring CMS through backcross progenies (Peterka *et al.*, 2002) chromosome composition varied and it was found that the BC_1 progeny plants had eight chromosomes from onion and 30-33 from leek. In the following generations *viz.*, BC_2 and BC_3, the transmission rate of univalent onion chromosome was decreased (varied from 0-8 in BC2) which resulted into high percent of alloplasmic leek plants. In another study, biased transmission of organelle DNA in somatic hybrids between leek and onion was reported (Buiteveld *et al.,* 1998). The somatic hybrids had rearranged Mt DNA that predominantly contained mtDNA and cpDNA of leek.

INTROGRESSION FROM *A. SATIVUM*

Majority of garlic cultivars are sterile, which preclude using them as a parent in sexual crossing program. However, occurrence of fertile garlic plants was also reported (Kononkov, 1953; Katarzhin and Katarzhin, 1978; Etoh, 1983), which opens the door for use of conventional breeding in garlic. Flowering in some garlic clones is often associated with presence of bulbils formed in the place of flowers over inflorescence (Kamenetsky and Rabinowitch; 2001). *Allium cepa* and *A. sativum* have not been hybridized sexually for a long time, because of the narrow genetic pool of fertile garlic and a large genetic distance (Kik, 2002; Keller *et al..,* 1996). A cross between *A. longicuspis* with fertile *A. sativum* as a pollen parent could able to produce few sterile F_1 progenies (Etoh, 1985). The first interspecific hybrids between onion and garlic were reported by Ohsumi *et al.,* (1993) and found allicin, a characteristic compound of garlic in the hybrid. Embryos were formed when onion is used as a seed parent and reciprocal crosses were not successful. Hybrids were regenerated in *in vitro* by embryo rescue which possessed 16 chromosomes in somatic cells, with clearly distinguishable two satellite chromosomes. Chromosome elimination was not observed in the study and pollen viability of the hybrids were very low (2%). A preliminary study of hybridization between diploid garlic and tetraploid leek has been attempted by Sugimoto *et al.* (1991) for the purpose of introducing fertility into garlic but characteristics of hybrid were not reported. Several triploid and aneuploids close to triploids hybrids were obtained. Worth mention is work of Yanagino *et al.* (2003) with aim of increasing fertility in garlic through sexual hybridization with leek. Success of this cross was accelerated, since both leek and garlic belong to the subgenus *Allium*. Leek was used as the female parent. Interspecific hybrids were obtained with the use of the embryo-rescue technique. Hybrids were triploids and possessed 24 chromosomes (16 from onion and 8 from garlic), however meiosis in PMC was not examined. Obtained hybrids were confirmed by molecular, cytological and morphological methods but all are almost completely sterile. Protoplast fusion was used to produce somatic hybrids of onion and sterile, garlic (Yamashita *et al.,* 2002). Onion and garlic have sixteen chromosomes but the two lines of somatic hybrids possessed 40 and 41 chromosome with three of each chimeric chromosomes. The chimeric chromosomes in somatic hybrids could have originated from chromosomal fusion or translocation between chromosomes of onion and Garlic.

INTROGRESSION FROM *A.GALANTHUM*

Introgression between *A.cepa* and *A.galanthum* has been attempted by many (Saini and Davis, 1967; McCollum, 1971). Majority of the generated hybrids were found sterile, but in few crosses seed set was observed and fertility of the progenies in later generation were restored by back cross (Yamashita and Tashiro, 1999).

A. galanthum contributed CMS trait in shallot (Yamashita and Tashiro, 1999), in onion (Havey, 1999) and in *A. fistulosum* (Yamashta *et al.*, 1999a). In *A. fistulosum*, alloplasmic line having *A. galanthum* cytoplasm was developed by continuous backcrossing (Yamashita *et al.*, 1999a). Male sterile populations were obtained by selection from BC_1 to BC_5. But the male fertile populations segregated into fertile and sterile in equal numbers. It indicates the fertility restorer gene was also from *A. galanthum* in the nucleus (Yamashita *et al.*, 1999b) segregating in the progenies. The location of the restorer loci was located on 5f region on *A. galanthum* chromosome (Yamashita *et al.*,2005). Studies in onion sterile could not able to locate on the same region of *A. galanthum* (Yamashita *et al.*, 2000; Havey, 1999), but on 2A in onion (Shigyo *et al.*, 2005).

Table: 1. Some of the efforts made in *Allium* interspecific hybridization

Sr. no	Crosses has been made between	Author & year
1	*A.cepa X A. fistulosum*	Emsweller & Jones 1935;Van der Meer & Van Benekon 1978; Currah & Ockendon 1988, Bark *et al.*.,1994
2	*A.Galanthum X A.cepa*	Mc Collum (1980)
3	*A.fistulosum X A roylei*	Mc Collum (1982), Khrustaleva and Kik (1998)
4	*(A.fistulosum X A. cepa) X A.cepa*	Van der Valk *et al.* 1991a, 1991b
5	*A.cepa X A.sativum*	Ohsumi *et al.*, 1993
6	*A.chinense X A. fistulosum*	Nomura *et al.* 1994 ; Nomura & Makara 1993
7	*A.fistulosum X A. tuberosum*	Amagai *et al.* 1995
8	*A.cepa x (A.fistulosum X A. roylei)*	Khrustaleva & Kik 1998
9	*A.cepa X A. ampeloprasum* (Leek)	Peterka *et al.* 1997
10	*A. cepa X A.sphaerocephalon*	Bino *et al.* 1989
11	*A.ampeloprasum X A. sativum*	Yanagino *et al.*, 2003
12	*A.fistulosum X A.schoenoprasum*	Umehara *et al.*, 2007
13	*A.cepa X A.roylei*	Chuda & Adamus (2012)
14	*A.roylei X A.cepa*	Van der Meer & de Vries 1990

WILD ONION NATIONAL STATUS

Negi and Pant (1992) reported the occurrence of lesser known wild species in mountainous regions and discussed information on their habitat, local uses, *etc.* The prominent wild species included in the study were *A. carolinianum* D.C. (syn. *A. bladum* Wall.), *A. chinense* G. Don (syn. *A. bakeri* Regel), *A. consanguineum* Kunth, *A. humile* Kunth (syn. *A. govanianum* Wall ex Baker), *A. przewalskianum* Regal (syn. *A. jacquemontii* Regel, *A. stolczkii* Regel), *A. steacheyi* Baker, *A. victorialis* L. and *A. wallichii* Kunth. The National Bureau of Plant Genetic Resources (NBPGR) has conducted extensive plant exploration in different allium-growing states/regions in India and collected wild relatives also, namely, *A. ampeloprasum*, *A. auriculatum*, *A. ascalonicum*, *A. carolinianum*, *A. chinensis*, *A. wallachi*, *A. tuberosum*, and *A. rubellum*.

The bureau has also introduced over 1100 accessions of *Allium* germplasm which include related *Allium* species from over 40 countries. (Singh and Rana, 1994).

The Himalayan *Allium* species comprise a rather diverse assortment of floristic elements. They include important cultivated food plants and taxa having so called secondary gene pooles, which are of potential value as plant genetic resources for future breeding programs (Xu, 1990; Gohil, 1973, 1992; Nayar *et al.*, 1992; Negi *et al.*, 1991; Chandel & Pandey, 1992). Most of them are potential and already been introduced as semi-domesticated cash crop into the backyards, dooryards and garden of natives of the Himalayan region (Negi, 2006, Negi *et al.*, 1995).

Pathak *et al.* (2001) studied that Stemphylium leaf blight (SLB) (causal agent *Stemphylium vesicarium*) is one of the major onion (*Allium cepa*) diseases of the tropics. Five Welsh onion (*Allium fistulosum*) and 106 *A. cepa* lines were screened in field and laboratory experiments. All the *A. fistulosum* lines were resistant or moderately resistant, whereas all the *A. cepa* lines were susceptible. Crosses were successfully made between five *A. fistulosum* and 29 *A. cepa* lines to introgress SLB resistance into onion lines. These crosses were further evaluated for disease resistance and fertility in the F_2 and F_3 generations. These progenies are now being used in breeding program to develop SLB resistant onion lines but, yet to get desirable success. Anitha *et al.* (2011) reported need for introducing *Allium ampeloprasum, A. bouddhae, A. obliquum, A. nutans* and *A. odorum* possessing traits of resistance to different pests.

Murthy *et al.* (2011) studied about wild onions. The white bulbs of 'Wild Onion, *Urginea indica*' samples were collected from the hillocks near Vaddagere village of Tumkur district, a southern part of Karnataka. The presence of primary and secondary metabolites such as *carbohydrate, proteins, alkaloids, phenolic compounds, saponins* was confirmed through preliminary phyto-chemical analysis. The extract was found to possess anti-bacterial activity in *E. coli, S. aureus and P. aeruginosa* isolated from infected patients. The Minimum inhibitory concentration (MIC) was found to be considerably effective against selected pathogenic bacteria. Such an effect might contribute to explaining the traditional use of wild onion sps, *Urginea indica* in the treatment of wound healing. Wild Onion sps was capable of reducing oxidants and scavenging free radicals. This also indicates that, tubers of wild onion, *Urginea indica*' are of therapeutic potential used in traditional medicine treatment. Hence, the formulation of extract of *Urginea indica*' needs to be purified using biophysical techniques towards development of a potential drug/ lead molecule against microbial infection, inflammation and wound healing respectively.

Underutilized Alliums (*Allium* and its allies) are among the plant genetic resources that have been valued for food, feed, medicines, spices and condiments *etc.* of about 700 species scattered all over the world, 34 species are reported mainly from the alpine and temperate zones of Himalaya, India. Besides cultivated species (*Allium cepa, A. sativum, A. ascalonicum* and *A. ampeloprasum*) some species are semi-domesticated/lesser-known types (*Allium carolineanum, A. chinense, A. c onsanguineum, A. humile, A. victorialis, A. wallichii* and *A. semenovii*) under sporadic cultivation. Large scale collection and utilization of some of the wild species for edible (leaves and bulbs) and flavouring purpose has resulted in loss of the natural diversity reported by Negi, 2011. There is urgent need to conserve these wild species.

EFFORTS AT THE DIRECTORATE OF ONION AND GARLIC RESEARCH

Wild species collection and conservation at DOGR

Due to changes of climate, urbanization, road construction, heavy rainfall, landslides many primitive cultivars and wild relatives of *Allium* spp. were on the verge of extinction. Hence, its conservation, evaluation and utilization is more important. Keeping above facts in mind efforts were to collect different wild and underutilized *Allium* species from different parts of the country as well as introduced from abroad through NBPGR from USDA, USA, CGN Netherlands, IPK, Germany *etc*. Till now at DOGR we have 25 number of *Allium* species with 116 number of lines. These species are successfully maintained under natural conditions. In North Eastern parts foliage is used as a substitute of onion and garlic in their day to day life. No work has been done earlier in this direction. Hence, identification of species for palatable foliage is being undertaken. Status of flowering and its season is also recorded. Four crosses were between wild *Allium* species with the cultivated onions and further being multiplied using embryo rescue technique. Nutritive value, cytology, molecular characterization, morphological characterization is being carried out in these species besides diseases and pest resistance for further utilization in breeding as well as direct consumption.

Exploration trip in Arunachal Pradesh state including Assam area was undertaken for *Allium* spp. (wild and cultivated) in collaboration with NBPGR Regional Station, Bhowali during 17 to 29 October, 2013. Efforts were made to collect primitive cultivars including *Allium spp.* (cultivated and wild) germplasm from seven districts i.e., West Kameng, East Kameng, Kurung, Kumey, Lower Subansiri, Upper Subansiri, West Siang and Pappum Pare of Arunachal Pradesh and a part of Sumitpur and Lakhimpur of Assam state covering overall about 2000 km including remote area before reaching their vulnerable stage. In most of the villages wild Alliums i.e., *Allium fasciculatum, A. tuberosum, A. hookerii, A. chinense* and *A. macranthum* are being semi domesticated for day to day use as spices and condiments including vegetables. Variation in foliage, bulb/clove size, shape, colour and taste were observed. Foliage of most of the *Allium* species are used for consumption by tribal populations in place of common onion and garlic for cooking purpose, making chatany and even for medicinal use. *Allium fasciculatum, A. tuberosum, A. hookerii* and *A. macranthum* is being cultivated in natives kitchen garden/ dooryards/ backyards. Two accessions of wild *Allium* yet to be identified are found where its roots/ rhizomes were highly prized for spices and condiments, locally known as *Modi dite* (Mountain Garlic/ Chives) and also used for medicinal purpose. In local market leaves of *A. tuberosum*, locally known as *Dite* is largely being sold. We hardly found common onion (*Allium cepa*) germplasm from Arunachal Pradesh. Onion is replaced by *Allium cepa var. aggegratum* (Multiplier onion), known as *Punyoo/ Punroo* in local parlance were noticed for their frequent cultivation in remote place or inaccessible areas in Assam. *Allium ampeloprasum*, leek known as *Lam* is also being used as medicinal purpose in limited area. A total of 49 germplasm collections were made. Due to changes of climate, urbanization, road construction, heavy rainfall, landslides many primitive cultivars and wild relatives of *Allium* spp. were on the verge of extinction. Foliage of these species can be used as an alternative to onion and garlic in crises. Hence, its conservation, evaluation and utilization is more important. The detail of the collections is given here under :

Allium hookeri grown in dooryards/ backyards

Foliage of *Allium tuberosum* sold in market

Roots of wild onion (Modi dite)

Young plant of modi dite

Allium chinense

Allium macranthum

Table: 2. Germplasm collections made during 2013 from Arunachal Pradesh and part of Assam

S. No.	Common/Local name	Botanical name	Cultivation status	No. of acc. collected	Area of collection	Use
1	Garlic, Lohru, Lahsoon	*Allium sativum*	Cultivated	07	Assam	Vegetable spices for cooking
2	Leek, Lam	*Allium porrum* syn. *A. ampeloprasum* var. *porrum*	Semi-domesticated	01	Arunanchal Pradesh	vegetable, pickle & condiment / flavor and as medicinal purpose
3	Multiplier onion, Punyoo/ Pyaz, Punroo	*Allium cepa.* var *aggregatum*	Semi-domesticated	05	Assam	Vegetable cooking
4	Chiense chives, Lapta, Lapsar, Narang, Dite	*Allium tuberosum*	Semi-domesticated	06	Arunanchal Pradesh	Vegetable for flavor/ condiment
5	Taley, Talap	*Allium hookerii*	Semi-domesticated/ Wild	04	Arunanchal Pradesh	Vegetable (fibrous roots), pickle (pseudostem), foliage for flavor and for making chatany
6	Modi dite	*Allium* spp.	Wild	02	Arunanchal Pradesh	Roots are used for medicinal purpose for blood pressure and cold diseases besides cooking as flavor
7	Lasun, Lapsa, Modi Byke, Modi Byako	*Allium macranthum*	Semi-domesticated	14	Arunanchal Pradesh	Foliage are used for flavor in vegetable and for making chatany
8	Rakkyo, Talap, Adi Talap	*Allium chinense*	Semi-domesticated	02	Arunanchal Pradesh	Used as vegetable, bulbs for pickle, condiment
9	Zap, Leppi, Lipee, Laptap	*Allium fasciculatum*	Semi-domesticated/ Wild	08	Arunanchal Pradesh	Vegetable and salad
	Total			49		

EFFORTS FOR INTERSPECIFIC HYBRIDIZATION

Interspecific crosses were attempted to facilitate the introgression of desirable traits of *Allium fistulosum* into *Allium cepa* var. Bhima Shubhra genome. For that purpose, crossing was done in 16 umbels, ovule enlargement was observed after pollination. But ovary shrivelled and ovules degenerated about 10 days after pollination and no seed development was observed. Ovule culture technique was tried to obtain crossed progeny. Ovules were cultured after fifth day of pollination on ovary culture media (BDS +7.5% sucrose +2mg/2-4-D +2mg/lit BAP). Ovaries enlarged normally on the media and plantlet germination was observed after a 40 days of inoculation. Germinated plantlets were shifted to basal media (B5 +2% Sucrose) for proper establishment. Two plants with proper root and shoot were obtained and established on basal media. After three months plants were shifted to green house for hardening. The interspcific hybrids showed a vigorous growth habit like *Allium fistulosum*; foliage were deep green in colour with waxy coating and no flowering were observed in last one year. The hybrid could not form a bulb and, showed tillering habit. The generated hybrids will be used for backcrossing.

MOLECULAR CHARACTERIZATION OF SOME *ALLIUM* SPECIES

Genetic diversity studies in twenty different *Allium* species was carried out using SRAP markers. A total of 11 SRAP primers were used, out of which 8 primers amplifeid. All the 8 amplified primers showed polymorphism. Analysis of SRAP markers showed that genetic similarity (GS) value among the 20 samples ranged from 0.44 to 1. The UPGMA dendrogram was prepared using NTSys pc 2.1 programme. The cluster analysis divided the 20 Allium species samples into two distinct clusters. In cluster 1 *Allium alticum* (CGN-14770), *Allium alticum* (CGN-14769) and *Allium alticum* (CGN-16417) has shown maximum similarity, while Allium fitulosum (Taiwan) grouped with *Allium alticum*(ALL-233). Pran-1in a subcluster grouped with *A.cepa aggrigatum* and *A. schenoprasum*. In cluster 2 comprised of 3 wild alliums *A. tuberosum, A. fistulosum* and *A. sativum*. Among these *A. sativum* and *A. tuberosum* were grouped together. Bhima Super was out rooted and was not showing much similarity with the wild *Alliums*.

Table: 3. *Allium* species used for molecular characterization

Sr. No.	Name		
1	Bhima Super	11	*A. cepa* var. aggrigattum (-5)
2	*A. alticum* (CGN-14770)	12	*A. tuberosum* (CGN-20779)
3	*A. fistulosum* (Taiwan)	13	*A.griffithianum* (IC-250414)
4	PRAN-1	14	*A. sativum*
5	*A. ampeloprasum* (EC-609483)	15	*A. tuberosum*
6	*A. alticum* (CGN-14769)	16	*A. aschaninii* (EC-328495)
7	*A. alticum* (ALL-233)	17	*A. schenoprasum*
8	*A. alticum* (CGN-16417)	18	*A. fistulosum* (ALL-646)
9	*A. ampeloprasum* (ALL-697)	19	*A. guttatum* (CGN-16418)
10	*A. cepa* (Uzbekistan)	20	*A.cepa* (Eshing Tilong)

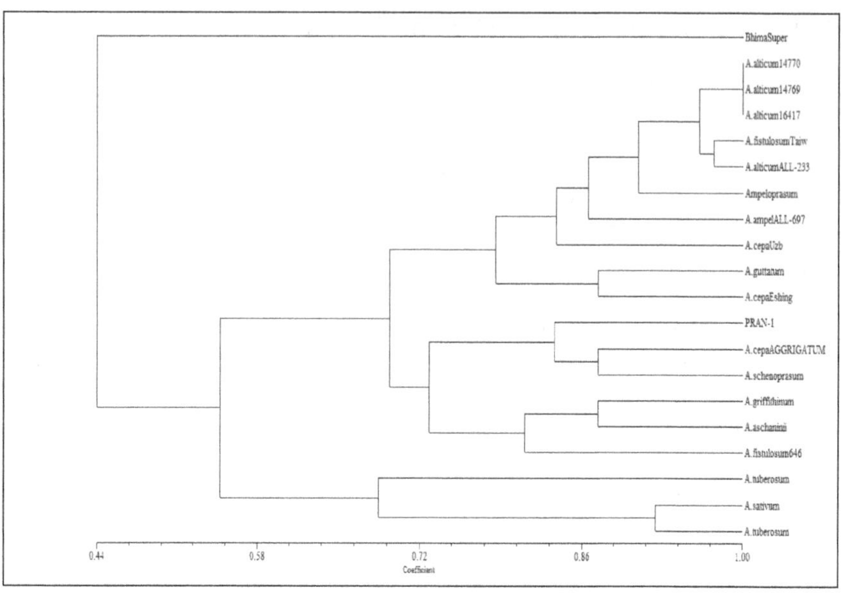

Fig. 1: UPGMA dendrogram prepared using NTSys pc 2.1 programme

CONCLUSION

For producing a interspecific hybrid successfully, overcome pre and post fertilization barriers is challenging. Pre- fertilization barriers in distance introgression , such as failure in pollen germination, slow pollen tube growth, or foreign pollen tube arrest in the style have been reported in alliums by Gonzalez & Ford-Lloyd, 1987; Ohsumi *et al.*, 1993; van der Valk *et al.*, 1991a and different species by Kiełkowska & Adamus, 2006; Manickam & Sarkar, 1999. Post fertilization barriers are associated with abnormalities in the development of the zygote, absence or abnormal development of the endosperm which causes embryo starvation and ultimately abortion (Zenkteller, 1990). Within the discussed species, the embryos developed regularly in the hybrids with introgression from *A. fistulosum*, or had to be rescued on the early stages of the development in hybrids with the leek and garlic introgression. Possibly, restoration of fertility of interspecific hybrids can be achieved by chromosome doubling (Jones, 1983). Chromosomes are doubled by treating the plant tissue with colchicine as antimitotic agent (Blakslee, 1939; Blakslee & Avery, 1937). Mc Collum (1980) reported fertile amphidiploid of *A. cepa* x *A. galanthum* obtained by colchicines treatment. Autotetraploids in *A. cepa* and *A. fistulosum* were obtained by soaking germinating seeds in aqueous colchicine solution, but the recovery of tetraploids was very low *i.e* 2% (Toole & Clarke, 1994). Song *et al.* (1997) treated calli form F1 hybrids of *A. fistulosum* x *A. cepa* with 0.1 and 0.2% colchicine and regenerated tetraploids. It seems that the potential is promising and the recovery of fertile plants from colchicine-treated F_1 hybrids has been reported in other species (Eigsti & Dustin 1955; Orton & Steidl, 1980). Wide taxonomic distance between *Allium* species does not hinder fertilization between them. On the contrary, our preliminary attempts to cross *A. sphaerocephalon* with leek, which belongs to the same subgenus, indicated strong barriers early after pollination Peterka *et al.* (1997). There is no close correlation between the level of relationship and crossability in the genus *Allium* (van Raamsdonk *et al.*. 1992). Further characterization of all the collected *Allium* species at ICAR-DOGR is being done both at morphological as well as molecular level. Studies are in progress for their nutritive value and biochemical constituents, diseases and pest resistance and cytological studies in Indian and exotic collections. Flowering time and behavior is also being studied. Non flowering species will be sent to CITH under long day conditions for flowering and utilize in crossing programme with cultivated *Allium* species including *A. cepa*. Efforts on flowering in garlic are also in progress. Further efforts will be made to flower useful species creating artificial conditions in plains with the aim to develop interspecific hybrids or use to transfer using biotechnological tools. There is need to collect or introduce some more *Allium* species for maintenance and its utilization which is being done in collaboration with NBPGR through ICAR-IIHR. Work on exploitation of Allium species has been already taken up at ICAR-DOGR, hence, these wild species can be exploited in improvement of onion and garlic with thorough evaluation and possible use through crossing or using biotechnological tools in future.

REFERENCES

Aiyar, A. K. and Yaganarayana (1956). The Antiquity of Some Field and Forest Flora of India. Bangalore Printing and Publishing Co. Ltd.

Amagai, M., Ohashi K. and Kimura S. (1995). Breeding of interspecies hybrid between *Allium fistulosum* and *A. tuberosum* by embryo culture. *Bull Tochigi Agr Exp Stn*, Vol. 43, pp.87–94.

Anitha K., Venkateswaran K., Chakrabarty S. K., Kumar G.S., Babu B. S., and Varaprasad K.S. (2011). Onion Genetic Resources and Pest Resistance: The Indian Scenario. *Indian Journal of Plant Protection* 39 (2): 81-92.

Bark H. O., Havey M. J., Joe N. Corgan (1994). Restriction Fragment Length Polymorphism (RFLP) Analysis of Progeny from an *Allium fistulosum* x *A.cepa* Hybrid. *J. Amer. Soc. Hort. Sci.* 119(5):1046-1049.

Bino, R.J., Janssen M.G., Franken, J. and de Vries, J.N. (1989). Enhanced seed development in the interspecific cross *Allium cepa* x *A. sphaerocephalon* through ovary culture. *Plant Cell Tiss and Org Cult*, Vol. 16, pp. 135-142.

Blakeslee, A. and Avery A. (1937). Methods of inducing doubling of chromosomes in plants by treatment with colchicine. *J Hered,* Vol. 28, pp. 393–411.

Blakeslee, A. (1939). The present and potential service of chemistry to plant breeding. *Am J Bot*, Vol. 26, pp. 163–172.

Brewster, J.L. (1994). The genetics and Plant breeding of *Allium* crops. In: *Onions and other vegetable alliums.* Brewster J.L., pp. 41-61, CAB International, ISBN 0581995101,UK.

Buiteveld, J., Suo Y., van Lookeren Campagne M.M. and Creemers-Molenaar J. (1998). Production and characterization of somatic hybrid plants between leek (*Allium ampeloprasum* L.) and onion (*Allium cepa* L.). *Theor Appl Genet*, Vol. 96, pp. 765-775.

Chandel, K.P.S. and Pandey, R. (1992). Distribution, diversity, uses and *in-vitro* conservation of cultivated and wild *Alliums* –a brief review. *Indian J. Pl. Genetic Resources* 5 (2): 7-36.

Chuda A. and Adamus A. (2012). Hybridization and molecular characterization of f1 a*llium cepa* x *allium roylei* plants. *Acta Biologica Cracoviensia* 54/2: 25-31

Currah, L. (1986). Leek breeding: a review. *J Hortic Sci,* Vol. 61, pp. 407-415.

Currah, L. and D.J. Ockendon, (1988). A hybrid index for *Allium cepa A. fistulosum crosses.* pp. 190 196. *Proc. 4th Eucarpia Allium Symposium.* Wellesbourne, United Kingdom.

de Vries, J.N., W.A. Wietsma, and T. de Vries. (1992). Introgression of leaf blight resistance from *Allium roylei* Stearn into onion (*A. cepa* L.). *Euphytica* 62:127–133.

Dole el, J.; Nowak F.J. and Lu ny J. (1980). Embryo development and *in vitro* culture of *Allium cepa* and its interspecific hybrids. *Z. Pflanzenzüchtg,* Vol. 85, pp. 177-184

Eigsti, D.I. and Dustin P. Jr (1955). Spindle and cytoplasm. In: *Colchicine in agriculture, medicine,biology and chemistry,* pp. 65-139, The Iowa State University College Press, ISBN 9780598806932, Ames.

Emsweller, S.L. and Jones H.A. (1935). An interspecific hybrid in *Allium. Hilgardia* Vol. 9, pp. 265-273.

Emsweller, S.L. and Jones H.A. (1945). Further studies on the chiasmata of the *Allium cepa* x *A. fistulosum* hybrid and its derivatives. *American Journal of Botany,* Vol. 32, pp. 370-379.

Etoh, T. 1983. Accomplishment of microsporogenesis in a garlic clone. *Mem. Fac. Agr. Kagoshima Univ.* 19:55–63.

Etoh, T. 1985. Studies on the sterility on garlic, *Allium sativum* L. Mem. Fac. Agr.

Kagoshima Univ. 21:77–132.

FAOSTAT. http://faostat.fao.org/. Accessed 02 August 2011.

Fritsch, R.M. and Friesen N. (2002). Evolution, domestication and taxonomy. In: *Allium Crop Science: recent advances.* Rabinowitch H.D. & Currah L., pp. 5-30, CABI Publishing, ISBN 0851995101, UK

Galvan GA, Wietsma WA, Putrasemedja S, Permadi ah, and Kik C. (1997). Screening for resistance to anthracnose (*Colletotrichum gloesporioides* Penz.) in *Allium cepa* and its relatives. *Euphytica* 95(2): 173–178.

Gohil, R.N. 1973. *Cytogenetic Studies of Some Indian. Alliums* Ph. D. Thesis, University of Kashmir.

Gohil, R.N. (1992) Himalayan representatives of Alliums. In: Hanelt P, Hammer K, Knüpffer H (ed.) *The genus Allium – taxonomic problems and genetic resources.* Gatersleben, Germany, pp. 335–340.

Gonzalez, L.G. and Ford-Lloyd B.V. (1987). Facilitation of wide crossing through embryo rescue and pollen storage in interspecific hybridization of cultured *Allium* species. *Plant Breed,* Vol. 98, pp. 318–322.

Hanelt P., 1990. Taxonomy, evolution and history. In: Rabinowitch H.D., Brewster J.L. (eds), Onions and Allied Crops. Volume 1. *Boca Raton*, CRC Press: 1–26.

Hanelt P, Schulze-Motel J, Fritsch R, Kruse J, Maass H, Ohle H and Pistrick K (1992). Infragenetic grouping of Allium – the Gatersleben approach. In: Hanelt P, Hammer K, Knu¨pffer H (eds) The genus Allium – taxonomic problems and genetic resources. Gatersleben, Germany, pp. 107–123.

Havey MJ (1999) Seed yield, floral morphology, and lack of malefertile restoration of male sterile onion (Allium cepa) population possessing the cytoplasm of Allium galanthum. J Am Soc Hort Sci 124:626–629.

Jones, H.A. and Mann L.K. (1963). Onions and Their Allies - Botany, Cultivation, and Utilization. Interscience Publishers Inc., ISBN 100249388731, New York3.

Jones, R.N. (1983). Cytogenetic evolution in the genus *Allium*. In: *Cytogenetic of crop plants*. Swaminathan M.S.; Gupta P.K. & Sinha V., pp. 516-554, Macmillan Press Ltd, ISBN 0333904230, India.

Jones, R.N. (1990). Cytogenics. In: *Onions and Allied crops. Vol I. Botany, physiology and genetics*, Rabinowitch H.D. & Brewster J.L., pp. 199-214, CRC Press, ISBN 0849363004, Florida.

Katarzhin, M. S. and I. M. Katarzhin (1978). Experimental on the Sexual Reproduction of Garlic. (in Japanese) *Japan. J. Breed*. 45 (2): 310.

Keller, R. J., Schubert I., Fuchs J. and Meister A. (1996). Interspecific crosses of onion with distant *Allium* species and characterization of the presumed hybrids by means of flow cytometry, karyotype analysis and genomic *in situ* hybridization. *Theor Appl Genet*, Vol. 92, pp. 417-424.

Khrustaleva L. I. and Kik C. (2000). Introgression of *Allium fistulosum* into *A. cepa* mediated by *A. roylei. Theor Appl Genet*, Vol. 100, pp. 17–26.

Khrustaleva L. I, and Kik C. (1998). Cytogenetical studies in the bridge cross *Allium cepa* (*A. fistulosum   A. roylei*). *Theoretical and Applied Genetics* 96: 8–14.

Kiełkowska, A. and Adamus A. (2006). Growth of pollen tubes from foreign species in carrot (*Daucus carota* L.) pistils. In: *Haploids and doubled haploids in genetics and plant breeding*. Adamski T. & Surma M., pp. 193-197, Prodruk, ISBN 8389887444, Poland.

Kik, C. (2002). Exploitation of wild relatives for the breeding of cultivated *Allium* species. In: Allium Crop Science: recent advances. Rabinowitch H.D. & Currah L., pp. 81-100, CABI Publishing, ISBN 0851995101, UK.

Kofoet A, and Zinkernagel V. (1990). Resistance to downy mildew (*Peronospora destructor* (Berk.) Casp.) in *Allium* species. *Journal of Plant Diseases and Protection* 97(1): 13–23.

Kofoet A, Kik C, Wietsma W. A. and de Vries (1990). Inheritance of resistance to downey mildew (Pernospora destructor (Berk.) Casp.) from *Allium roylei* Stearn in the back-cross *Allium cepa* L. 9 (A. roylei 9 A. cepa). *Plant Breed*105:144–149.

Kohli B, and Gohil R. N.(2009). Need to conserve *Allium roylei* Stearn: a potential gene reservoir. *Genetic Resources and Crop Evolution* 56: 891–893.

Kononkov, P.F.(1953). The question of Obtaining Garlic seed (in Russian) Sad i *Ogorod* 8:38-40.

Konvi ka, O. and Levan A. (1972). Chromosome studies in *Allium sativum*. Hereditas, Vol. 72, pp. 129-148.

Levan, A. (1941). The cytology of the species hybrid *Allium cepa* x *fistulosum* and its polyploidy derivatives. *Hereditas,* Vol. 27, pp. 253-272.

Havey M. J. (1991). Molecular Characterization of the Interspecific Origin of Viviparous *Onion J Hered* (1991) 82 (6): 501-503.

Maeda, T. (1937). Chiasma studies in *Allium fistulosum, Allium cepa* and their F_1, F_2 and backcross hybrids. *Jap J Genet,* Vol. 13, pp. 146-159.

Mangum PD, and Peffley EB.(2005). Central cell nuclear-cytoplasmic incongruity: a mechanism for segregation distortion in advanced backcross and selfed generations of [(*Allium cepa* L.   *A. fistulosum* L.)   *A. cepa*] interspecific hybrid derivatives. *Cytogenetic and Genome Research* 109: 400–407.

Manickam, S. & Sarkar K.R. (1999). Maize, pearl millet and sorghum pollen tube growth rate in maize silk. *Ann Agricult Research,* Vol. 20(2), pp. 216-219.

McCollum, G. D. 1982. Experimental hybrids between *A. fistulosum* and *A. roylei*. Bot. Gaz. 143:238–242.

McCollum, G.D. (1971). Sterility of some interspecific *Allium* hybrids. *J Am Soc Hort Sci,* Vol. 96, pp. 359-362.

Mc Collum, G.D. (1976). Onion and Allies In: Simmonds N. W. (ed.). *Evolution of Crop Plants.* London and New York, Longman, pp.186-190.

McCollum, G.D. (1980). Development of the amphidiploids of *A. galanthum x A. cepa, J Heredity,* Vol. 71, pp. 445-447.

Murthy, P. G., Mamtharani, D.R., Tejas, T.S and Niranjan M. Suarlikerimath (2011) Phytochemical analysis, in vitro anti-bacterial and antioxidant Activities of wild onion sps. International Journal of Pharma and Bio Sciences 2 (3): 230-237.

Nayar, N. M., Ahmedulla, M. & Singh, R. (1992). *Alliums* in South Asia. Importance, ehnobotanical uses, genetic resources, enumeration of species and distribution. In: *The Genus Allium- Taxonomic Problems and Genetic Resources* (Eds. Hanelt, P., Hammer, K. and Knupffer, H.). pp. 205-213.

Negi, K.S. (2006). *Allium* species in Himalayas and their uses with special reference to Uttaranchal. *Ethnobotany* 18, 53–66.

Negi, K.S. (2011). Scenario of underutilized Alliums in India. Souvenir: Nat. Symp. On Alliums: Current Scenario and Emerging Trends, 12-14, March, 2011: 29-42.

Negi, K.S, Gaur, R.D. & Pant, K.C. 1995. Biodiversity: untapped plant wealth in Uttarakhand. *Indian Hort.* 40 (2): Cover I, II, IV 32-40.

Negi, K.S. and Pant, K.C. (1992). Less-known wild species of *Allium* L. (Amaryllidaceae) from mountain region of India. *Economic Botany* 46(1), 112–114.

Negi, K.S., Pant, K.C., Koppar, M.N. & Thomas, T.A. 1991. Wild relatives of gsenus *Allium* L. in Himalayas. *Indian J. Pl. Genetic Resources.* 4 (1): 732-77.

Nomura, Y. and Makara K. (1993). Production of interspecific hybrids between Rakkyo (*Allium chinense*) and some other *Allium* species by embryo rescue. *Jpn J Breed*, Vol. 43, pp. 13–21.

Nomura, Y., Maeda M., Tsuchiya T. and Makara K. (1994). Efficient production of interspecific hybrids between *Allium chinense* and edible *Allium* spp. through ovary culture and pollen storage. *Breed Sci*, Vol. 44, pp. 151-155.

Ohsumi, C.A., Kojima K., Hinata K., Etoh T. and Hayashi T. (1993). Interspecific hybrid between *Allium cepa* and *Allium sativum*. *Theor Appl Genet*, Vol. 85, pp. 969-975.

Orton, T.J. and Steidl P.J. (1980). Cytogenetic analysis of plants regenerated from colchicinetreated callus cultures of an interspecific *Hordeum* hybrid. *Theor Appl Genet*, Vol. 57, pp. 89-95.

Pandey UB, Kumar Ashok, Pandey Ruchira, and Venkateshwaran K. (2005). Bulbous crops – cultivated alliums. **In:** Plant Genetic Resources: Horticultural Crops (Dhillon BS, Tyagi RK, Saxena).

S, and Randhawa GJ, eds.). Narosa Publishing House, New Delhi, India. pp. 108–120.

Pandey, A., Pandey, R, Negi, K.S. and Radhamani J. (2008). Realizing value of genetic resources of *Allium* in India. *Genet. Resources Crop Evol.* 55:985–994.

Pathak, C.S., Black, L.L., Cheng, S.J., Wang, T.C. and Ko, S.S. (2001). Breeding onions for stemphylium leaf blight resistance. *Acta Hort.* (ISHS) 555:77-81.

Peffley, E.B. (1986). Evidence for chromosomal differentiation of *A. fistulosum* and *A. cepa*. *J Am Soc Hort Sci*, Vol. 111, pp. 126-129.

Peffley, E.B.; Corgan J.N.; Horak H.G. & Tanksley S.D. (1985). Electrophoretic analysis of *Allium* alien addition lines. *Theor Appl Genet*, Vol. 71, pp. 176-184.

Peterka, H.; Budahn H. and Schrader O. (1997). Interspecific hybrids between onion (Allium cepa L.) with S-cytoplasm and leek (*Allium ampeloprasum* L.). *Theor Appl Genet*, Vol.94, pp. 383-389.

Peterka, H.; Budahn H.; Schrader O. and Havey M.J. (2002). Transfer of male-sterility-inducing cytoplasm from onion to leek (*Allium ampeloprasum*). *Theor Appl Genet*, Vol. 105, pp.173-181.

Rabinowitch HD and Brewster J (1990). Onion and allied crops –botany, physiology and genetics, vol I. CRC Press, *Boca Raton*, pp 1–26.

Rabinowitch, H.D. (1997). Breeding alliaceous crops for pest resistance. *Acta Hort*, Vol. 433, pp. 223-246.

Ray, P. and Gupta, H. N. (1980). "*Charaka Samhita*" *Scientific Synopsis*. Indian National Science Academy, New Delhi.

Ray, P., Gupta, H. N. and Roy, M. (1980). "*Sankruta Samhita*" *Scientific Snopsis*. Indian National ScienceAcademy, New Delhi.

Kamenetsky R. and H. D. Rabinowitch (2001). Floral development in bolting garlic *Sex plant Reprod* 13:235-241.

Saini SS and Davis GN (1967). Compatibility in some Allium species.Proc Am Soc *Hort Sci* 91:401–409.

Scholten OE, Van Heusden AW, Khrustaleva LI, Burgermeijer K, Mank RA, Antonise RGC, Harrewijn JL, Van Haecke W, Oost EH, Peters RJ, and Kik C. (2007). The long and winding road leading to the successful introgression of downy mildew resistance into onion. *Euphytica* 156: 345–353.

Shigyo M, and Kik C. (2008). Onion. In: Prohens J, Nuez F [eds], *Vegetables II: Fabaceae, Liliaceae, Solanaceae and Umbeliferae*, 121–159, Springer New York.

Shigyo M, Yamane N, Martin WJ, Yamauchi N, Havey MJ (2005).Assignment of onion RFLP-linkage map to chromosomes byusing alien monosomic additions. *Abstr J Jpn Soc Hort Sci* 74(Suppl 1):426.

Shigyo, M.; Imamura K. Iino M.; Yamashita K., & Tashiro Y. (1998). Identification of alien chromosomes in a series of *Allium fistulosum* – *A. cepa* monosomic addition lines by means of genomic *in situ* hybridization. *Genes Genet Syst*, Vol. 73, pp. 311-315.

Singh BP, Rana RS (1994). Collection and conservation of Allium genetic resources: an Indian perspective. *Acta Hort* (ISHS) 358:181–190.

Song, P.; Kang W. & Peffley E.B. (1997). Chromosome doubling of *Allium fistulosum x A. cepa* interspecific F_1 hybrids through colchicine treatment of regenerating callus. *Euphytica*, Vol. 93, pp. 257-262.

Sugimoto H, Tsuneyoshi T, Tsukamoto M, Uragami Y, Etoh T(1991) Embryocultured hbrids between garlic and leek. *Allium Improv News* 1: 67-68

Toole, M.G. & Clarke A.E. (1994). Chromosome behavior and fertility of colchicine-induced tetraploids in *A. cepa* and *A. fistulosum*. *Herbertia*, Vol. 11, pp. 295-303.

Ulloa, M.; Corgan J.N. & Dunford M. (1995). Evidence for nuclear-cytoplasmic incompatibility between *Allium fistulosum* and *A. cepa*. *Theor Appl Genet*, Vol. 90, pp. 746-754.

Umehara, M.; Sueyoshi T.; Shimomura K.; Hirashima K.; Shimoda M. & Nakahara T. (2007). Production and characterization of interspecific hybrids between *Allium fistulosum* L. and *Allium schoenoprasum* L. *Bull of the Fukuoka Agric Res Center*, Vol. 26, pp. 25-30.

Van Der Meer QP, and De Vries JN. (1990). An interspecific cross between *Allium roylei* Stearn and *Allium cepa* L.,and its backcross to *A. cepa*. *Euphytica* 47: 29–31.

Van der Meer, Q.P. & van Bennekom J.L. (1978). Improving the onion crop (*Allium cepa* L.) by transfer of characters from *A. fistulosum*. *Biuletyn Warzywniczy*, Vol. 22, pp. 87-91.

Van der Valk P.; de Vries S.E.; Everink J.T.; Verstappen F. & de Vries J.N. (1991a). Pre- and post-fertilization barriers to backcrossing the interspecific hybrid between *Allium fistulosum* L. and *A. cepa* L. with *A. cepa*. *Euphytica*, Vol. 53, pp. 201-209.

Van der Valk, P.; Kik C.; Verstappen F.; Evernik J.T. & de Vries J.N. (1991b). Independent segregation of two isozyme markers and inter-plant differences in nuclear DNA content in the interspecific cross (*Allium fistulosum* L. x *A. cepa* L.) x *A. cepa* L. *Euphytica*, Vol. 55, pp. 151-156.

Van Heusden AW, Shigyo M, Tashiro Y, Vrielink-Van Ginkel R, and Kik C. (2000b). AFLP linkage group assignment to the chromosomes of *Allium cepa* L. via monosomic addition lines. *Theoretical and Applied Genetics* 100: 480–486.

Van Heusden, A.W.; van Ooijen J.W.; Vrielink-van Ginkel M.; Verbeek W.H.J.; Wietsma W.A. & Kik C. (2000a). A genetic map of an interspecific cross in *allium* based on amplified fragment length polymorphism (AFLP) markers. *Theor Appl Genet*, Vol.100, pp. 118-126.

Van Raamsdonk LWD, Wietsma Wa, and De Vries JN. (1992). Crossing experiments in *Allium* L. section *Cepa. Botanical Journal of the Linnean Society* 109: 293–303.

Xu, J.M., Hanelt, P. and Chun-Lin, L. (1990). Key to the Alliums of China.*Herbertia* 46(2):pp.138–164.

Yamashita K, Arita H, Tashiro Y (1999a). Cytoplasm of a wild species, Allium galanthum Kar. et Kir., is useful for developing the male sterile line of A. fistulosum L. *J Jpn Soc Hort Sci* 68:788–797.

Yamashita, K.; Hisatsune Y.; Sakamoto T.; Ishizuka K. & Tashiro Y. (2002). Chromosome and cytoplasm analyses of somatic hybrids between onion (*Allium cepa* L.) and garlic (*A.sativum* L.). *Euphytica*, Vol. 125, pp. 163-167.

Yamashita K, Arita H, Tashiro Y (1999b). Isozyme and RAPD markers linked to fertility restoring gene for cytoplasmic male sterile Allium fistulosum L. with cytoplasm of A. galanthum Kar. et Kir. *J Jpn Soc Hort Sci* 68:954–959.

Yamashita K. and Tashiro Y (1999). Possibility of developing male sterile lines of shallot (*Allium cepa* L. Aggregatum group) with cytoplasm from *A. galanthum* Kar. et Kir. *J Jpn Soc Hort Sci* 68:256–262.

Yamashita K, Iino M, Shigyo M, Tashiro Y (2000). Visualization of nucleus substitution between *Allium galanthum* and shallot (*A. cepa*) by genomic in situ hybridization. *J Jpn Soc Hort Sci* 69:189–191.

Yamashita K, Takatori Y, Tashiro Y (2005). Chromosomal location of a pollen fertility-restoring gene, Rf, for CMS in Japanese bunching onion (Allium fistulosum L.) possessing the cytoplasm of A. galanthum Kar. et Kir. revealed by genomic in situ hybridization. *Theor. Appl Genet* (2005) 111: 15–22.

Yanagino, T.; Sugawara E. & Watanabe M. (2003). Production and characterization of an interspecific hybrid between leek and garlic. *Theor Appl Genet* Vol. 107, pp. 1–5.

Zenkteller, M. (1990). *In vitro* fertilization and wide hybridization in higher plants. *Plant Sci*, Vol. 9, pp. 267-279.

Distant Hybridization in Horticultural Crops   *Pages* **221–254**
Editors: **M.R. Dinesh & M. Sankaran**
Published by: **ASTRAL INTERNATIONAL PVT. LTD., NEW DELHI**

14

Distant Hybridization and Tropical Tuber Crops: Designing for Adaptive Food-nutrition and Livelihood

Archana Mukherjee

ICAR-Regional Centre of Central Tuber Crops Research Institute
Bhubaneswar-751019, Odisha, India

INTRODUCTION

Wide hybridization and domestication played pivotal role in development of agri-horticultural crops including tropical tuber crops. In fact scientific advancement of crop breeding had actually marched forward with Mendelism that too with vegetable crop. Such logistic approach was the key guidance in breeding of other food crops. Wide crosses which was earlier Intergeneric, Interspecific even Intervarietal often in case of incompatible combinations is now advanced to transgenerics.

Chromosome manipulation through polyploidization and mutations are the effective tools for wide crosses. Thereafter 'transgene' combinations became reality with the discovery of recombinant DNA technology. Such techniques advanced further with genomics, functional genomics aided with radiation breeding and tilling.

More recently 'reverse designing' of crops with 'integrated grid model' based on phenotypes, bio-molecular markers with system biological approach are found to be the best option for "model product". Such study will harness the harvest of 'precise crop product'. In the context of climate change, low input high energy tropical tuber crops like cassava, sweet potato, taro, yams and elephant foot yam have immense potential towards food security. These foods cum vegetable crops propagated vegetatively. Nature and domestication have played significant role to evolve present cultivars of many food crops. Tropical tuber crops are not the exception. Many of their present day cultivars are the product of wide crosses. To evolve need based products, distant hybridization was also attempted classically as well as through genetic engineering in sweet potato, cassava and other horticultural crops. Further to address the agri-horticultural issues, today's distant hybridization need to be de-emphasized to satisfy 'consumer needs' and to 'cope with adverse climate'. To achieve that, 'farmers-researchers participatory breeding' approach through National and International network would be the most logical to facilitate desirable wide combinations. European Union aided **'International Network for Edible Aroids (*INEA*)'** is being continued in twenty two different countries including India to develop 'taro' adaptive to climatic and commercial changes. Results are encouraging across the globe for adaptive food-nutrition and livelihood.

Distant Hybridization

Distant hybridization or wide crossing of crop is the mating between distantly related or unrelated individuals. Sexual or somatic cells may be involved in this fusion. When fusion takes place between somatic cells, it is called para sexual hybridization. Wide crosses or distant hybridization are generally used to improve crop varieties for disease resistance, pest resistance, stress resistance, quality, adaptation, yield *etc*. Wide hybridization in crop species including tuber crops [Fig. 1 (A-E)] and 'domestication' observed to play a significant role in evolving the varieties coupled with many natural happenings including land mass drift. Such incidents facilitates nature aided wide crosses, man made crosses, trimmed with spontaneous mutation and natural selection pressure. Progressive evolution of wide crosses in 19th to 20th century marched scientifically with 'Mendelian factors', 'chromosome manipulation', 'radiation rather mutation breeding'.

Fig. 1 Flowers-sources of wide hybridization in tuber crops; sweet potato (A&B), taro (C), elephant foot yam (D) and yam bean (E)

Later, crop breeding widened further for **'transgene combination'** aided with recombinant DNA or genetic engineering techniques. Genetic manipulation concerns the transfer of genetic information –in the form of DNA sequences-across sexual barriers between species, which under normal conditions would not exchange DNA. The resulting organisms are called genetically modified organisms (GMOs) or transgenics. To feed the world population which is expected to be 8.5 billions in 2020 with less per capita available water and land resources, the humankind would depend largely on **"precised technology"**. Use of agro-chemicals for fertigation, pest control and food storage is causing serious health hazards and polluting the scarce natural resources. There has been tremendous growth in transgenic research all over the world and development of *Bt* transgenic is emerging as effective crop improvement methodologies.

Sustainable food security can be achieved only through higher productivity per unit of land, water, energy and time which warrants integration of both traditional and modern eco-technologies. Thus, precision and use of need based tools [Fig. 2 (A & B)] are the demand of the day. Precision and expression of desired characters (genes) through functional genomics has encouraged the biotechnological research.

Fig. 2 (A-B), Precision (A) & need based use of tools (B)

The Gene Technology for crop improvement marched forward with Mendelism

This has been unveiled with the discovery of Mendelism in 1860.

Later 'gene movement' the base for wide hybridization was continued with newer tools and approaches [Fig. 3 (A-E)] as briefed below:

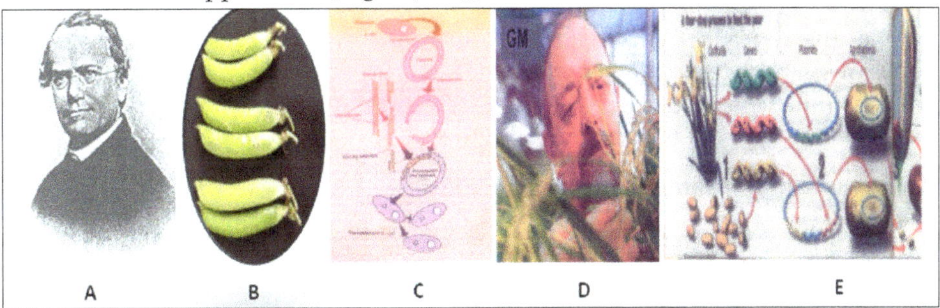

Fig. 3 March of genetechnology; Mendel (A), Pea (B), Recomninant DNA (C), Ingo (D) & Golden Rice (E)

Mendel : making crosses, introducing genes

1920 Discovery of hybrid vigor

1950 Inducing mutations

1960 Tissue culture and embryo rescue

1980 Plant transformation and GM crops

2000 onwards Genomics (study of all the genes), functional genomics, tilling *etc.*

A brief overview of domestication and distant hybridization, its progressive advancement aided with natural, classical and modern tools in different crop species including climate resilient tuber crops are presented as follows.

DOMESTICATION

The term 'domestication' refers to a plant genotype which has been adopted by human activity.

The term 'cultivar' represents a particular line or breed of a domesticated type. Wild types are associated with wild niche, on the other hand domesticates and cultivars are associated with the managed habitats. However vice versa is also possible.

The zones of domestication of major crops are as follows:

Mesoamerica: corn, common bean, squash, sweet potato, tomato, upland cotton

Andean region: potato, common bean, peppers, cassava,   pineapple, pima cotton

African Sahel + Ethiopia: sorghum, African rice, coffee,   okra, melon, cowpea, white yam

Near East: wheat, barley, pea, lentil, chickpea, grape,

Eastern India and South-East Asia: mungbean, cucumber, banana and plantain, eggplant, coconut, tea, taro, elephant foot yam

Southern China: Asian rice, soybean, citrus, tea, cabbage.

Land races

Relative reproductive isolation and differential agro climatic conditions created numerous landraces of our crops. Landraces do not have a fixed complement of genes (alleles) because subsistence farmers are always trading seeds and the conditions change over time span of hundreds of years.

Domestication of land races and speciation

All the traditional starch-rich staples (major and minor), most vegetables are vegetatively propagated. Over time, vegetative propagation is thought to have selected for parthenocarpic varieties in several species, including bananas and yam cultivars. Vegetative propagation, whether of parthenocarpic and sterile varieties or not, is a way of genetically isolating a cultivated or adapted plant from surrounding populations. It is also a mechanism through which plants can be moved beyond the natural range in which sexual reproduction is viable. Climate, absence of pollinators or low density in the landscape may all inhibit sexual reproduction in a plant. The history of plant domestication for example in New Guinea is complex and yet to be fully unravelled for most staples. Those are the product of intra and inter-specific hybridisation with Southeast Asian (SE) cultivars viz. bananas, sugarcane, taro and some yams, (Lebot 1999). Carol Lentifer (University of Queensland) outlined a multi-locus and multi-phased domestication process for two sections of Musa banana. Domestication of AA diploid varieties of 'Musa section' eventually gave rise to the most important groups of cultivated bananas in the world.

Many such horticultural crops are found to have evolved through either intergeneric, interspecific or inter-varietal crosses under nature aided circumstances. Whether it is natural, traditional or through advanced gene manipulation techniques 'distant hybridization' may be interspecific or intergeneric.

The theoretical and practical aspects of wide hybridization, problems associated with it and the techniques for its success are briefed prior to its implications and development in tropical tuber crops.

Interspecific Hybridization

Hybridization between two species of the same genus usually takes place by sexual fusion. It is usually practiced to transfer desirable genes from wild species of plants to cultivated species. Interspecific crosses may be fully fertile [tomato (*L.esculentum* x *L.peruvianum*)], partially fertile or sterile (wheat 6X 4X).

Interspecific crosses help in introgressive hybridization which is the transfer of some genes from one species into the genome of another species. Fertility level of interspecific crosses depends on the homology of chromosomes in the parental species. In the case of sterile crosses, amphidiploidy is induced with colchicine and the fertility is restored.

Intergeneric Hybridization

This refers to crosses between two different genera of the same family. Such crosses are not commonly used in crop improvement. However, such crosses may become desirable in a number of situations. Intergeneric crosses can be used when the desirable genes are not present in the same genus, but they are present in allied genera. F_1 hybrids of this type of crosses are always sterile. However, they can be made fertile by chromosome doubling. Intergeneric hybridization has been used successfully in the development of the synthetic cereal, for example-triticale.

The major problems associated with wide crosses are

- Cross Incompatibility
- Hybrid Inviability
- Hybrid Sterility

Cross incompatibility

This is the inability of the pollen grains of one species or genus to affect fertilization in another species or genes. This can be overcome by adapting different techniques like reciprocal crosses, bridge crosses, using pollen mixtures, pistil manipulations or use of growth regulators *etc.*

Hybrid inviability

This refers to the inviability of the hybrid zygote or embryo. In some cases, zygote formation occurs but further development of the zygote is arrested. In some other cases, after the completion of the initial stages of development, the embryo gets aborted due to either unfavorable interactions between the chromosomes of the two species or unfavorable interaction of the endosperm with the embryo or reciprocal crosses as have been evident in case of tuber crops (Mukherjee *et al.* 1991, Mukherjee 1999) reciprocal crosses, application of growth hormones and embryo rescue are the techniques that can be used to overcome these problems.

Hybrid sterility

This refers to the inability of a hybrid to produce viable offspring. This is more prominent in the case of intergeneric crosses. The major reason for hybrid sterility is the lack of structural homology between the chromosomes of the two species. This may lead to meiotic abnormalities like chromosome scattering, chromosome extension, lagging of chromosome in the anaphase, formation of anaphase bridge, development of chromosome rings, chains, and irregular as well as unequal anaphase separations. These irregularities may lead to aberrations in chromosome structure. Sterility caused by structural differences between the chromosomes of two species can be overcome by amphidiploidization. Hence Techniques like alien addition and alien substitution through manipulation of chromosomes can be effective tool for wide cross.

Techniques to make wide crosses successful

Selection of plants

The most compatible parents available should be selected for the crosses.

Reciprocal crosses

Reciprocal cross may be attempted when one parental combination fails.

Manipulation of ploidy

Diploidization of solitary genomes to make them paired will be helpful to make the cross fertile.

Bridge crosses

When two parents are incompatible, a third parent that is compatible with both the parents can be used for bridge crosses and thus it becomes possible to perform cross between the original parents.

Use of pollen mixture

Unfavorable interaction between pollen and pistil in the case of wide crosses can be overcome to some extent by using pollen mixture.

Manipulation of pistil

Decapitation of the style will sometimes prove helpful in overcoming incompatibility.

Use of growth regulators

Pollen tube growth can be accelerated by using growth hormones like IAA, NAA, 2,4-D and Gibberellic acid.

Manipulation of chromosomes

Biotechnological tools

Chromosome manipulation and biotechnological tools for crop improvement are discussed as follows.

Manipulation of chromosomes

Chromosome manipulation for crop improvement has been widely used as an important tool in case of wide or distant hybridization. The chromosome behaviors in F1 hybrids provide the essential genetic basis for chromosome manipulation. The induction of homologous pairing in F_1 hybrid plants followed by the incorporation of a single-chromosome fragment from an alien or a wild species into an existing crop species by translocating chromosomes has been used in the production of translocation lines (Figs. 4 & 5). Most efforts to transfer a beneficial trait from wild plants into crops so far have bridged the species gap via alien chromosome translocation lines. Chromosome doubling in somatic cells or gametes of F1 hybrids followed by the incorporation of all alien chromosomes has been used in the production of amphidiploids. Amphi-diploidy can be used as a bridge to move a single chromosome from one species to another or for the development of new crops. Chromosome elimination of a uni-parental genome during the development of F_1 hybrid embryos has been used in the production of haploids. Haploids are very useful in double-haploid breeding of a true-breeding crop such as wheat and rice since this method can quickly replace genetic recombination while enhancing breeding efficiency or facilitating genetic analysis (Qi *et al.* 2007; Liu *et al.* 2014).

Wild species provides a vast gene pool for crop improvement. Most pioneering efforts in chromosome engineering have involved the *Triticum* species in *Triticeae*, with the greatest emphasis on improving common wheat (*T. aestivum* L., 2n = 6x = 42, AABBDD) (Qi *et al.* 2007; Reynolds *et al.* 2009).

Albinism

Albinism is a common problem encountered in interspecific crosses and tissue culture experiments including anther culture and generation of doubled haploids. It is characterized by partial or complete loss of chlorophyll pigments and incomplete differentiation of chloroplast membranes.

Albinism has been attributed to deletions in ptDNA in addition to nuclear-plastid genome incompatibility in wide hybrids.

2   Distant Hybridization: A Tool for Interspecific Manipulation of Chromosomes   27

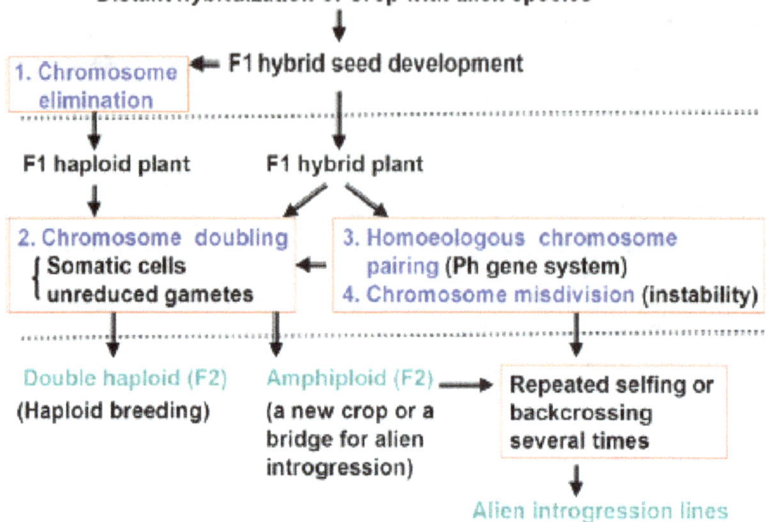

Fig. 4. Manipulation of chromosomes tool for interspecific crosses (Source Liu *et al.* **2014)**

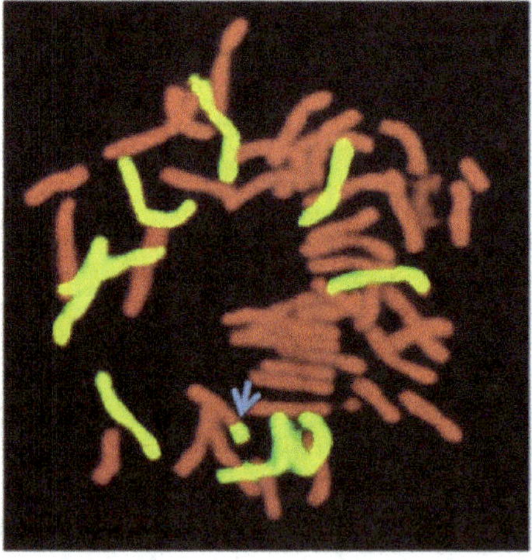

Fig. 5. Wheat-rye translocation chromosone (arrow) observed at mitotic metaphase in root-tip cells of F2 plants (Source: Liu *et al* **2014)**

Approaches to overcome albinism at a cellular level

Manipulation of ploidy in hybrids- albinism is difficult to eliminate, but in some cases it can be overcome through development of doubled haploids.

Somatic hybridization and cybrid development- through somatic hybridization using cotyledon-derived protoplast culture techniques in albino mutant and wild species as demonstrated by Bordas *et al.* (1998) in Cucumis.

Biotechnological tools for crop improvement

- Tissue culture techniques like *in vitro* fertilization, embryo rescue & protoplast fusion enabled breeders to recover useful hybrids as an aid to widen hybridization.

- Genetic engineering has further advanced the improvement program through *Agrobacterium* mediated transformation, electroporation and particle bombardment techniques.

- Production of transgenics are underway to develop stress tolerant plants, to extend shelf-life of fruits and vegetables as well as to enrich staple food crops with vitamins such as

- Transgenic soyabeans with higher protein

- Potatoes with more starch and amino acid

- Golden rice with β-carotene

- GM fresh produce of Broccoli, Carrots, Melons

- GM flavoured Pepper and Melons *etc.*

- Disease resistant transgenics such as papaya resistant to ring spot virus, transgenic cassava against common mosaic virus, transgenic sweet potato against sweet potato feathery mottle virus.

To precise the GMOs or GE products, concept of functional genomics, proteomics and more recently tilling (Slad *et al.* 2005) have been launched. Genomics is a branch of biotechnology that has developed rapidly over the last decade. It involves the large scale sequencing of DNA including entire genome. The science of bio informatics has enriched the genomic research. A detailed knowledge of plant and animal genomes will speed up the wide hybridization for 'precised product'.

The march of crop improvement since the discovery of Mendelian factors no doubt commendable. Further the recombinant DNA technology opened a new avenue to gear up through trans-hybridization. Despite all those advancements the GHI (Global Hunger Index) of our country is alarming. Even after completing 154 years of Mendelism, 34 years of GE technology, India with GHI at 24.1 ranked 67th position out of 84 countries (GHI, 2010). The proportion of underweight children is even inferior to Bangladesh, Pakistan.

In fact India harbors 42% of under weight children   of the world (Fig. 6). To tackle this, we have to think deeply about the food habits of people live in disadvantageous sections. Further a serious threat of food insecurity is haunting especially in the context of climate change. Therefore, climate resilient alternate food and vegetable crops need to be of prime importance (Mukherjee 2013; Mukherjee *et al.* 2013a).

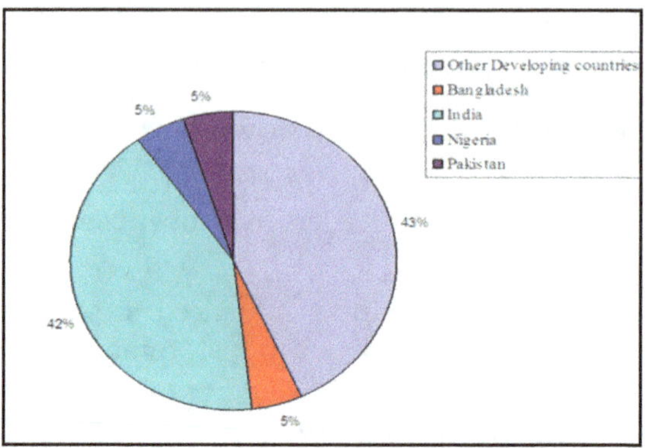

Fig. 6. Share of under weight children under five years of age. Source: UNICEF 2009

Fig.7. (A-H). High energy tuber crops. Cassava (A&B), Sweet Potato (C&D), Taro (E&F) and Elephant foot yam (G&H)

Climate change vs. tropical tuber crops and its improvement

Climate change and the threat of food insecurity is the burning agri-horticultural issue. Thus a paradigm shift has taken place to develop climate proof crops with valued traits. The tropical tuber crops viz. cassava, sweet potato, taro and yams highly responsive to organics are the staple food of people live in Islands and fragile environment. These crops [Fig. 7 (A-H)] are gaining importance for their high productivity (20-50 t ha^{-1}), calorie and ability to thrive on harsh climate. The minerals and fibre contents in most of those crops are 1.5 to 5 times higher than rice and wheat (Table 1). The climatic resilience of sweet potato and taro as life support species was the reality during super cyclone in 1999 in Odisha, Tsunami in 2004. Such fact led to think the potential of these third world crops for livelihood security under awakening threats of global warming. R & D work at ICAR-Central Tuber Crops Research Institute (CTCRI) and its AICRP Centres resulted in developing sweet potato tolerant to biotic (weevil) and abiotic (salt) stress having high yield, starch (18-20%), β-carotene (6-12 mg/ 100g) and anthocyanin (85 mg/100g). Likewise improved varieties of elephant foot yam and other tuber crops have also been developed.

Developed and developing countries have shown greater concern about the loss of genetic diversity of such climatic resilient valued tuber crops genotypes. Conservation of gene source with valued traits is thus essential for its adaptive advancements.

The BPGR and NBA-PPVFRA based system in India supports the conservation and resolve the various issues related to biodiversity. Awareness on conservation especially community based conservation starting from grass root level is the need of the hour. Similarly distant hybridization through farmers- researcher participatory approach is most logical towards wholesome benefits. The International network for edible aroids for climatic and commercial changes through 'Farmers–researcher participatory breeding' can unearth the paradox to realize the potential of these crops in evolving "desirable products" for food and nutrition security against adverse climatic conditions.

The results on evolving improved varieties of tropical tuber crops and the mode of dissemination (Fig. 8) of improved technologies towards food-nutrition security are briefed crop wise as follows.

Sweet potato

Sweet potato (*Ipomoea batatus* (L.) Lam.) belongs to the family *Convolvulaceae*. It is grown more extensively in tropical and sub tropical countries. Sweet potato, the fifth most important food crop is a short duration creeper (90-110 days). This crop seems to be most suitable to grow and check soil erosion in degrading and fragile lands as ecofriendly crop to cater food (194 MJ/ha/day), feed, nutritional [vitamin C (23 mg/100g), + vitamin E (4.56 mg/100g)] and industrial demands (16-20% starch). Of the various *Ipomoea* species, sweet potato botanically *Ipomoea batatus* is the only edible species. It is hexaploid with basic chromosomes of x =15. (Mukherjee *et al.* 1998a; b & 2001) It has many wild relatives, pentaploid sweet potato was developed through distant hybridization using wild relatives of *Ipomoea* species. To recover desirable wide hybrids, embryo rescue technique developed in sweet potato and its wild relatives (Mukherjee *et al.* 1991; Mukherjee and Vimala 1994).

Relationship among Novel Ipomoea Interspecific Hybrids

Interspecific hybridization can be used to broaden the genetic base to generate novel species and to understand genetic relationships through introgression of elite alien genes. However, interspecific hybridizations using wild parents outside the *Ipomoea* section '*Batatas*' are difficult due to self -cross incompatibility and sterility coupled with ploidy variability. Three novel interspecific hybrids were generated (Qing-He *et al.* 2014) by crossing *Ipomoea batatas* (L.) Lam. *I. hederacea* Jacq., *I. batatas* (L.) Lam. *I. muricata* (L.) Jacq., and *I. batatas* (L.) Lam. *I. lonchophylla* J.M. Black. The ploidy level of the interspecific hybrids was determined by flow cytometry. The cross, *I. batatas I. hederacea*, yielded the artificial pentaploid *Ipomoea* hybrid. The other two hybrids, *I. batatas I. hederacea* and *I. batatas I. muricata* were tetraploid. The hybrids showed storage roots in interspecific *Ipomoea*. AFLP (Amplified Fragment Length Polymorphism) molecular markers based studies shown the genetic relationship of interspecific hybrids with three other natural diploid, tetraploid, and hexaploid species of the *Ipomoea* section 'Batatas'. Cluster analysis of AFLP bands showed that these three evolved interspecific hybrids were closely related to cultivated sweet potato (*I. batatas* L. Lam.). Results indicate that those hybrids can be used as an interspecific bridge to transfer alien genes from wild to cultivated sweet potato.

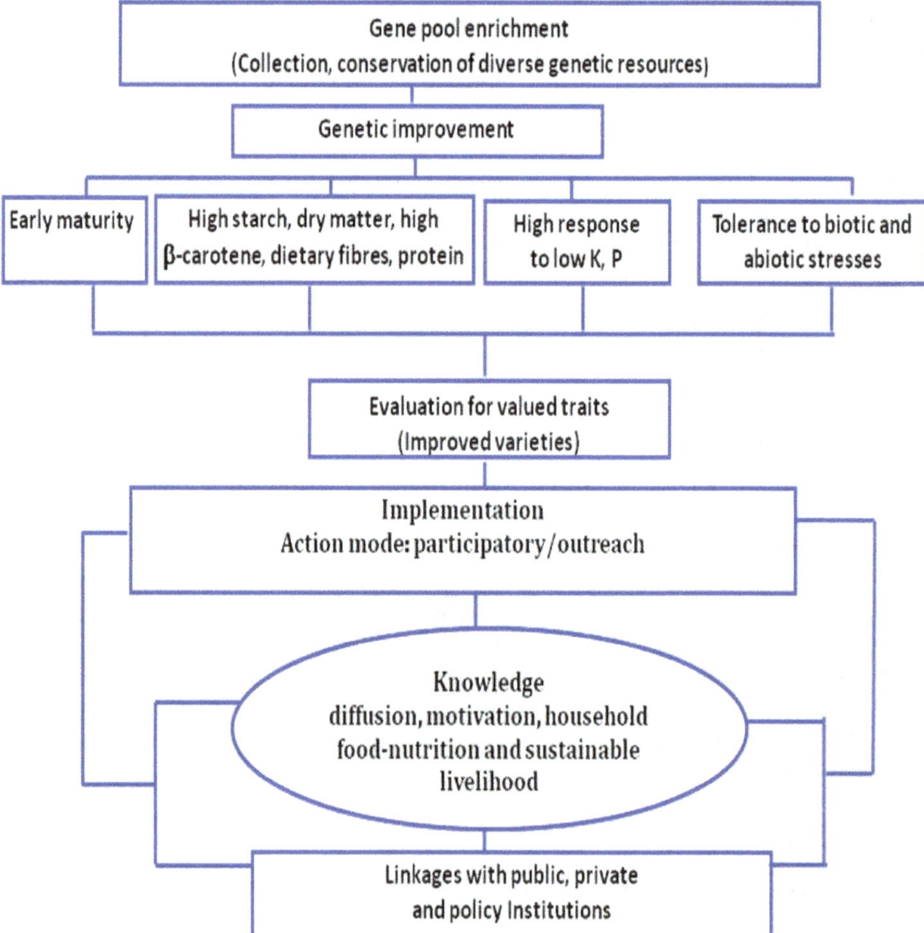

Fig.8. Thte road map for development and dissemination of tuber crops technologies

The amplified AFLP bands in the novel synthesized tetraploid interspecific hybrids were similar in number, while the old hybrid *I. tabascana* had fewer bands, indicating that some DNA bands might have been lost during speciation. These were explained by the fact that polyploidization is usually followed by a genome-wide loss of some of the redundant genomic material. Such differential gene loss following polyploidization among closely related plants was also explained in cereals.

Sweet potato breeding in India

Most of the improvement work is confined to ICAR-Central Tuber Crops Research Institute (CTCRI), India. Wide crosses using wild relatives [Fig. 9 (A-D)] resulted in developing hypo, hyper ploids sweet potato. Such chromosomal variability was also noticed in callus cultures of sweet potato [Fig. 10 (A-D); Mukherjee et al 1998 a & b]. However to satisfy consumer demands inter varietal hybridization was conducted with widened genetic base. Progressive breeding and evaluation resulted to evolve sweet potato tolerant to biotic (weevil), abiotic salt stress (6-8dSm-1) packed with high yield (>15tha–1), starch (18-20%), beta carotene (6-10mg/100g) and anthocyanin (85mg/100g). Such high valued sweet potato [Fig. 11 (A-F) & Fig. 12] are recommended for release and are registered at NBPGR, New Delhi (Mukherjee and Naskar 2012; Mukherjee 2013; Naskar and Mukherjee 2013).

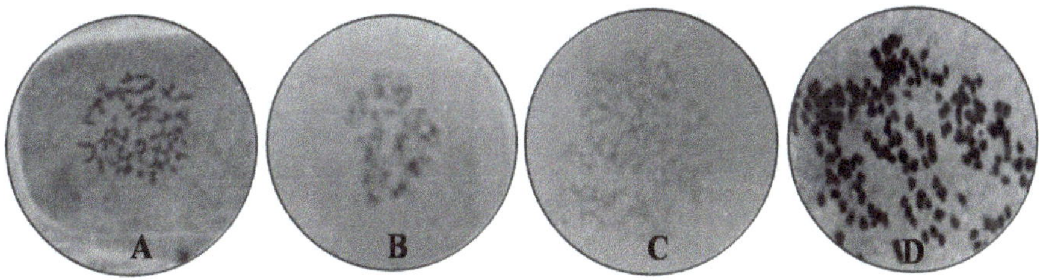

Fig. 9. (A-D). I trifida (A,B) wild species and edible sweet potato, I. Btatus (C,D)

Fig. 10. (A-D). Hypo-(A,B) and hyperploidy (C,D) in sweet potato

Transgenic sweet potato

Transgenics have developed in sweet potato (Mukherjee 1999, Mukherjee *et al.* 2012). Transgenics for fungal resistance was also developed by genetic transformation with chitinase and glucanase genes. An efficient transformation system using *A. tumefaciens* strain containing marker for SPFMV coat protein genes has also been reported (Mukherjee *et al.* 2012). In India micropropagation including somatic embryogenesis, synthetic seeds [Fig. 13 (A-F)] as well as *Agrobacterium*-madiated putative transformation has also been reported in sweet potato (Mukherjee 2002; Mukherjee *et al.* 2012).

Fig. 11.(A-F). White (A&B), Orange (C&D) & Purple flesh (E&F) Sweet potato

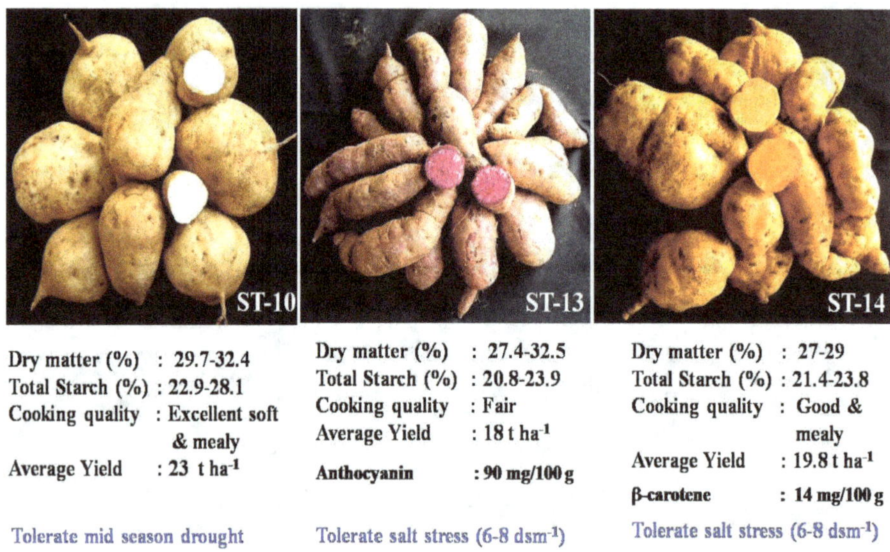

Dry matter (%)   : 29.7-32.4
Total Starch (%) : 22.9-28.1
Cooking quality   : Excellent soft
                              & mealy
Average Yield   : 23 t ha⁻¹

Tolerate mid season drought

Dry matter (%)   : 27.4-32.5
Total Starch (%) : 20.8-23.9
Cooking quality   : Fair
Average Yield   : 18 t ha⁻¹

Anthocyanin   : 90 mg/100 g

Tolerate salt stress (6-8 dsm⁻¹)

Dry matter (%)   : 27-29
Total Starch (%) : 21.4-23.8
Cooking quality   : Good &
                              mealy
Average Yield   : 19.8 t ha⁻¹

β-carotene   : 14 mg/100 g

Tolerate salt stress (6-8 dsm⁻¹)

Fig. 12. Valued traits of stress tolerant white (ST10), Purple (ST-13) & Orange (ST-14) flesh sweet potato

Fig.13. Micropropagation of sweet potato, somatic embryogenesis (A-C), syn seeds (D), hardening (E) & RAPD profile of source and regenerants (F)

Cassava

Cassava or tapioca (*Manihot esculenta* Crantz) which belongs to the family *Euphorbiaceae* is a native of Brazil (South America), widely cultivated in the tropics. The adventitious roots of cassava store carbohydrates and form the tubers and the tubers being rich in starch, form an important subsidiary food. In coastal and tribal areas it is a staple food. It is also the raw material for starch & sago industry and a component of animal, fish and poultry feeds in many developing nations including India.

Wide hybridization in cassava-the synthesis of a new cassava-derived species *Manihot vieiri* Nassar.

The species was synthesized artificially by chromosome doubling of an interspecific hybrid. The ensuing polyploid type exhibited apomictic nature and reflected its morphological characteristics in the progeny. It showed a frequency of 29% of multiembryonic sacs of its examined ovules whereas the multiembryonic sacs were absent in the diploid type.

The diploid hybrid of *M. anomala* with cassava is one of the few that showed complete chromosome pairing. Hence, its tetraploid type tends to have less fertility due to quadrivalent chromosome configuration. In other cases of polyploidizing interspecific hybrids of *Manihot* species the pairing was extremely different as exhibited in the interspecific hybrid of cassava with *M. glaziovii*. The viability of the polyploidized type was 92% compared to 13% in the diploid type. Apparently phylogenic affinity was reported more between *M. esculenta* and *M.anomala*.than that between *M. glaziovii* and *M. esculenta*. Hybridity and polyploidy confer variability to a certain population, then apomixis fixes and perpetuates genotypes adapt in certain environment (Nassar *et al.* 1998).

Progress in India

In India, cassava is cultivated predominantly in Kerala and Tamil Nadu. It is also grown in Andhra Pradesh, Assam, Karnataka, Madhya Pradesh, Pondicherry, Nagaland, Tripura, Mizoram and the Andaman & Nicobar group of islands. Breeding thrust in cassava resulted in developing and release of varieties like Sree Visakham, Sree Jaya, Sree Vijaya (Fig. 14), Sree Prabha, Sree Rekha, Sree Padmanabha, Sree Prakash, Sree Sahya and more recently high starch yielding triploids like 5-3(3X) and 4-2. Among these, Sree Padmanabha (MNga-1), (Fig. 15) is the product of wide cross of East and West African cassava gene sources.

Fig. 14. Different varieties of cassava

Fig. 15. MNga-1 (Sree Padmanabha: Field resistance to CMD)

Transgenic Cassava

Tissue culture techniques especially micropropagation and somatic embryogenesis have well documented [Fig. 16 (A-E)] in cassava (Mukherjee 1999). Disease free propagules can be produced in cassava through syn-seeds (Fig. 17). Transgenic research is underway in India. Transgenic resistances for common mosaic virus (ACMV) with disfunctional ACMV genes are identified. Four different research groups have developed transgenic cassava plants. The CIAT, Cali, Colombia, is the first laboratory to claim it. To circumvent the problem of natural *A. tumefaciens*, Swiss Federal Institute of Technology, Switzerland used a modified strain of *A. tumefaciens* which does not transfer disease causing genes.

Fig.16 (A-E). Micropropagation of cassava. Regeneration (A-D), RAPD profile of source and regenerants (E)

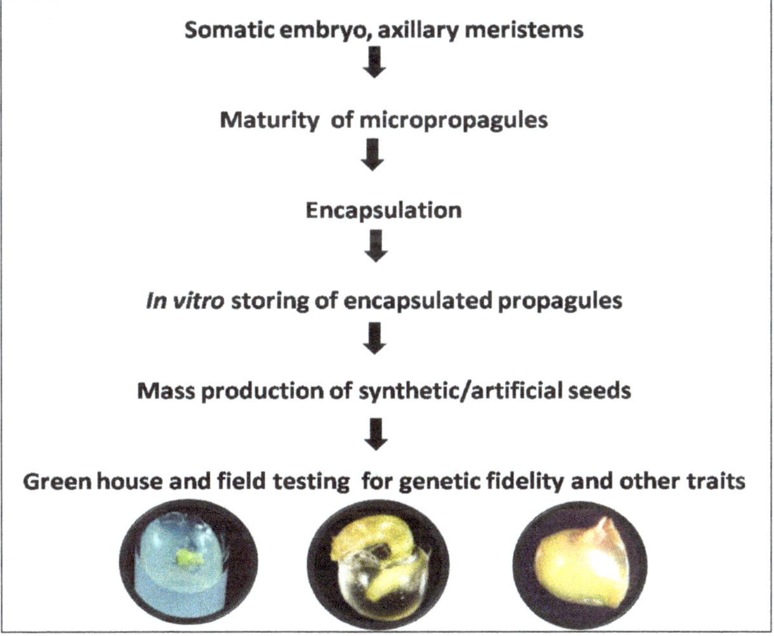

Fig. 17. Stages of synthetic seed mediated propagation in cassava

Yams

Edible yams constitute a group of *Dioscorea* species cultivated widely in the tropics, of which the most important ones are greater yam (*D. alata*), lesser yam (*D. esculenta*) and white yam (*D. rotundata*).

The former two are Asiatic in origin and the later African. Yam belongs to the family Dioscoreaceae under monocotyledons. Yams are herbaceous climbers characterized by weak stems that twine on supports. *D. alata* tubers are large, varying in shape and are usually 1-3 per plant. Yam tubers are mainly consumed as vegetable. They are also useful for processed product like fries and chips. In India, they are grown practically in all the states but the major yam producing states are Kerala, West Bengal, Bihar, Odisha, Assam, Gujarat, Maharashtra and Rajasthan.

Genetic nature of yams-wide hybridization

Such study has shown that through domestication farmers influenced and increased the genetic diversity in yam by using sexual reproduction of wild and possibly cultivated yams through wide hybridization (Scarcelli *et al.* 2006).

Wild species are the main progenitors of the cultivated complex [Fig. 18 (A-C)]. Wild yams mainly reproduce sexually while cultivated yams are vegetatively propagated. However, many yam cultivars have limited flowering potential and thus hybridisation between varieties has been used in varietal improvement.

Fig. 18. (A-C). Yam plantation (A), wild (B) & cultivated varieties (C)

The phylogenetic tree showed three clustered groups that corresponded overall to the three species *D. abyssinica, D. praehensilis and D. cayenensis–D. rotundata.* This classification was well supported by AMOVA results on the genetic distance matrix of *D. abyssinica, D. praehensilis and D. cayenensis–D. rotundata* individuals. Its consistency with those of the NJ-tree corresponding to the three species was also reported (Scarcelli *et al.* 2006). Two hypotheses could be put forward to explain these phenotypes which deviate from the typical *D. praehensilis* phenotype. First, these yams could be interspecific hybrids between *D. praehensilis* and *D. cayenensis–D. rotundata.* AFLP analysis showed that half of the predomesticated plants belonged to *D. abyssinica and D. praehensilis* species. In their study, three spontaneous plants morphologically identified as *D. praehensilis* were clustered genetically with the

D. cayenensis– D. rotundata group. Results indicate that some cultivated genotypes might have grown outside the fields and that farmers might have collected those spontaneously. On the contrary, some cultivated plants might have escaped into the wild environment. Secondly, these plants could be the result of natural hybridization (Baco 2000).

Thus the domestication practices involve production of present day cultivars by sexual reproduction of wild yams and probably cultivated yams. Through the domestication practice, sexual reproduction contributes to the evolutionary dynamics of yam, a vegetatively propagated plant. Domestication is still practiced in Benin, but this trend is declining. Moreover, yam genetic diversity is threatened by the introduction of improved varieties and other cultivated species. The domestication practice should thus be taken into account for on-farm conservation of plans for gene introgression though wide hybridization. In India, the clonal selection as well as hybridization   resulted in developing improved yam varieties viz. Sree Roopa, Sree Silpa, Orissa Elite, Da-25 *etc.* (Fig. 19) to cater the need of greater Eastern India.

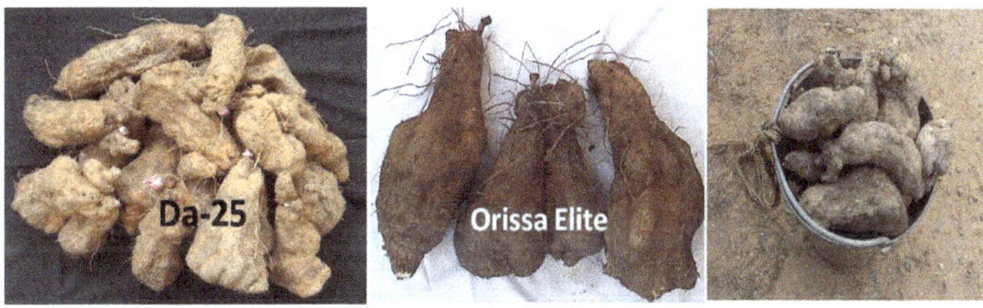

Fig. 19. Different improved varieties of yams

Transgenic research in yam

Application of biotechnological tools including tissue culture is most important in yam (Mukherjee 1999, Mukherjee *et al.* 2013b) to develop desirable hybrids in overcoming the bottlenecks of conventical   and non-conventional breeding. Micropropagation techniques developed in edible yams (Fig. 20; Mukherjee *et al.* 2013b).

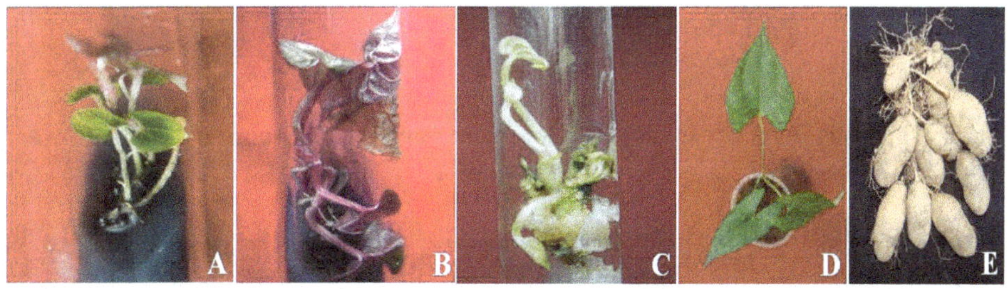

Fig. 20. Micropropagation in yam. Stages of regeneration (A-C), hardening (D) & minitubers (E)

Information on genetic transformation in yams is still limited. Being a monocotyledonous plant, *Agrobacterium* mediated transformation was difficult in *Dioscorea* sp. (Achmus and Brinks 1996). However, Schaefer *et al.* (1987) described a dicot tissue mediated technique to overcome difficulties associated with *Agrobactrium* mediated transformation. Use of cell layers of tubers of *D. bulbifera* treated with *Agrobacterium* strain which was pretreated with the exudates of *Solanum tuberosum* (dicot) tissue culture. Those yam plants produced crown gall.

Elephant foot yam, its species diversity

Elephant foot yam (*Amorphophallus paeoniifolius* (Dennst) Nicolson is the major one in aroid groups (family *Araceae*). The genus *Amorphophallus* with 90 to 170 species primarily occurs in Tropical Asia and Africa, (Bogner *et al.* 1985) Ten species are endemic to India viz. *A. carnosus, A. longistylus* and *A. oncophyllus* to Andaman Islands. *A. arunachalensis* and *A. bognerianus* to Arunachal Pradesh, *A. mysorensis* to Karnataka, *A. longiconnectivus* to Madhya Pradesh, *A. bonaccordensis, A. nicolsonianus* and *A. smithsonianus* to Kerala. *A. paeoniifolius, A. hirsutus, A. brooksi, A. jijas* and *A. prainii* are species that occur in Sumatra. *A. konjac* is seen in Korea and China. The tall type with largest inflorescence (10 feet height) *Amorphophallus* titanium occurs in Indonesia *A. konjac* and *A. jaijas* (*A. brooksi*) are also having tall spadices (4 feet to 10 feet height) *A. bulbifer, A. kachinensis, A. hirsutus, A. nepalansis, A. paeoniifolius* and *A. sylvaticus* are the species with wide distribution in India, Myanmar, Thailand, Laos, China, Sumatra, Bhutan, Nepal, Srilanka and Pacific Islands. The edible species are *A. paeoniifolius* (Dennst) Nicols *var. campanulatus* (Decne) Sivad, *A. Konjac* K. Koch, *A. mulleri* Bl, and *A. variabilis* Bl. (Bailey, 1950).

It is cultivated in many states of India. Breeding efforts in elephant foot yam resulted in developing improved varieties with high yield, acridity free as well as good culinary qualities viz. Gajendra, Sree Padma, Bidhan Kusum and the first hybrid variety Sree Athira. Elephant foot yam is common vegetable for rural and traditional dishes. It is used both for its food-nutrition and medicinal values. Its spp. Diversity [Fig. 21 (A-F)] offers immense scope for wide crosses.

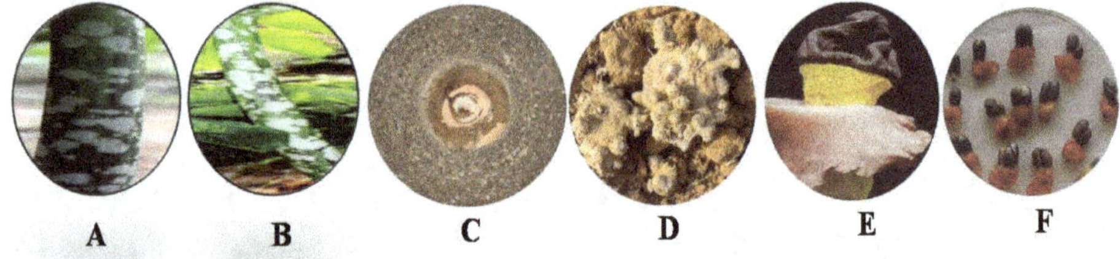

Fig. 21. Species diversity in elephant foot yam. Stem (A&B), Tubers (C & D), Flower (E) and Seed (F)

Cytological and histological studies

Cytological and histological studies along with biochemical parameters indicated the relative performance of a species and provides a mean to understand interspecific, intraspecific or in some cases intergeneric variability.

Cytological variation have been observed in the species, with 26, 28 and 39(2n) numbers in *Amorphophallus paeoniifolius* indicating the presence of triploids and hyperploids.   A *bulbifer* has 2n = 36 (Abraham *et al.* 1976). Natural seed set was observed among the wild species although the genus is burdened with extreme protogyny and irregular flowering (Rajendra *et al.* 1977, Jos & Bai, Mukherjee *et al.* 2010).Chromosome number and karyotype analysis of 13 species of *Amorphophallus* revealed close and straightforward evolutionary trend among them.

Molecular marker based studies

Phylogenetic studies involving molecular markers, DNA sequencing and protein characterization not only overcome the exsiting confusion in the classification of *Amorphophallus* but also provides information on molecular level for genetic variability among species along with taxonomic and evolutionary studies.

RAPD studies and cluster analysis were employed to analyze 22 species of *Amorphophallus* genus. 9 of the 40 decamer nucleotide primer showed consistent banding pattern. After amplification, 84 bands were produced, of which 77 were polymorphic. The 22 accessions were clustered into 5 groups by unweighted pair group average. *A. konjac* and *A. albus* were found to be genetically close, and *A. tokinensis* is genetically most distant from *A. konjac* followed by *A. dunii.* (Zhang, YuJin, 2001). Molecular variation at interspecific and intraspecific level have indicated an evolutionary trend of increasing insertion /deletion events (Chiang *et al.* 1998).

Transgenic research

Being monocot, transgenic research has not progressed in elephant foot yam. However micropropagation through tissue culture has already succeeded through axillary shoot proliferation, organogenesis and somatic embryogenesis [Fig. 22 (A-D), Mukherjee *et al.* 2010]. Such results are encouraging to overcome the barriers of wide hybridization in elephant foot yam.

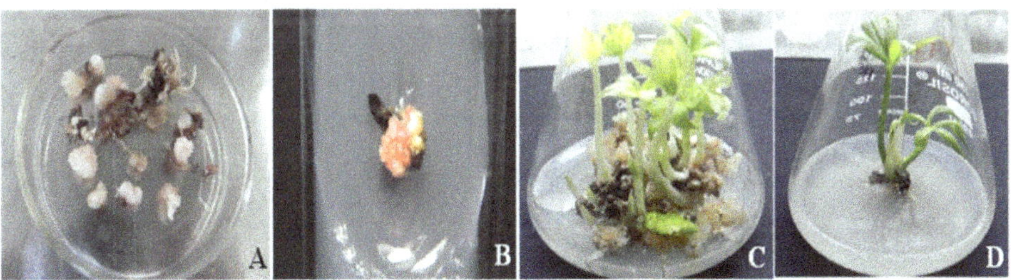

Fig. 22. Micopropagation 0f elephant foot yam. Sprouting (A & B), somatic embryogenesis (C), whole plant (D)

Taro

Taro [*Colocasia esculenta* (L.) Schott] belonging to the family Araceae (Aroideae). It ranks fourteenth among staple/vegetable crops worldwide. It is most popular food cum vegetable crop in many parts of the humid tropics and sub tropics. This crop responsive to organic farming is the staple of people live in islands and fragile environment around the world. It has wide adaptability can fit well into agro forestry systems as well as in challenged soil conditions like swamps. Further the origin, domestication, botany as well as the challenges and opportunities of taro farming reflect the immense potentiality of conserving its valued genetic resources for food nutrition and livelihood.

Dispersal, origin, domestication and botany

A commonly accepted speculation is that the centre of origin and domestication of taro in Southeast Asia. Matthews (2004) explained that domestication occurred in many places across the natural distribution through wild progenitor. Two gene pools appeared with domestication occurring in Southeast Asia and with separation of the land masses of Sunda and Sahul overlapping in Indonesia (Lebot *et al.* 2004). Based on these gene pools, two botanical varieties of taro have been designated *C. esculenta* var. *esculenta*, commonly known as dasheen and *C. esculenta* var. *antiquorum*, commonly known as eddoe with different cytotypes [Fig. 23 (A-E)].

Fig. 23. (A-E), Diversity in tato. Eddoe & dasheen types (A-C), diploids (D) and triploids (f)

In general, cytotypes (triploids, diploids) varied at high altitude and latitude environments suggesting that such conditions promote the occurrence of unreduced gametes. The genetic diversity of cultivated forms of taro (*C. esculenta*) was initially assessed based on morphological and cytological characters (Yen and Wheeler 1968; Kuruvilla and Singh 1981). There are diploid (2n=2x=28) and triploid (2n=3x=42) cytotypes (Mukherjee et al 1998a). Isozymes studies were also conducted on major collections of five Southeast Asian countries and two Pacific countries (Indonesia, Malaysia, Papua New Guinea, Philippines, Thailand, Vanuatu and Vietnam) under TANSAO, "the Taro Network for Southeast Asia and Oceania". The results from these studies reflected-

- Two distinct gene pools, one in Southeast Asia and the other from Southwest Pacific.

- The allelic diversity of the wild taros was similar to that of cultivated forms.

Integrated participatory breeding approach in India to adapt harsh conditions

India is said to be the secondary centre of origin of taro endowed with diverse genetic resources. That has been further conformed with wide variations in isozyme profiles of Asian taro from India, Indonesia and Japan (Lebot and Aradhya 1992).

Recently under *INEA* programme a sample core germplasm of India was evaluated, DNA analysis of which revealed wide diversity of taro genetic resources in India (Unnikrishnan *et al.* 2013). It is grown in most of the states of India in a wide range of agro ecological conditions. Both tubers and leaves of these crops are an alternative source of dietary energy. Taro starch is easily digested and used in baby food. Some of the land races of this ancient crop found to have desirable stress tolerant traits. To make taro more adaptive to climate changes, an extensive study was taken up to evaluate the genetic resources integrating conventional and non conventional methods to tap the vast potential of genetic diversity in isolating biotic (blight) and abiotic (drought, salinity) stress tolerant taro with following major objectives.

- Collection and maintenance of diverse genetic resources.

- Screening of genetic resources with high yield, dry matter, high starch with non acrid, good cooking quality and palatability.

- Screening of genetic resources for tolerance to biotic (leaf blight) and abiotic stresses like salinity, drought and water logging.

- Characterization for yield and biochemical responses of selected genotypes viable to commercial and climatic changes.

- Breeding taro to adapt under changing environment

Such studies resulted in identifying biotic (blight) and abiotic (salt, drought, submergence) stress tolerant taro [Fig. 24 (A-E) & 25] with good yield (12-15 t/ha) [Mukherjee *et al.* 2011]. Higher antioxidant and pronounced isozymes activities were reported in tolerant genotypes under stress (Mukherjee *et al.* 2004; Sahoo *et al.* 2007, 2009 & 2010). Cluster analysis of tolerant and sensitive lines [Fig. 24(D & E)] revealed that tolerant lines share the same node (Mukherjee 2004; Mukherjee *et al.* 2011).

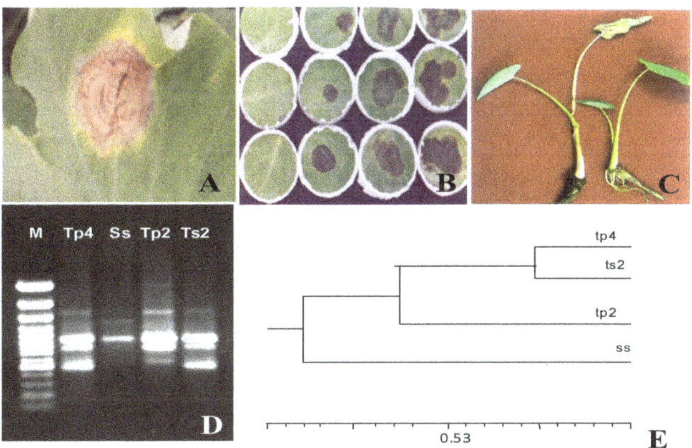

Fig.27. (A-E). Biotic (A&B), Abitic (C), stress tolerance in taro, DNA and cluster anlysis of tolerant and sensitive

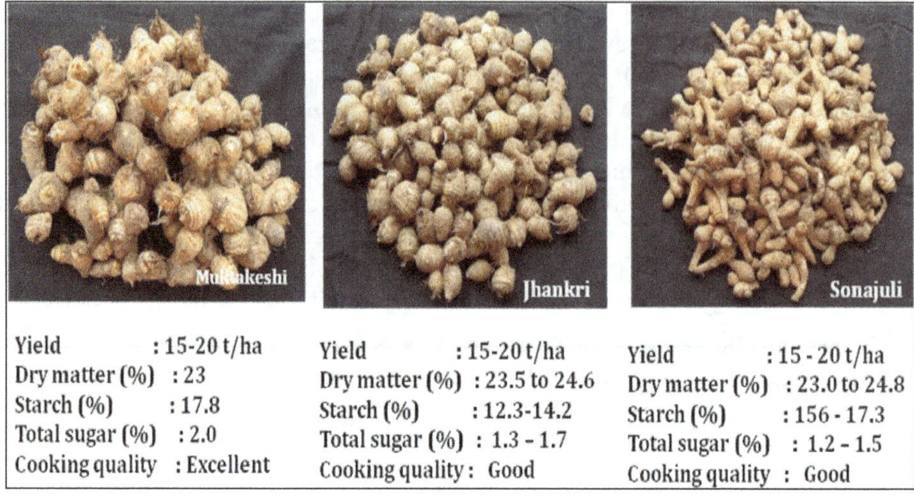

Fig. 25. Developd taro verities (Muktakeshi, Khnakri & Sonajuli) tolerant to both biotic (blight) and abiotic stresses (drought and salinity)

Irrespective of taro farming in different nations and islands around the world, to ascent with this ethnic crop, its inherent challenges and opportunities need to be reviewed.

Challenges -Genetic erosion

Genetic erosion owing to continuous vegetative propagation resulting in sexual degeneracy in most of taro growing areas.

Genetic vulnerability

The genetic base of taro, outside of Southeast Asia is narrow and vulnerable to biotic and abiotic stresses. In the Pacific, and in relatively new taro cultivation regions, narrow genetic base has made the crop vulnerable to a range of very damaging biotic stresses such as taro beetles (Papuana spp.), taro viruses such as Alomae-Bobone virus complex (ABVC), and the most serious one taro leaf blight (TLB) caused by *Phytophthora colocasiae* Racib.

Bottleneck in taro breeding

- Lack of flowering and erratic flowering, Protogynous nature
- Cytogenetical anomalies, triploidy impedes conventional hybridization thereby gene flow.

Opportunities for taro farming and launching of International Participatory breeding programme

Taro is an environment friendly unique crop. It is able to grow in harsh ecological conditions wherein other crops fail to grow. It is highly productive (15-20 t/h), short duration and shade loving crop. Shade tolerance fits the crop as a profitable inter-crop and it is also with associated with ethnic culture and revenue generation. Taro fits well as an alternative crop in the rice-based cropping systems as land that has been prepared for flooded rice is equally suitable for flooded taro. Considering the opportunities across the globe, an International network project on taro was launched in 2011 involving 22 different countries (Fig. 26) with the **'breeding thrust' of widening the genetic base and to breed through farmers-researchers participatory approach to adapt taro in climatic and commercial changes.** As the programme is through National Agricultural Research System (NARS), farmers across the globe have easy access to wide genetic resources rather gene sources of different regions. The source for international transfer of genetic resources was *in vitro* plants (Fig. 27) through NARS. The steps involve for inland programme presented in Fig. 28.

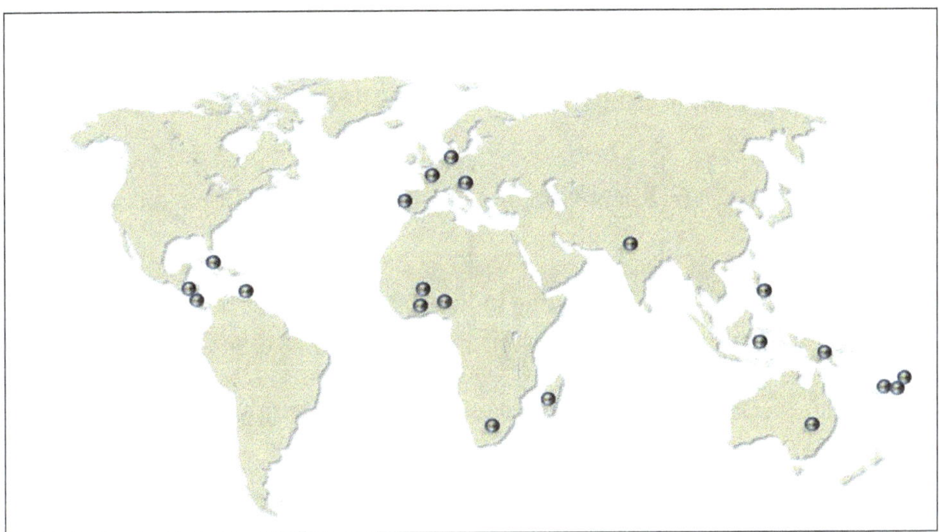

Fig. 26. International Network for Edible Aroids (INEA)

Fig. 27. Stages of adaptation (A-D) of in vitro exotic taro plants

Adapting taro for climatic and commercial changes

↓

Gene pool enrichment

↓

In vitro multiplication, hardening, field multiplication

↓

Selection (blight resistance, high yield, no acridity, high mineral contents)
especially to combat **"hidden hunger"** through multilocation trial

↓

Breeding (Distant hybridization) through participatory approach

↓

Continuing in India & other countries
(Trivandrum, Bhubaneswar and different villages in Odisha, India)

[Participatory taro breeding under *INEA*-taro, India]

Fig. 28. Stages of distant hybridization in taro through participatory breeding

Biochemical and DNA analysis of Indian taro

Starch and mineral contents of Indian taro (Table 2) revealed very high micro nutrients especially contents of Ca, Fe, Zn. Such high valued crops can address the issues on 'hidden hunger' apart from its calorie value. DNA analysis revealed two distinct gene pools with wide genetic diversity (Fig. 29).

Gene pool 1 (26 accessions), Gene pool 2 (18 accessions)

- Gene pool 1 contained accessions from Southern States as well as those from North, East and North-Eastern States.

- Gene pool 2 contained accessions from only North and North-Eastern parts of India parts of India.

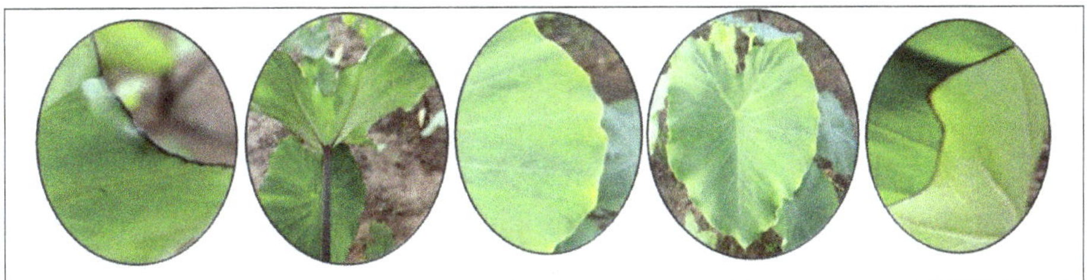

Fig. 29. Genetic diversity in taro.

Biochemical and molecular markers for phylogeny of wide crosses

RFLP analysis of chloroplast DNA was also used to study phylogenetic affinities between taro species and in genus *Alocasia* (Yoshino 1994; Yoshino *et al.* 1998). These studies indicated the presence of *Alocasia* chloroplast DNA in *Colocasia* samples, suggesting possible hybridization between the two genera. Lebot & Aradhya (1991) studied isozyme variation in 1417 cultivars and wild type taros from Asia and Oceania. Asian cultivars showed greater variation than the Oceanic cultivars. Hence wide hybridization between indigenous and exotic taro [Fig. 30 (A-F)] can strengthen the adaptive food-nutrition and livelihood programme.

Fig.30. Distant hybridization in taro for adaptive food-nutrition. Taro plant eddoe type (A), dasheen type (B), eddoe type tubers (C), hybridized flower (D) Fruiting (E) & seeds (F)

In this context, a thorough breeding integrating molecular dissection of the field tolerant and susceptible lines, their progenies and their character association studies through bulk segregate analysis would help to identify ESTs for stress resistance to recover the desirable one (Mukherjee *et al.* 2011) towards climatic and commercial demands.

The developed stress tolerant, nutrient rich sweet potato and taro are now being popularized in tribal, backward areas through National and International programme (Figs. 31 & 32). In this context, LANSA (Leveraging Agriculture for Nutrition in South Asia) and FSN (Farming System for Nutrition) programme led by Prof. M. S. Swaminathan needs special mention.

Fig. 31. Training and participatory nutrie farming in Malkangiri, Odisha

Cross section

- Nature and natural happenings played major role to offer desirable crop product through distant hybridization

- Molecular tools with integrated grid of morpho-phylogenetic studies of "desirable natural product" can open up a new avenue to analyze its origin

- Synthesis of new 'crop products' adaptive to climatic and consumer needs

- National, International Network programme can further accelerate the process

Fig. 32.

Women at Work	Former making sure that his wife is working	Agronomist monitoring the farmer	Sociologist obeserving the agronomists behaviour	Anthropologist observing relation between observer and obeserved	Political scientist oberving the observation process	Donor funding the whole thing	Tax payer ignoring what's being done with his money

Fig. 33. Togetherness? Is it with belongingness?

CONCLUSION

To address the issues of 'environmental protection', 'food security', 'climate change', 'genetic erosion' and 'gene sources' integrating all techniques with region specific agro techniques is the demand. Success through the integration of modern and traditional tools depends on proper planning, political will, support and public awareness with the real spirot of belongingness. World today needs an evergreen, sustainable growth. Certain issues like "Farmer's right", "intellectual property right" and "socio-economic ethical chaos" advocated with modern tools need to be resolved sensibly.

Distant hybridization, whether traditional or advanced, ultimately what matters is **"Goals"-** to design the crop for its 'product' with and 'Most of the cases, the hypothecated programme is found to be reality. The genesis of yams itself is a burning example. Similarly the development of taro, elephant foot yam and other tuber crops and their domestication from wild relatives and land races are also the product of nature aided distant hybridization.

ACKNOWLEDGEMENT

I take this opportunity to express my sincere gratitude to Dr. N.K. Krishnakumar, Honourable Deputy Director General (Hort.), ICAR, India and Dr. S.K. Chakrabarti, Director, ICAR-CTCRI, Trivandrum for their constant guidance and encouragement. I also acknowledge the financial support of EU-aided *INEA* taro project and the encouragement of Dr. Vincent Lebot, Global Leader of *INEA* taro. Sincere thanks to all my *INEA* team members across the globe and my colleagues of CTCRI for their support.

Table 1. Comparative proximate composition of tuber crops to combat hidden hunger

Tuber crops	Grams per 100 g on dry weight basis					
	Protein	Fat	Minerals	Fibre	Carbohydrates	Calories
Potato	7.3	0.4	2.4	1.6	89.0	382
Sweet potato	3.8	0.9	3.1	2.5	88.5	377
Cassava	1.7	4.9	2.5	1.5	84.9	386
Yam	4.7	0.3	5.3	3.3	86.6	370
Taro	11.6	0.4	6.3	3.7	78.5	361
Elephant foot yam	5.6	0.5	3.8	3.8	86.3	371
Rice	7.8	0.8	0.7	0.2	89.9	397
Wheat	13.5	1.7	1.7	1.4	81.2	393

Table 2. Starch and mineral contents in taro, India

Sample	Starch (%)	Elemental Concentration							
		(mg/kg)						(%)	
CEIC-Acc. No.		Ca	Cu	Zn	Mn	Fe	Mg	K	P
264484	53	3775	3.53	50	43	64	2227	3.24	0.472
419621	62	1327	9.66	94	30	67	1947	1.61	0.465
089560	55	3134	13.1	101	90	38	1532	1.79	0.277
211584	64	2357	11.3	73	24.0	46	1102	1.62	0.279
204239	52	2230	11.5	131	49	74	2036	2.59	0.451
089624	62	1366	9.14	63	20.5	41	1426	1.44	0.425
204336	55	2976	12.5	118	82	41	1899	2.33	0.261
12459	64	1727	13.7	82	20.0	61	1100	2.04	0.362

Source: *INEA*

REFERENCES

Abraham A, Ninan CA, Nair PNC, Philomena K and Pillai PG (1976). An inventory of Germplasm of plants of Economic Importance in South India. Department of Botany, University of Kerala, Kariyavattom, Trivandrum, India. pp. 268.

Baco MN (2000). La domestication désignâtes sauvages dans la sous-prefecture de Sinende´: savoirs locaux et pratiques endoge`nes d'ame´lioration ge´ne´tique des Dioscorea abyssinica Hochst. The`sed'Ingenieur Agronome, Universite´Nationale du Be´nin, Benin.

Bailey (1950). The standard cyclopedia of Horticulture. The Macmillan Co. Newyork 1: 276-277.

Bogner J, Mayo S and Sivadasan M (1985). New species and changing concepts in *Amorphophallus*. *Aroideana* 8: 14-25.

Bordas M, Gonza'lez-Candelas, L, Dabauza M, Ramo'n D and Moreno V (1998). Somatic hybridization between an albino *Cucumis melo* L. mutant and *Cucumis myriocarpus* Naud. *Plant Science* 132: 179–190.

Jos JS and Vijaya Bai K (1986). Seed set and polyembryony in *Amorphophallus* campanulatus. *Journal of Indian Botanical Society* 65(2): 178-184.

Kuruvilla KM, Singh A (1981). Karyotpic and electrophoretic studies on taro and its origin. Euphytica 30: 405-413.

Lebot V (1999). Biomolecular evidence for plant domestication in Sahul. Genetic Resources and Crop Evolution 46: 619–628.

Lebot V, Aradhya KM (1991). Isozyme variation in taro (*Colocasia esculenta* (L.) Schott) from Asia and Oceania. Euphytica 56: 55-66.

Lebot V, Gunua T, Pardales JR, Prana MS, Thongjiem M, Viet NV (2004). Characterisation of taro [*Colocasia esculenta* (L.) Schott] genetic resources in Southeast Asia and Oceania. Genetic Resources and Crop Evolution 51: 381–392.

Matthews PJ (2004). Genetic diversity in taro and the preservation of culinary knowledge. Ethnobotany Research and application 2: 55-71.

Mukherjee A (1999). Tuber Crops.In: Biotechnology and its application in Horticulture. Edited by Ghosh SP, Narosa Publishing House, New Delhi. pp. 267-294.

Mukherjee A (2002). Effect of NaCl on *in vitro* propagation of sweet potato. (*Ipomoea batatas*). Applied Biochemistry and Biotechnology 102: 431-441.

Mukherjee A (2013). Climatic resilience and salinity cum submergence stresses tolerance of sweet potato (*Ipomoea batatas L.)* genotypes. **In:** Climate change and Environment. Eds. J. Sundaresan, S. Sreekesh, A.L. Ramanathan, Leonard Sonnenschein, Ramboojh. Scientific publisher, Jodhpur, India, pp: 89-99.

Mukherjee A and Vimala B (1994). *In vitro* seed germination and plantlet establishment of two wild relatives of sweet potato (*Ipomoea trifida* and *I. Amnicola*). Current Science 66(12): 941-942.

Mukherjee A, Debata BK and Naskar SK (1998b). Cytology of callus and regenerated plants of sweet potato (*Ipomoea batatas* L.). *Cytobios* 96: 109-118.

Mukherjee A, Debata BK and Naskar SK (1998a). Somatic embryogenesis and chromosomal stability of regenerants in sweet potato and taro. *J Sci Ind Res* 57: 709-715.

Mukherjee A, Debata BK, Mukherjee PS and Mallick SK (2001). Morpho-histo biochemical characteristics of embryogenic and non embryogenic callus cultures of sweet potato (*Ipomoea batatas* L). *Cytobios* 106: 113-124.

Mukherjee A, Naskar SK (2012). Performance of orange and purple fleshed sweet potato genotypes of coastal locations of Odisha. *J Root Crops* 38 (1): 26-31.

Mukherjee A, Naskar SK and Misra RS (2011). Biotic and abiotic stress tolerant sweet potato and taro vs. 'the paradox' of food insecurity under changing environment. **In:** *National Seminar on "Climate Change and Food Security: Challenges and Opportunities for Tuber Crops"* (NSCFT2011), 20-22, CTCRI, Thiruvananthapuram, pp. 38.

Mukherjee A, Naskar SK, Nedunchezhiyan M and Rao KR (2010). *In vitro* propagation of elephant foot yam. *Indian J. Hort.* 66(4): 530-533.

Mukherjee A, Naskar SK, Pati K, Misra RS, Sahoo BK, Chakrabarti SK, George Vimala JB, Unnikrishnana M and Sreekumari MT (2013a). Valued traits in genetic resources of tuber crops: ascent to food and nutrition security under changing climate, ICTRT-2013, Trivandurm, pp.19.

Mukherjee A, Naskar SK, Rao KR, Ray RC (2012). Sweet Potato: Gains through Biotechnology In: *Fruit, Vegetable and Cereal Science and Biotechnology* 6 (spl. 1) *Global Science Books* x-y (1-13).

Mukherjee A, Poddar A and Pati K (2013b). Propagation of edible *Dioscorea* species *in vitro. Int J Innovative Horticulture* 2(2): 111-116

Mukherjee A, Unnikrishnan M and Nair N G (1991). Growth and morphogenesis of immature embryos of sweet potato (*Ipomoea batatas* L.) *in vitro. Plant Cell Tissue and Organ Culture* 26: 97-99.

Mukherjee A, Unnikrishnan M, Nair NG and Easwari Amma CS (1989). Influence of genotypes and media in pollen embryogenesis in cassava. *J Root Crops* 15(2): 133-138.

Naskar SK and Mukherjee A (2001). Evaluation of stability in sweet potato using different methods. *Annals of Tropical Research* 23(1): 11-20.

Nassar NMA, Vieira C and Grattapaglia D (1998). Molecular and embrionic evidence of apomixis in cassava interspecific hybrids of *Manihot* sp. Canad. *J. Plant. Sci.* 78: 348-352.

Qi L, Friebe B, Zhang P, Gill BS (2007). Homoeologous recombination, chromosome engineering and crop improvement. *Chromosome Res* 15: 3–19.

Qing-He Cao, Jun Tang, Ang LI, Wolfgang Gruneberg, Kelvin Humani and Daifu Ma (2014). Ploidy Level and Molecular Phylogenic Relationship among Novel *Ipomoea* Interspecific Hybrids. Czech *J. Genet. Plant Breed* 50(1): 32–38.

Reynolds M, Foulkes MJ, Slafer GA, Berry P, Parry MA, Snape JW, Angus WJ (2009). Raising yield potential in wheat. *J Exp Bot* 60: 1899–1918.

Sahoo MR, Dasgupta M, Kole PC and Mukherjee A (2010). Biochemical changes in leaf tissues of taro [*Colocasia esculenta* L. (Schott)] infected with *Phytophthora colocasiae J. Phytopathol.* 158:154-159.

Sahoo MR, Dasgupta M, Kole PC, Bhat JS and Mukherjee A (2007). Antioxidative enzymes and isozymes analysis of taro genotypes and their implications in Phytophthora blight disease resistance. *Mycopathologia* 163: 241–248.

Sahoo MR, Kole PC, Dasgupta M, Mukherjee A (2009). Changes in Phenolics, Polyphenol Oxidase and its Isoenzyme Patterns in Relation to Resistance in Taro against *Phytophthora colocasiae. J Phytopathol* 157: 145–153.

Scarcelli Nora, Serge Tostain, Ce´dric Mariac, Cle´ment Agbangla1, Ogoubi Da1, Julien Berthaud and Jean-Louis Pham (2006). Genetic nature of yams (*Dioscorea sp.*) domesticated by farmers in Benin (West Africa) *Genetic Resources and Crop Evolution* 53: 121–130.

Slad AJ *et al.* (2005) A reverse genetic, non-transgenic approach to wheat crop improvement by tilling. *Nature Biotechnology* 23, 439-442.

Unnikrishnan M, Mukherjee A, Srinivas T, Naskar SK, Pradhan DMP & Sharma T (2013). Valued traits in taro: influence of cytotypes. ICTRT-2013, CTCRI Trivandrum, Abst. pp. 52.

Yen DE, Wheeler JM (1968). Introduction of taro the Pacific. The indications of the chromosome numbers. *Ethnology* 7: 259-267.

Yoshino H (1994). Studies on phylogenetic differentiation in taro, *Colocasia esculenta* (L.) Schott. Kyoto University, Japan.

Yoshino H, Ochiai T, Tahara M (1998). An artificial inter-generic hybrid between *Colocasia esculenta* (L.) Schott and *Alocasia macrorrhiza* G. Don. In: Proceeding of the Monocots II and Grass III, pp. 340. Sydney.

15

Hybridizing Genetically Distant Wild and Cultivated Types for Crop Improvement in Ashwagandha (*Withania somnifera* Dunal)

D.H Sukanya

Section of Medicinal Crops
ICAR-Indian Institute of Horticultural Research,,
Hessaraghatta Lake post, Bengaluru-560089

INTRODUCTION

A large germplasm (102 wild and 88 cultivated) collection were assessed over years for qualitative (23) and quantitative (30) traits at   different stages to assess the extent and nature of differentiation. The cluster analysis revealed three major clusters, two for wild and one for cultivated indicating their distinct morphology and existence of two morphological syndromes in the wild type. Analysis of molecular variance based on RAPD data, also grouped the wild and cultivated accessions in to two distinct clusters indicating distinct genetic grouping. Twenty one crosses, including both direct (Cultivated x Wild) and reciprocal (Wild x Cultivated) were generated. The F_1 hybrids were not economically useful. Plants combining the traits of importance were selected and the progeny showed delayed segregation, but continued selection in to F6 generation resulted in uniform progeny. All these results indicate that the cultivated and wild type belong to two distinct groups and supports other studies which strongly view that the two should be placed in separate species /sub-species and they can significantly contribute for further improvement of the crop through hybridization between them.

W. somnifera, commonly known as Ashwagandha, is an important medicinal plant that has been used in Ayurvedic and indigenous medicine for over 3,000 years. In view of its varied therapeutic potential, it has also been the subject of considerable modern scientific attention. Ashwagandha roots are a constituent of over 200 formulations in Ayruvedha, Siddha and Unani medicine, which are used in the treatment of various physiological disorders. *Withania* appears in WHO monographs on Selected Medicinal Plants and an American Herbal Pharmacopoeia monograph is also forthcoming . Roots and leaf of *W. somnifera* are the source of drug withanolides possessing anti-inflammatory, antitumor , immunomodulatory, antistress, antioxidant properties, prescribed for arthritis and rheumatism and to improve overall health and longevity .

The genus *Withania* (Solanaceae) consists of 26 species distributed in South Asia and the Eastern Mediterranean area (Kaul 1957; Kumar *et al.* 2007). Whereas only two species (*W. somnifera* and *W. coagulans*) have been reported from India. *Withania somnifera* is widely distributed throughout the drier and subtropical parts of India (Hooker, 1885) and is well represented in Bombay, Gujarat, Rajasthan, Madhya Pradesh, and Uttar Pradesh, Punjab plains extending to the mountainous regions of Himachal Pradesh and Jammu & Kashmir where it ascends up to an elevation of 1800 m above sea level (Nigam and Kandalkar, 1995). The crop is also commercially cultivated in Manasa, Madhya Pradesh, and in some parts of Rajasthan, Andhra Pradesh and Uttar Pradesh (Anon 1976; Kothari *et al.* 2003). The species dwells in a variety of phyto-geographic regions differing from each other in climate and edaphic characters (Singh and Kumar, 1998). It has also been reported from Pakistan, Afghanistan, Palestine, Egypt, Jordan, Morocco, Spain, Canary Island, Eastern Africa, Congo, Madagascar and South Africa and occupies areas, which differ in their soil, rainfall, temperature and altitudinal profiles (Dymock *et al.*, 1981).

According to an estimate, the annual requirement of the drug at about 9127.5 tons far exceeds the annual production of about 5905.1 tons under cultivation (Ministry of Health and Family Welfare, 2002). The herb   identified by National Medicinal Plant Board of India as one of the thirty-two selected priority medicinal plants, is in great demand in both domestic and international markets (Prajapati *et al.*, 2003). Lack of post harvest storage technology for roots and adequate information on the genetic basis of yield contributing traits (Govil *et al.*, 1993; Vitali *et al.*, 1996; Singh and Kumar, 1998) are other unfavourable factors. Long time gap between planting and harvesting, excessive exploitation of natural resources, non-availability of procedures for synthetic production of withanolides are the reasons for ever increasing demand-supply ratio for the drug in Ashwagandha. The species because of high "off-take" from the natural resources has been listed as "endangered" (Govil *et al.*, 1993) and "on verge of extinction" (Vitali *et al.*, 1996).

Vast amount of intra-specific morphological variability characterizes *W. somnifera* and it calls for a systematic assessment of the different morphological and geographical variants. Stewart (1869) recorded two varieties (red and white), in Punjab. Kaul (1957) listed morphological differences between the cultivated and wild varieties and named the former as *W. ashwagandha* and the later as *W. somnifera*. Incidentally, most of the species in *Withania* are differentiated on the basis of apparently trivial characters (Singh and Kumar, 1998). For example, *Withania obtusifolia* var. tackholm bears membranous and obtuse leaves in contrast to somewhat thicker and acute leaves of *W. somnifera*. Atal and Schwarting (1962) made an extensive study of *W. somnifera* herbarium specimens, representing different bio-geographical regions, from Forest Research Institute, Dehradun, Uttaranchal (India) and Gray Herbarium, Harvard University, Cambridge, Massachusetts (USA) and observed a great deal of intra-specific morphological variation along its distribution range.

Significant levels of genotypic and phenotypic variability between the wild and cultivated morphotypes, have been reported by several researchers (Kaul 1957; Atal and Schwarting 1962; Dhalla *et al.* 1961; Atal *et al.* 1975) and they have also discussed the species diversification and differentiation. Realising the lack of consistencies in the extent and nature of differentiation in *Withania somnifera* genetic resources in India, Arun Kumar et al (2011) have carried out morphological, cytological, phytochemical and molecular studies on a representative set of samples. All the above studies are of the opinion that two types have to be given the rank of distinct species or delineated into subspecies. Kaul et al (2005) and Arun Kumar et al (2011) observed very low success in experimental hybridization between wild and cultivated accessions, and very low seed set in the hybrids. With this background the present study with large number of germplasm with representation of wild and cultivated types along with a good number of crosses taken through advanced generations was studied to throw more light on the differences between cultivated and wild types in morphological , phytochemical and molecular level and behaviour during and after crossing, and in various advanced generations to support the fact that the cultivated and wild are quite distinct and they have to be either placed as subspecies or separate species.

MATERIALS AND METHODS

The study comprised a large collection of 190 germplasm which included 88 cultivated and 102 wild types. The collections were from different geographical regions in India (Fig 1). The morphological data on 30 quantitative and 23 qualitative traits were recorded at vegetative, reproductive, harvest and post-harvest stages over two years of evaluation (Table 1).

For molecular differentiation 141 random decamer Operon primers were tested for their amplification with representative samples. On the basis of robustness and repeatability of the amplification, clarity and scorability of banding patterns finally a set of 40 primers were employed.

To combine   distinct differences in plant type, root yield and quality and disease reaction between cultivated and wild types, 21 crosses, 14 direct crosses (Cultivated x Wild) and 7 reciprocal crosses (Wild x Cultivated) were generated and taken through advanced generations. Data was analyzed using appropriate statistical packages.

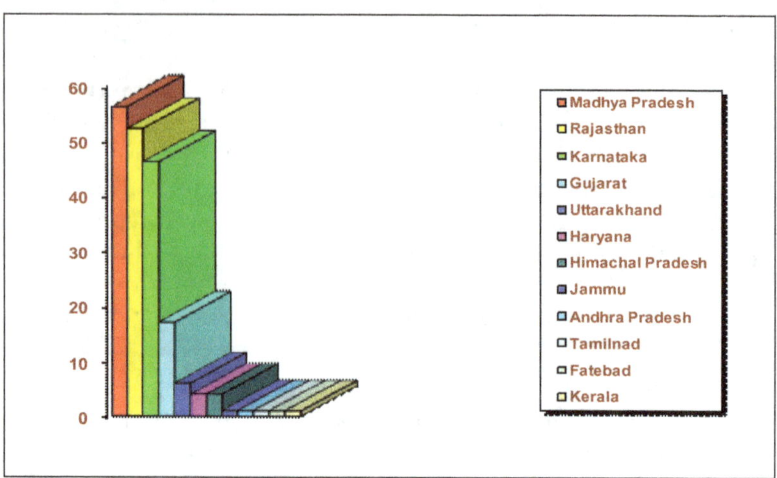

Fig. 1: Geographical representation of germplasm assessed in the study

Table 1: List of morphological traits assessed on the germplasm

Quantitative traits	Qualitative traits
Plant height (cm)	Plant type
Plant spread No of primary & secondary branches secondary and tertiary branches	Stem colour Length of primary, Stem pubescence
No of leaves /branch,	Calyx colour
No of flowers /branch,	Mature fruit colour
Stem thickness,	Immature fruit colour
No of fruiting node /branch	Leaf shape
No of fruits per node and per branch	Leaf tip shape
Leaf length and width,	Leaf base shape
No of leaf axils present,	Leaf margin leaf colour
Length of leaf axil	Upper leaf pubescence
Calyx length and width,	Lower leaf pubescence
Inter nodal length of primary, secondary and tertiary branches	Leaf orientation
Fruit length and width	Anther length
No of roots, root length	Filament length
Root diameter	Stigma length
Fresh root weight	Leaf margin curve
Dry root weight	Secondary branching
Pests and diseases	Calyx shape
	Open/closed calyx
	Plant growth
	Root colour

RESULTS AND DISCUSSION

Morphological differentiation

The results revealed significant genetic variation among 190 accessions, as well as among 88 annuals and 102 perennials as indicated by variance components. This variation was substantial in individual years and also in pooled data over years. This indicates that the population exhibits consistent genetic variability across years revealing the efficiency of morphological criteria in differentiating the accessions.

The Shannon and Weaver diversity index (H) is used as a measure of phenotypic diversity of each trait. This index was estimated for each of the 30 quantitative and 23 qualitative characters over all 190 entries. The grouping of similar genotypes depends on the level of dissimilarity among them, which can be determined by the phenotypic diversity index. Among quantitative traits except for No of secondary branches all the 29 traits recorded highest index values ranging from 0.489 to 0.636 indicating very high phenotypic diversity. In qualitative characters leaf pubescence, tertiary branching pattern, anther length and calyx openness recorded low index values revealing low phenotypic variability for these traits. While the stem pubescence, stem colour, calyx colour, immature and mature fruit color, leaf shapes, stigma and filament length had higher index values (0.301 to 0.815). This clearly indicates that the collection represents substantial phenotypic diversity for most of the traits observed. As the traits with higher phenotypic diversity index contribute to maximum diversity, these traits could be potentially employed for morphological characterization.

The genetic similarity between 190 genotypes for both 30 quantitative and 23 qualitative traits was assessed by Gower's similarity coefficient which ranged from 0.10 to 1.00 indicating abundant variability. The hierarchical cluster analysis revealed three clusters, two major clusters for wild types and one major cluster for annuals cultivated type revealing that the cultivated and wild have distinct morphology and wild have two morphological syndromes (Fig 2).

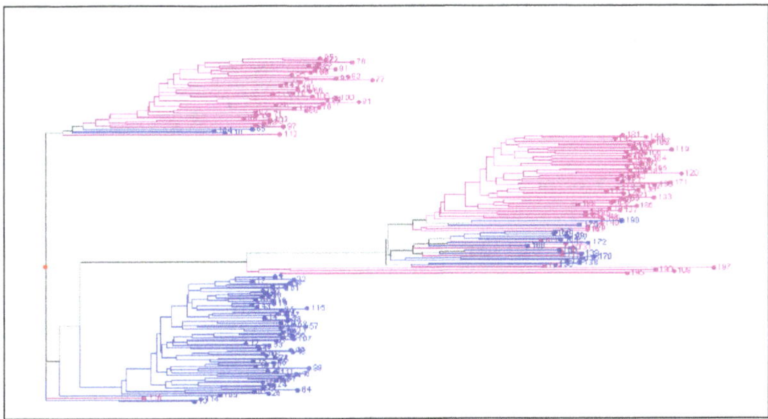

Fig. 2 Tree diagram based on gower's distance matrix of 23 qualitative and 30 quantitative traits of 190 Accnos (Blue is annual and Pink is perennials)

This is very much true as cultivated types were annuals, with <60 cm height, leaves with wavy margin and   pubescence , 4-6 fruits /node , elongated calyx with yellow to orange fruits, with bigger seed size and light yellow seeds. While wild types/ perennials were >60 cm to 150cm height , sparsely pubescent entire leaves, with more (6-10)fruits /bunch , calyx was round with red fruits and dark yellow colour oily seeds(Fig .3).

The   two morphological syndromes in wild / perennial is mainly due to existence of very tall types (120-150 cm height) with very high root yield, higher number of leaves etc in one of the groups. The results obtained here are in tune with earlier reports (Kaul 1957; Atal and Schwarting 1961, 1962; Atal *et al.* 1975; Kumar *et al.* 2007, Arun Kumar et al, 2011) that cultivated and wild have distinct morphology and the present study with very large collections of cultivated and wild germplasm from all over the country, confirms the same in a bigger scale.

Fig. 3 Morphology of cultivated and wild types of Ashwagandha

Analysis of Molecular variance (AMOVA) and Cluster analysis

AMOVA of cultivated and wild indicated substantial variation (69%) within group   as compared to between groups (31%). The clustering of the accessions resulted into distinct distribution of cultivated / annuals and wild / perennials indicating their independent genetic grouping (Fig 4). This clearly indicated that at DNA level also the cultivated and wild differed. And many a times molecular markers are used to solve many taxonomical problems. The present study is based on RAPD data, while Arun Kumar, 2011 have proved the taxonomic differences using AFLP and ITS species specific marker. In the process of screening, one putative

RAPD marker **(OPE-2 $_{650B\,P}$)** has been identified that differentiated the cultivated and wild types (Fig. 5). This marker could be developed as sequence specific / SCAR marker which will be of potential use in the identification of cultivated and wild types in the segregating population.

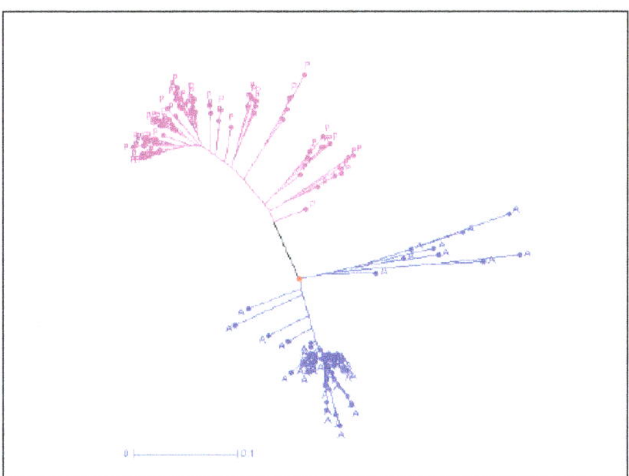

Fig. 4 Tree diagram based on RAPD data of 148 accessions (>50% data points) Pink is perennual blue annual

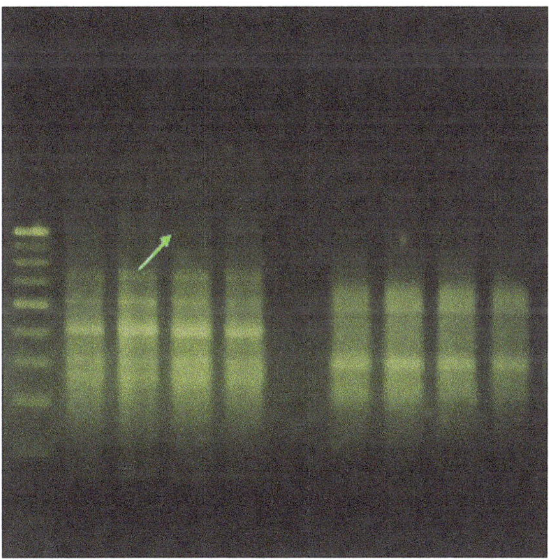

Fig 5. A putative RAPD marker (OPE-2 $_{650B\,P}$) for plant type Cultivated /Wild

Hybridization to combine traits

The cultivated and wild types differed for various economic traits (Table) , And there is a strong need to combine the traits for the improvement of the crop through hybridization.

Distribution of Economic Traits	
Cultivated	**Wild**
Annual in habit	Perennial in habit
Short Duration	Long Duration
Low in root biomass	High in root biomass
High in active ingredient	Low in active ingredient
Susceptible to diseases	Resistant to diseases
Resistant to pests	Susceptible to pests

To combine the distinct differences in plant type, root yield and quality and disease reaction between cultivated and wild types,   21 crosses representing 14 direct crosses (Cultivated x Wild) and 7 reciprocal crosses (Wild x Cultivated) were made. In each case, a large no of crosses were attempted as in each of the crosses less than 30% viable seeds were obtained. This was in tune with the observations of Kaul *et al.* (2005) and, Kumar et al 2011 who observed low success in crossing, lower seed set and germination of crossed seeds. These results on crossability and fertility indicate that the crosses behaved like inter-specific hybrids. Further, the F1 hybrids raised from all 21 crosses exhibited luxuriant plant growth with limited root yield indicating that hybrids are economically not useful. This gigas nature of F1s also indicated distant genetic relationship between the two types. The material was then advanced to further generations to study their behaviour. In F_2 generation wide segregation was observed with expression of transgressive segregants in both directions with extreme phenotypes. Plants combining the traits of importance were selected in F_2 , F_3 , F_4 and F_5 . The progenies in these generation exhibited segregation i.e delayed segregation. The continued selection in to F6 generation resulted in uniform progeny. All these observations in advanced generations also strongly confirm the fact that cultivated and wild are genetically distinct and their crosses and advanced generations behave like inter-specific or distant hybrids. .

To conclude all these results clearly indicate that cultivated and wild type of ashwagandha are distinctly different with respect to genotype and phenotype and are not easily crossable with segregants taking many generations to attain uniformity. This clearly revealed different lineage of cultivated and wild types and crossing between them as distant hybridisation with potential for crop improvement.

Crossing and advanced generation behaviour of crosses between wild and cultivated

REFERENCES

Atal CK, Schwarting AE (1961). Ashwagandha—an ancient Indian drug. *Eco Bot* 15:256–263.

Atal CK, Schwarting AE (1962). Intraspecific variability in Withania somnifera. I. A preliminary survey. Llyodia 25:78–87.

Kaul KN (1957). The origin, distribution and cultivation of Ashwagandha. The so called *Withania somnifera* of Indian Literature. In:Symposium on the Utilization of Indian Medicinal Plants.*Council of Scientific & Industrial Research*, New Delhi, pp. 7–8.

Atal CK, Gupta OP, Raghunathan K, Dhar KL (1975). Pharmacognosy and phytochemistry of *Withania somnifera* (L.) Dunal (Ashwagandha). Central Council for Research in Indian Medicine & Homeopathy, New Delhi.

Kumar A, Kaul MK, Bhan MK, Khanna PK, Suri KA (2007). Morphological and chemical variation in 25 collections of the Indian medicinal plant *Withania somnifera* (L.) Dunal (Solanaceae). *Genet Resour Crop Evol* 54:655–660.

Anon (1976). The wealth of India, raw materials vol. X: SpW. Publications and Information Directorate, CSIR, New Delhi, pp. 581–585.

Kothari SK, Singh CP, Kumar YV, Singh K (2003). Morphology, yield and quality of ashwagandha (*Withania somnifera* L. Dunal) roots and its cultivation economics as influenced by tillage depth and plant population density. *J Hort Sci Biotech* 78:422–425.

Arun Kumar , Bilal A. Mir ,Deepmala Sehgal ,Tanvir H. Dar , Sushma Koul , Maharaj K. Kaul,•Soom N. Raina ,Ghulam N. Qazi Utility of a multidisciplinary approach for genome diagnostics of cultivated and wild germplasm resources of medicinal *Withania somnifera*, and the status of new species, *W. ashwagandha*,in the cultivated taxon *ArPlant Syst Evol* (2011) 291:141–151.

Hooker JD (1885) Flora of British India. Vol.4. Reeve and Co, London, pp 228.

Nigam KB and Kandalkar VS (1995). Ashwagandha, In: Chadha KL, Rajendra G (eds) Advances in Horticulture Vol.11-Medicinal and Aromatic Plants. Malhotra Publishing House, New Delhi, India, pp. 337-344.

Anon 1976;

Kothari SK, Singh CP, Kumar YV and Singh K (2003). Morphology, yield and quality of Ashwagandha (*Withania somnifera* L. Dunal) roots and its cultivation economics as influenced by tillage depth and plant population density. *J. Hortic. Sci. Biotechnol.* 78: 422-425.

Singh S and Kumar S (1998). In *Withania somnifera*: *The Indian Ginseng, Ashwagandha*. Central Institute of Medicinal and Aromatic Plants, Lucknow, India.

Dymock W, Warden CJH and Hoeper D (1981) Pharmacographica Indica, Vol. 2.

Govil JN, Singh VK, Shamima H (1993). **In:** Glimpses in Plant Research Vol. X Medicinal Plants: New Vistas of Research. (Part I). Today and Tomorrows Printers and Publishers, New Delhi, India, pp. 232-253.

Misra HO, Sharma JR and Lal RK (1998a). Genetic divergence in Ashwagandha (*Withania somnifera*). Journal of Medicinal and Aromatic Plant Sciences 20: 1018-1021.580 *Recent Advances in Plant Biotechnology and its Applications*

Misra HO, Sharma JR, Lal RK and Sharma S (1998b). Genetic variability and path analysis in asgandh (*Withania somnifera*). *Journal of Medicinal and Aromatic Plant Sciences* 20: 753-756. Ministry of Health and Family Welfare, 2002.

Vitali *et al.*, 1996; Vitali G, Conte L and Nicoletti M (1996). Withanolide composition and *in vitro* culture of Italian *Withania somnifera*. *Planta Medica* 62: 287-288.

Prajapati *et al.*, 2003 Prajapati ND, Purohit SS, Sharma AK and Kumar T (2003). A handbook of medicinal plants – A complete source book. Agrobios, Jodhpur, India.

Distant Hybridization in Horticultural Crops   *Pages* **265–273**
Editors: **M.R. Dinesh & M. Sankaran**
Published by: **ASTRAL INTERNATIONAL PVT. LTD., NEW DELHI**

16

Pollen Cryopreservation in Distant Hybridization

Rajasekharan, P.E and S. Ganeshan

Division of Plant Genetic Resources,
ICAR-Indian Institute of Horticultural Research, Bangalore 560 089

INTRODUCTION

DISTANT HYBRIDATION involves crossing plants belonging to different species or genera, to effectively combine diverse genetic material into one nuclear genome. Most efforts to transfer a beneficial trait from wild plants into crops so far have bridged the species gap via alien species genetic material. Such hybridization attempts to break the species or generic barrier enabling transfer of desired trait specific genes from one species to another, results in genotypic changes in the progenies. The male gametophyte has a vital role to play, provided there is cross compatibility between the species and genera in question.

A 'Pollen Cryobank' which conserves nuclear genetic diversity, can collect and maintain genetically diverse stocks collected from distantly related plant species and provide the same as male parent in a viable form for primary and supplementary pollination needs, to accomplish wide hybridization. Extended availability of the nuclear gene pool through application of cryogenic technology can help redesign crop-breeding strategies and improve breeding efficiency, integrating conventional and modern techniques for crop improvement. Trans-boundary exchange of cryopreserved pollen pose relatively less quarantine problems, providing a genetic snapshot of the G X E variability captured for traits expressed in the source ecosystem. Besides overcoming asynchrony in flowering, wide inter-specific or even inter-generic hybridization can be attempted through shuttle breeding and developing trait specific genetic stocks through pre-breeding, leading to developing core / mini-core germplasm sets. The presentation will make an attempt to review success stories on distant hybridization in horticultural crops using cryostored pollen

Wide or distant hybridization, a mating between individuals of different species or genera, provides a way to combine diverged genomes into one nucleus. Wide hybridization breaks what is known as the species barrier for gene transfer and thus makes it possible to transfer the genome of one species to another, which results in changes in genotypes and phenotypes of the progenies. Interspecific and intergeneric crosses are made to introduce new genetic variation into cultivated plants. Is there any technological interventions possible to achieve wide hybridization in plant species ? Pollen cryopreservation offers such an option for carrying out distant hybridization in horticultural species/crops. Cryopreservation is defined as conservation of viable cells, tissues, organs and organisms at ultra-low temperatures, usually in liquid nitrogen to a minimum temperature of -196 C. This long-term conservation method is increasingly used in the management of crop plant genetic resources and is also an important component of many plant biotechnology programs. Cryogenic storage is also used to conserve germplasm derived from wild relatives, ancient and modern cultivars and biotechnologically derived genotypes. A pollen cryobank for a given crop can provide a constant supply of viable and fertile pollen and can also allow supplementary pollinations for improving seed set (Ganeshan and Rajasekharan,2000).

Cryopreservation is thus now an accessible conservation and use option for a wide range of users and it has the potential to support both small- and large-scale laboratories and conservation centers. Cryopreservation of pollen is the method of choice for which no elaborate tissue culture methods are needed. It is cheaper than field maintenance of accessions, circumventing the risk of loss caused by diseases, climate changes and other forms of genetic erosion at the haploid stage. The biological material kept in a cryogenic state is said to be in a state of suspended animation with minimal metabolic changes affecting the conserved material. For details see reviews(Ganeshan and Rajasekharan,2000 Ganeshan et al 2008, Rajasekharan et al,2013). How pollen cryopreservation technology helps in distant hybridization in horticultural crops discussed in details in this chapter

CRYOPRESERVATION RESEARCH AT IIHR

With the support of a ICAR sponsored FAO UNDP project during the early eighties, IIHR was able to establish facilities, train manpower and initiate research programs for carrying out research in the area of cryopreservation in horticultural crops. During the last 33 years technology has been developed for long-term cryopreservation of Nuclear Genetic diversity (NGD). Long term cryopreserved pollen in citrus, papaya, grape, mango, tomato, eggplant, onion, capsicum, rose, gladiolus, gerbera, carnation and RET species of medicinal plants are maintained in liquid nitrogen. The pollen cryobank is managed by periodic replenishment of the cryogen, for maintaining a constant cryogenic temperature throughout the storage duration. The technology developed is globally acceptable and will be useful for production of hybrid seeds, besides its use in gene banks and biotechnological interventions.

At the pollen cryobank of IIHR, which was featured in the LIMCA BOOK OF RECORDS in 2001 as the first of its kind in India, information pertaining to 650 pollen samples of more than 45 species belonging to 15 families is stored in a database. The software support provided for this database was FoxPro version –2. For easy management of data one main database and three sub-databases were created. Genus, species and cultivar information is available in a separate database. Information pertaining to longevity, viability and fertility of pollen after cryopreservation, media used for pollen germination, year wise listing of collected samples, year-wise list of samples cryopreserved in different years are available within the database, which is menu driven and user friendly.

Use of pollen cryopreservation in distant hybridization

Redesigning crop breeding strategies

Crossing desirable genotypes involves multiple and staggered plantings in order to synchronize flowering. This can be avoided when cryopreserved viable pollen is available, facilitating hybrids between genera, species and genotypes. This could effectively conserve field and greenhouse space. A pollen cryobank for a given crop can provide a constant supply of viable and fertile pollen and can also allow supplementary pollinations for improving seed set .Male sterile populations can be perpetuated by cryostored maintainer pollen, thus avoiding frequent planting of maintainer line. Large-scale consolidation of potential pollen from male parents will ensure an uninterrupted supply of the male gametophyte for hybrid seed production at a given location, and pollen can be transported to different locations where seed parents are grown for crossing (Ganeshan and Rajasekharan, 1995).

Pollen as a component of genepool for national and international exchange of germplasm

The international transfer of germplasm in the form of dry pollen is not generally restricted (Hoekstra 1995). Moreover, this will eliminate the need for growing plant populations to produce pollen. Pollen is subjected to less stringent quarantine restrictions. So, it can be easily shipped and used. Through exchange of pollen nationally and internationally , desired crosses can be made directly on the seed parent, allowing introgression of characters at a much faster rate. This would find favour especially in breeding of tree species with a long juvenile phase.

Extent of value addition contributed by cryopreserved pollen in genetic enhancement:

The availability of pollen with good quality in a pollen cryobank will provide a constant supply of the same for extended durations. Pollen of such a state can be termed "value added" by virtue of its potentially extended life, for having been able to kept viable and fertile for extended durations to perform its natural function of fertilization, leading to formation of fruit and seed set.

Cryopreservation offers a safe reliable technology for collection, conservation ad sustained maintenance of viable pollen. Methods exist for the cryopreservation of pollen from many crops and species (Ganeshan and Rajasekharan, 2001, 2010 and 2013). This particularly true with many horticultural species. The uses of pollen cyobank are usually supplemented to the established national or international bank's role in supplying seed or vegetative propagules. Supplying pollen to the user community is viewed as an additional service and the costs associated with the user harvesting the pollen from the plant are transferred to the genebank. The challenge for a pollen cryobank, then, is not so much in the development of pollen cryopreservation, but in the development for a system that utilizes the advances made in pollen biology and able to fulfill the goal of providing pollen samples to the user community.

Other uses of cryopreseved pollen in distant hybridisation

Crossing with wild relatives

As you aware the wild relatives of different horticultural species will not be growing well in different seasons. It is ideal to collect and store pollen during the growing season and it can be used whenever and wherever required.

Pollen cryopreservation may also help to overcome genotypic differences and to overcome environmental specificity. Grow adequate male parents in ideal environmental conditions and collect and store pollen in liquid nitrogen.

Sometimes male parent is difficult to grow or is susceptible for many diseases. These male parent could be grown in ideal conditions and pollen could be collected and stored in liquid nitrogen for further use.

Asynchronous flowering can also overcome by storing pollen liquid nitrogen. Generally male parent is sown earlier than the female.

In cross pollinated crops genetic drift and splitting of adaptive complexes can be reduced by storing pollen.

Security of male parent is ensured if you cryopreserved male parent in the form of pollen.

Study supply of pollen at the appropriate time is ensured if you collect and store pollen.

Uses of cryopreservation in breeding programmes

1. General crosses: a male parent repeated over time and space in a number of crosses

2. Shuttle breeding: to speed up variety development, depending on seasonal advantages, the generations are forwarded in different locations and the common male can be used in crosses.

3. Maintenace of B line in male sterile based seed production

4. During unfavourable weather conditions as in the rainy season, pollen from the common or frequently used parent can be collected, stored and used.

5. Appropriate male to female ratio is to be maintained and this could be labour intensive.

6. Normally 15% time is required for pollen collection and storage.

Uses of cryopreserved pollen in production fields:

1. Poor pollen producers: in case of poor pollen producers male parents can be grown in a favorable environment, pollen collected pooled and stored in liquid nitrogen in advance so that in production plots there is no constraint on availability of pollen.

2. When male parent is difficult to grow or is susceptible to some disease or is poor in vigour, is technology can be used.

3. Genetic stability for male parent is ensured if you are able to collect and store pollen in lquid nitrogen.

4. There can be more number of female plants in production plot ratio of 3:1 if you are able to collect the male parent in the form of pollen.

5. When MS line is used, parental increase A line can be made easier by storing the pollen of B line.

6. Wherever the production areas are localed cryolobiological systems could be located and cryopreserved pollen can be used for hybrid seed production. This will save time, resources *etc.*

Protocols for pollen cryopreservation in horticultural crops

Ganeshan and Rajasekharan (2000) reviewed the current status of pollen cryopreservation research and its relevance to tropical horticulture. Protocols for pollen collection, viability assessment (pre and post storage), processing for cryopreservation, retrieval, and fertility assessment followed at IIHR have been described in detail by Ganeshan etal (2008) and Rajasekharan etal,2013

Flow chart of activities for pollen collection and cryopreservation

Cryopreservation of pollen

Anthesis

$\downarrow$

Anther and pollen collection	Quality and quantity is important

$\downarrow$

Viability assessment	By cellophane or hanging drop

$\downarrow$

Processing for storage capsules	Pollen is transferred in to gelatin and inturn to laminated alumiinium pouches and sealed airtighlty

SUMMARY

Pollen cryopreservation could be used as a technological intervention for helping distant hybridization in many horticultural crops and species In the last 32years technology has been developed for long-term cryopreservation of Nuclear Genetic diversity. Long term cryopreserved pollen in the form of nuclear genetic diversity (NGD) of citrus, papaya, grape, mango, tomato, eggplant, onion, capsicum, rose, gladiolus, gerbera, carnation and RET species of medicinal plants are maintained in liquid nitrogen. This technology developed is globally acceptable and will be useful for production of hybrid seeds in the above-mentioned crops, besides its use in gene banks. At the pollen cryobank of IIHR, information pertaining to 650 pollen samples of more than 45 species belonging to 15 families is stored using this database. a breeder involved in genetic enhancement of a given crop can have access to a Pollen Cryobank facility, for nuclear genetic diversity (NGD) inputs in his amelioration program. Many problems encountered by the breeder could be successfully overcome by using this technology.

Table 1: list of crops, species and varieties of pollen available at Pollen cryobank of IIHR , Bangalore

Crop	Genus	Species	Variety	No. of accns	Date cyp	Last ViabilityTest Date
Fruit						
	Magnifera	*indica*	Alphonso	29	12/23/1988	1/25/1991
			Totapuri	7	12/23/1988	12/23/1988
			Totapuri T2	2	1/16/1985	1/16/1985
			Langda	4	3/8/1986	3/8/1986
			Mulgoa	2	1/10/1991	1/10/1991
			Anfas	1	1/19/1993	1/19/1993
			Maharjah Pasand	2	1/23/1993	3/2/1994
			Suvarnarekha	1	2/3/1993	2/3/1993
			Hyder Saheb	1	1/27/1993	1/27/1993
			Lazzath Bakh	1	2/4/1993	2/4/1993
			Dwarf local	2	3/14/1988	3/14/1988
	Vitis	*vinifera*	Anab-E-Shahi	24	12/14/1982	12/23/1982
			Anab-E-Shahi		12/7/1988	12/23/1988
			Arka Hans (BBxTS)	2	11/3/1988	11/3/1988
			Arka Kanchan (BCxTS)	20	12/17/1988	12/19/1988
			Beauty Seedless	2	12/14/1984	12/14/1984
			Black Champa	10	12/11/1984	12/10/1985
			Convent Large Black	3	12/20/1984	12/10/1985
			Kishmish Beli	2	12/14/1984	12/14/1984
			Parlette	4	12/20/1984	12/27/1984
			Pusa Seedless	2	12/11/1984	12/11/1984
			Queen of Vineyards	8	12/7/1982	12/4/1992
			Sonaka	2	11/3/1988	1/3/1988

Contd...

Crop	Genus	Species	Variety	No. of accns	Date cyp	Last ViabilityTest Date
			Thompson Seedless	37	8/26/1985	11/11/1988
			Bangalore Blue	4	11/29/1982	12/27/1989
			Bangalore Purple	5	11/29/1982	12/10/1985
			Concord	3	12/10/1985	12/10/1985
			Red Prince	2	12/10/1985	12/10/1985
			Pusa Ruby	5	2/9/1994	1/25/1994
	Carica	*papaya*	CO-1	2	12/20/1989	12/20/1989
			Coorg Honey Dew	12	7/18/1983	9/30/1987
			Washington	69	4/2/1983	11/13/1991
			Washington Hybrid 5h	1	10/17/1988	1/17/1988
			Washington (Pro. frozen)	2	8/11/1983	8/11/1983
			Washington Des Hybrid 5h	3	1/17/1988	10/17/1988
	Citrus	*limon*	Italian Sap & pap	1	9/19/1986	9/19/1986
			Nepali round	1	12/28/1991	12/28/1991
			Row No. 1 Seville	2	11/7/1991	11/7/1991
			Row No. 2 & 3 Italian	3	6/3/1985	6/18/1985
			Seville	13	6/3/1984	8/4/1989
			Seville Sap & pap	6	8/21/1989	6/13/1984
			Rangpur Lime	4	3/4/1991	11/14/1991
			Hill Lemon	3	6/13/1985	6/18/1985
			Eureka	1	2/17/1992	2/17/1992
			Nepali Oblonga	--	--	--
	Genus	*species*	variety	no. of accns	date cyp	Last Viability Test Date
	Allium	*cepa*	Arka Kalyan	14	3/27/1985	3/27/1985
			Arka Niketan	30	3/27/1985	3/21/1989
	Alium	*fistu*				
	Solanum	*melangona*	Arka Kusumakar	12	10/21/1988	5/26/1992
			Arka Navaneeth	4	10/3/1988	10/18/1988
			Arka Shirish	27	12/29/1988	2/14/1991

Contd...

Crop	Genus	Species	Variety	No. of accns	Date cyp	Last ViabilityTest Date
			Arka Shirish x A . K	4	12/21/1989	9/19/1991
	Lycopersicon	*esculentum*	Arka Ahuti	5	1/29/1994	3/2/1994
VEGETABLES			Arka Vikas	44	2/16/1984	2/1/1993
			Arka Saurabh	25	1/11/1984	3/2/1994
	Capsicum	*annum*	Arka Gaurav	20	8/27/1987	9/6/1993
			Arka Mohini	2	9/1/1989	9/1/1989
ORNAMENTAL	*Gladiolus*		Aarti	2	10/23/1991	10/23/1991
	Rosa	*indica*	Queen Elizabeth	11	9/5/1991	10/28/1992
			Free Flowering Edwards	3	11/3/1992	8/27/1992
			Dr. Fleming	2	10/19/1992	10/19/1991
			Paradise	6	9/3/1991	12/4/1992
			Happiness	12	9/10/1991	10/21/1993
			Bullsred	7	9/17/1991	11/3/1993
			Cherry Blossom	2	3/5/1983	3/5/1983
MEDICINAL(RET)						
	Oroxylum indicum			1	2/3/2011	2/3/2011
	Celastrus paniculatus			1	2/3/2011	2/3/2011
	Cayratia pedata			1	2/3/2011	2/3/2011
	Holostemma adakodien			1	2/3/2011	2/3/2011

REFERENCES

Alexander,M.P. and Ganeshan,S.1993.Pollen Storage In:Advances in Horticulture Vol.I.Fruit Crops Part-I(Eds.K.L.Chadha and O.P.Pareek) Malhotra Publishing House,New Delhi India 110 064.481-496.

Ganeshan, S. and P. E. Rajasekharan.2000.Current status of pollen cryopreservation research: relevance to tropical horticulture. Cryopreservation of tropical plant germplasm. current research and application. JIRCAS, Tsukuba, and IPGRI, Rome, Italy F. Engelmann and H. Takagi (eds.)pp.360-365.

Ganeshan ,S. and P.E.Rajasekharan,1995.Genetic conservation through pollen storage in ornamental crops.In:Advances in Horticulture Vol-12-Part-I. Ornamental Crops(Eds:K.L.Chadha and S.K.Bhattacharjee)Malhotra Publishing House,New Delhi.India. 87-108.

Ganeshan, S. P.E.Rajasekharan, S.Shashikumar.W.Decruze 2008. Cryopreservation of Pollen In: Plant Cryopreservation: A practical Guide Springer (USA).

Hoekstra FA (1995). Collecting pollen for genetic resources conservation. In: Collecting Plant Genetic Diversity. Technical Guidelines (L. Guarino, V. Ramanatha Rao and R. Ried, (Eds.) pp. 527-550, CAB International, Wallingford, U.K.

Rajasekharan, P.E. B.S. Ravish, T. Vasantha Kumar, and S. Ganeshan 2013. Pollen Cryobanking for Tropical Plant Species.in M.N. Normah H.F. Chin Barbara M. Reed Editors Conservation of Tropical Plant Species pp. 65-76 Springer (USA).

Distant Hybridization in Horticultural Crops   *Pages* **275–282**
Editors: **M.R. Dinesh & M. Sankaran**
Published by: **ASTRAL INTERNATIONAL PVT. LTD., NEW DELHI**

17

Using Bees for Crop Improvement

V.V. Belavadi

Department of Entomology,
University of Agricultural Sciences, Bangalore560024

INTRODUCTION

Pollination is an important ecosystem service that is often taken for granted. Economic valuation of pollination service at global level is estimated to be around € 153 billion annually. In addition to being important in enhancing production of over 75% of our major food crops, pollinators can be useful in seed production. Our studies on foraging behaviour of pollinators in small cardamom show that bees exhibit patch fidelity and hence can be used for quality seed production by having compact blocks separated by a relatively small isolation distance of about five metres.   We also hypothesise that since pollinators are pollen vectors, their service can be utilized in transferring pollen between two distinct groups of plants – a pollen donor and a pollen receiver. The problems involved in such an exercise and certain observations and methodologies to be followed in using bees for crop improvement are discussed.

Global population has crossed 7.3 billion and is expected to reach 9 billion by 2050. The ever increasing population has dramatically enhanced the demand for food. It is generally considered that crop production will have to increase by more than 50 per cent in the next 25 years to meet the demand (Le Buanec, 2009). The only way to meet these challenges is to increase productivity significantly. Seeds and plant genetic resources are central to the biological basis of agriculture and improved crop varieties in turn are crucial for global food security.   Availability of quality seed is the foundation for food production and a precursor to crop and food diversification in any region. Efforts to improve performance of agricultural sector should therefore include seed production and delivary systems (Setimela *et al.*, 2004). Seed is a product of pollination. The present paper attempts to answer some basic questions on the utility of pollinators in seed production. The term bee is used here as referring to pollinators and not necessarily only honey bees. In cross pollinated crops, though bees are involved in transfer of pollen and are responsible for seed set, the following questions are raised with reference to seed production:

1.   Bees being agents of pollen transfer, do they alter purity of seeds?

2.   Do they restrict their visits to one floral type? If so can they be used in quality seed production?

3.   Can we determine isolation distance between seed blocks based on foraging behaviour of bees?

4.   If they do transfer pollen between varieties what are the prerequisites for using them in crop improvement?

To answer the above questions, we recorded detailed observations on the foraging behaviour of bees on small cardamom (*Elettaria cardamomum* L. Maton).

METHODS

Studies on small cardamom

The study was conducted at the Zonal Agricultural Research Station, Mudigere (13°13'N& 75°64'E; 980 m above MSL), in the Western Ghats of Karnataka. The system comprised of cultivated cardamom variety M-1 and its flower visitors including two main species of *Apis* viz., *A. cerana* and *A. dorsata*. All observations were recorded between 2004 and 2007 and some additional information needed were collected through sporadic visits to the observation sites in Mudigere in 2010, 2012 and in 2013.

We marked a total 613 foraging bees (474 *A. cerana* foragers and 139 *A. dorsata* foragers) using quick drying paints (Testor®)over a period of three years between 2005 and 2007 and recorded observations on the following aspects: a) Time of activity; b) Number of flowers visited per trip; c) Number of trips made/day; d) Movement pattern between flowers; e) Distance between flowers visited; and f) Foraging distance from the clump where the bee was marked. Additional observations on the flowering pattern in cardamom, flower duration, etc were recorded.

RESULTS AND DISCUSSION

a.   Time of foraging activity: Though cardamom flowering commenced from last week of April and continued till third week of Septermber, peak flowering was in July – August (Fig.1). Bees (*A. cerana* and *A. dorsata*) foraged on cardamom flowers almost through out the day, with a peak foraging activity between 11 am and 1 pm (Fig. 2). Belavadi and Parvathi (2000) observed the bees foraging on cardamom through out the day and found that in the morning hours the bees foraged for pollen while they shifted to nectar foraging in the afternoon hours. Similarly Sinu and

Shivanna (2007) observed pollen collectors to be more effective pollinators of cardamom.

b.   Number of trips by a single forager and number of flowers visited per trip: A single forager made a maximum of five trips in a day. The number of flowers visited by a single forager increased exponentially in the beginning but leveled off by late afternoon. The number of flowers visited by a single bee ranged from 42 ± 4.45 in the morning hours to 129.6 ± 7.66 in the late afternoon (Fig. 3).

c.   Movement pattern between flowers: Individually marked foragers of *A. cerana* were tracked for their movement. We found that the foragers did not move between flowers randomly but followed a set pattern. They either moved in clock-wise or anti-clock-wise pattern, and almost always visited the nearest neighboring flowers. On an average a single bee 127.4 ± 9.04 flowers per trip.

d.   Distance between flowers visited: Observations on marked bees showed that individual bees covered a total distance of 1515.9 ± 157 cm in one foraging trip. The marked bees returned to the same clumps on successive trips indicating a patch fidelity in foraging (Fig.4). When bees were marked at clumps in the border of a block, they did not cross the empty space between the blocks, when there was no cue of cardamom flowers. They continued to forage on the clump at which they were marked or moved to the nearest clumps.

Observations on the foraging behaviour of pollinators in cardamom clearly indicated that they exhibit patch fidelity. Being faithful to visiting flowers on only a few clumps of cardamom, it is evident that the bees are in fact bringing about self pollination or at the most transferring pollen between nearby flowers. Sinu and Shivanna (2007) pointed out pollen collection by bees from cardamom flowers. While collecting pollen the bees may bring about selfing, transferring pollen to the stigma of the same flower, as cardamom is self compatible. This brings about two important questions – whether bees can be used in seed production? and whether bees can be used for crop improvement?

a) Utility of bees in seed production: For utilizing pollinators in crop improvement, it is important to consider some of the behavioural aspects of the pollinator. One important criteria for a good pollinator is its floral constancy. Honey bees exhibit this behaviour mainly due to the fact that the foragers are *recruited* to harvest nectar or pollen, and due to memory constraints. This means a foraging bee remains faithful to the species or variety of plant as long as reward is available. Hence, in most cases bees are involved in selfing.

It has been established that pollinator foraging patterns strongly influence selfing rates within and among populations (Karron *et al.*, 1995, 2004; Harder and Barrett, 1995), and may therefore play an important role in the evolutionary stability of mixed-mating systems, a topic of considerable recent theoretical research (e.g. Goodwillie *et al.*, 2005; Johnston *et al.*, 2009). Several workers have recently shown that selfing rates are influenced by spatial and temporal variation in the composition and abundance of the local pollinating fauna (Brunet and Sweet, 2006; Kameyama and Kudo, 2009). Selfing rates may even vary on much finer spatial scales, due to the effects of the composition of co-flowering species competing for pollination (Bell *et al.*, 2005; Mitchell *et al.*, 2009), variation in floral morphology among neighboring plants (Karron *et al.*, 1997; Medrano *et al.*, 2005), and variation in the order of pollinator probes on individual floral displays (Karron *et al.*, 2009).

Keeping the above points in view, it appears reasonable to make an attempt to utilize honey bees in seed production. The basic condition is that the target crop is cross pollinated or often cross pollinated and/or self-compatible. Bees can also be used in hybrid seed production. This may require some preliminary studies on standardizing the design or planting architecture as to what should be the ratio between male and female lines, *etc.*

b) Using bees for crop improvement: In a preliminary study, several promising entries from a germplasm collection may be planted in a single block, a bee colony may be placed close by to effect pollination between entries to get more variations in the offsprings.

In a much advanced trial, male and female parent lines may be identified for crossing with the aid of bees. Some of the prerequisites would be the following:

a. The floral characters of parent lines do not differ significantly in terms of flower colour, structure, odour, anthesis and longevity

b. Initial studies on compatibility and viable seed production with these lines have been done

c. Detailed studies on the floral biology is available

d. Detailed observations on the foraging behaviour of honey bees on the target plant have been made.

Selections can also be made from a highly variable population with reference to the time of anthesis and/or nectar production. Such plants that fall to the extremes of distributions may be separated and see if different species of pollinators visit these and whether there are other advantages in selecting these.

Floral biology in relation to pollination: Though published information on floral biology of the target crop may be available, it is always better to take a fresh look keeping in mind the pollinator. From the pollinator's point of view information regarding the exact time of anthesis, pollen dehiscence, nectar quality and quantity, time of nectar production and longevity of the flower becomes important. The activity of the pollinator should coincide with all these parameters including pollen viability and stigma receptivity.

Foraging behaviour of the pollinator: In most cases, there will not be any information available on the foraging behaviour of the pollinator (honey bee in this context) with reference to the target crop we are working on. Hence, it is necessary to have first hand information on this aspect of pollination biology of the crop. Some key questions on foraging behaviour include, the time during which the bees forage on the target flowers. Is there a peak time of foraging during the day? Whether the peak foraging time corresponds with the pollen viability and peak stigma receptivity? It is also necessary to work out a couple of indices like the pollen removal efficiency index (PREi) (Freitas and Paxton, 1998) and Pollination efficiency index (PEi) (Spears, 1983).

PREi is computed as   PREi = $Ri - N / V - N$

Where Ri is the mean number of pollen grains removed per flower with a single visit by species i; N is the mean number of pollen grains removed per flower that received no visits and V is the mean number of pollen grains removed per flower receiving unrestricted visitation by bees.

Similarly, PEi is computed as PEi = $Pi - Z / U - Z$

Where Pi is the mean number of seed in the fruit that received only one visit by species i; Z is the mean number of seeds in the fruit that received no visits and U is the mean number of seeds in the fruits that received unrestricted visits.

The values of both the indices range from 0 to 1. A value closer to 0 indicates that the flower visitor is a poor or useless pollinator, while a value closer to 1 would indicate that the flower visitor is the potential pollinator.

While recording observations on foraging behaviour, it is also useful to record data on the number of flowers visited by a single bee per trip, the pattern of movement, number of visits a single flower gets in a day (or during its entire period of longevity); whether the same bee visits the same flowers repeatedly and is there a patch fidelity observed by foraging bees. If the crop in question is monoecious or a dioecious (say a cucurbit), it is all the more important to observe the frequency of visits by a single bee to male and female flowers and the pattern of movement. Many of these observations can be very easily made by marking individual bees using quick drying paints or markers and tracking them.

These observations help us to plan our layout for breeding trials. A simple layout for a pollinator mediated crop improvement trial could be two rows of female parent alternated with one row of male parent. If the female parent is male sterile or self incompatible, it is even better. If not, care should be taken to emasculate the flowers before the buds open. Studies on utilizing bees as vectors of pollen from allied varieties are very few. Some attempts have been made in hybrid seed production in radish (Evans *et al.*, 2011) and in brassicas (Sushil *et al.*, 2013).

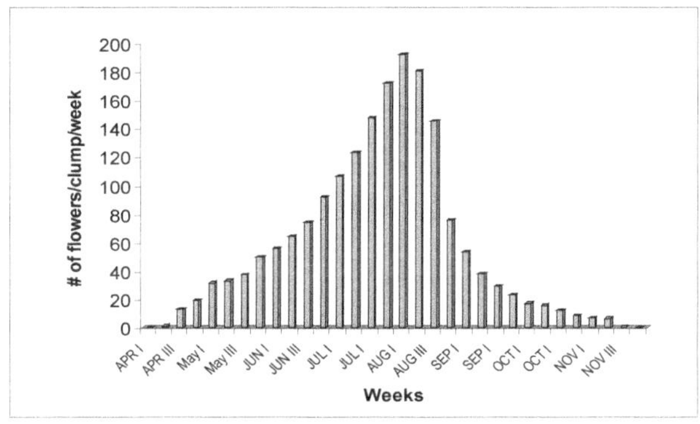

Fig. 1 Flowering pheology of cardamom

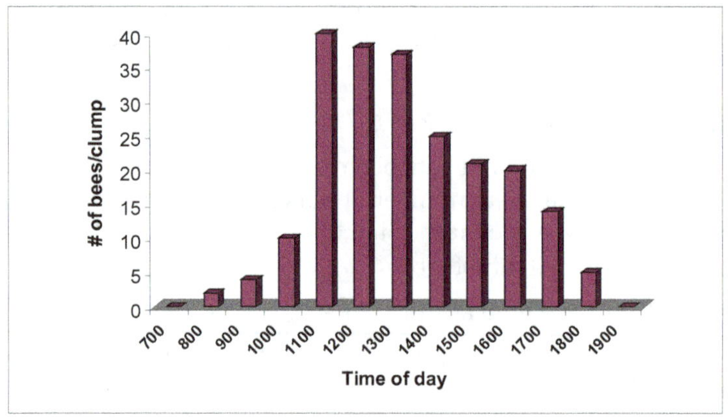

Fig. 2 Foraging activity of bees in a day

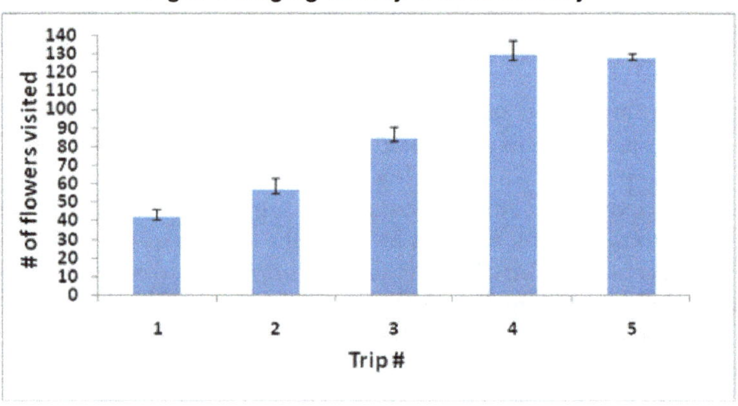

Fig. 3 Number of flowers visited on successive trips by a single forager (n = 10; error bars indicate standard error of mean)

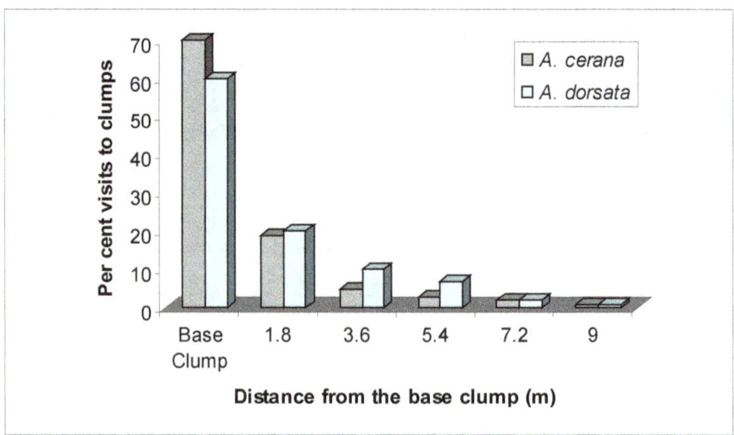

Fig. 4 Patch fidelity in bees foraging on cardamom. Base clump refers to the clump at which the bee was marked (n= 73 marked bees)

REFERENCES

Belavadi, V.V. and Parvathi, C., 2000. Pollination in small cardamom (*Elettaria cardamomum* Maton). *Journal of Palynology*, **35-36**: 141-153.

Bell, J.M., Karron, J.D. and Mitchell, R.J., 2005. Interspecific competition for pollination lowers seed production and outcrossing rate in *Mimulus ringens*. *Ecology* **86**: 762–771.

Brunet, J. and Sweet, H.R., 2006. Impact of insect pollinator group and floral display size on outcrossing rate. *Evolution*, **60**: 234–246.

Evans, L.J. , Goodwin, R.M. , Walker, M.K. and Howlett, B.G., 2011. Honey bee (*Apis mellifera*) distribution and behaviour on hybrid radish (*Raphanus sativus* L.) crops. *New Zealand Plant Protection*, **64**: 32-36.

Freitas, B.M. and Paxton, R.J., 1998. A comparison of two pollinators: the introduced honey bee *Apis mellifera* and an indigenous bee *Centris tarsata* on cashew *Anacardium occidentale* in its native range of NE Brazil. *J. Appl. Ecol.*, **35**: 109-121.

Goodwillie, C., Kalisz, S. and Eckert, C.G., 2005. The evolutionary enigma of mixed mating systems in plants: occurrence, theoretical explanations, and empirical evidence. *Annual Review of Ecology and Systematics*, **36**: 47–79.

Harder, L.D. and Barrett, S.C.H., 1995. Mating costs of large floral displays in hermaphrodite plants. *Nature*, **373**: 512–514.

Johnston, M.O., Porcher, E. and Cheptou, P.O., 2009. Correlations among fertility components can maintain mixed mating in plants. *American Naturalist*, **173**:1-11.

Kameyama, Y. and Kudo, G., 2009. Flowering phenology influences seed production and outcrossing rate in populations of an alpine snowbed shrub, *Phyllodoce aleutica*: effects of pollinators and self-incompatibility. *Annals of Botany*, **103**:1385-1394.

Karron, J.D., Thumser, N.N., Tucker, R. and Hessenauer, A.J., 1995. The influence of population density on outcrossing rates in *Mimulus ringens*. *Heredity*, **75**: 175–180.

Karron, J.D., Jackson, R.T., Thumser, N.N. and Schlicht, S.L., 1997. Outcrossing rates of individual *Mimulus ringens* genets are correlated with anther–stigma separation. *Heredity*, **79**: 365–370.

Karron, J.D., Mitchell, R.J., Holmquist, K.G., Bell, J.M. and Funk, B., 2004. The influence of floral display size on selfing rates in *Mimulus ringens*. *Heredity*, **92**: 242–248.

Karron, J.D., Holmquist, K.G., Flanagan, R.J. and Mitchell, R.J., 2009. Pollinator visitation patterns strongly influence among-flower variation in selfing rate. *Annals of Botany*, **103**: 1379–1383.

Le Buanec, B., 2009. Responding to the challenges of a changing world: The role of new plant varieties and high quality seed in agriculture, *Proceedings of the second world seed conference*, FAO, Rome.

Medrano, M., Herrera, C.M. and Barrett, S.C.H., 2005. Herkogamy and mating patterns in the self-compatible daffodil *Narcissus longispathus*. *Annals of Botany,* **95**: 1105–1111.

Mitchell, R.J., Irwin, R.E., Flanagan, R.J. and Karron, J.D., 2009. Ecology and evolution of plant–pollinator interactions. *Annals of Botany,* **103**: 1355– 1363.

Setimela, P.S., Monyo, E. and Banziger, M (eds.), 2004. *Successful community based seed production strategies*, Mexico, DF: CIMMYT. p. 75.

Sinu, P. A. and Shivanna K. R. . 2007. Pollination ecology of cardamom (Elettaria cardamomum) in Western Ghats, India. *Journal of Tropical Ecology* 23: 1-4.

Spears, E.E., 1983. A direct measure of pollinator effectiveness. *Oecologia,* **57**: 196-199.

Sushil, S.N., Stanley, J., Hedau, N.K. and Bhat, J.C., 2013. Enhancing Seed Production of Three *Brassica* vegetables by Honey Bee Pollination in North-western Himalayas of India. *Universal Journal of Agricultural Research,* **1**(3): 49-53.

Distant Hybridization in Horticultural Crops　　　　　*Pages* **283–289**
Editors: **M.R. Dinesh & M. Sankaran**
Published by: **ASTRAL INTERNATIONAL PVT. LTD., NEW DELHI**

18

Prospects, Opportunities & Concerns for Commercialization of Hybrids

Sudha Mysore

*Division of Social Sciences and Training,
ICAR-IIHR, Hesaraghatta, Bengaluru-560089*

The focus of plant breeding programmes in agriculture show a paradigm shift in the recent times from sheer productivity enhancement towards 'crop loss aversion' through resistance breeding. 'Multiple disease resistance' is the most sought after focus, while the scope and means for achieving this goal also have been expanded through recent developments in 'breeding' either through traditional or modern marker assisted &through allele mining techniques. Besides the techniques, the focus of research also has shifted in sourcing of such 'traits' cutting across the boundaries of traditional breeding approaches of sourcing from within the species. Several new opportunities have been thrown open for the modern day plant breeders aiming at crop productivity enhancement through hybridization, especially, the interspecific hybridization.

With Indian agriculture falling in line with the global order of Intellectual Property Rights regime, ICAR and the State agricultural universities today have huge potential for technology commercialization through technology licensing and through varied forms of Public Private Partnerships. Although ICAR adopts a policy of 'non-exclusive licensing' the scope for exclusive research collaborations also gets channelized through different models of PPP. This paper attempts at reviewing the prospects for commercialization of 'hybrids' with special reference to horticulture and deliberates on the opportunities and concerns.

TECHNOLOGY COMMERCIALIZATION, STATUS AND PROSPECTS

The publicly funded R&D institutions have been the main sources of technology innovation, demonstration, and transfer across countries, especially in the field of agriculture in emerging economies. Governed largely by the policy of providing easy access to technology, these institutions adopted efforts similar to other science and technology fields and helped technology dissemination primarily through front line demonstrations to producers and other public or private sector research institutions and seed companies. This approach, though successful, often undermines the contributions of publicly funded R&D towards improving technologies and enhancing agricultural productivity. Although these efforts to improve knowledge included an innovative process, it tended to be at 'proof of concept' level needing further improvement prior to being applied at commercial production level. In the absence of adequate financial assistance and infrastructure, the R&D institutions have not been geared up for up-scaling their innovations to a commercially viable scale. Hence, the technological advancement and innovative contributions of public R&D institutions such as elite germplasm in agriculture, or an innovative technical knowhow, seldom reached the ultimate consumer. In addition, the inventor is often underrepresented as the product is commercialized under a different brand name or by a different organization. This in the long run could lead to complacency among researchers from emerging economies resulting in the loss of competitiveness in work culture that is demonstrated through their reluctance to share innovations. Literature categorizes such technology policy among research institutes and universities into three paradigms viz., the market failure paradigm, the mission paradigm and the co-operative technology paradigm (Bozeman, 2000). Accordingly, the public sector R&D across countries adopt mission paradigm behavior, where in the public sector is only expected to play a complementary role to the private sector and not expected to compete with it. Studies in the past do point out to a continued use of other forms of knowledge sharing between universities and industry through research publications as well (Wu, 2010).

The concept of Technology Transfer Offices (TTOs) as incubation facilities that promulgate entrepreneurship development for technology commercialization is well adopted and accepted across developed countries. A plethora of such efforts, models and success stories of innovation lead societal welfare have all been documented world over (Sudha, et al, 2012). The enactment of the Bayh-Dole act 1980 in the US is often considered a milestone that changed the university - industry linkage in technology transfer and commercialization. Evidence from the US universities and research organizations suggests strengthening of university industry linkage post Bay-Dole act with technology transfer and commercialization efforts generated a fourfold increase in numbers of licenses and royalty incomes between 1980 and 1990. The share of total R&D support to US colleges and universities had risen steadily from 5.3% in 1953, to 10.0% in 1975, to 11.0% in 2000. Of the $42,431 billion in research performed by the US academic sector in 2004, 61.5% was provided by the federal government, 19.3% from the institutions' own funds, 9.0% by private non-pro t organizations, and approximately 5% each by industry and special state government programs [Arlington, VA 2006], suggesting that innovation lead incubation and technology transfer directly or indirectly encouraged Public Private Partnership (PPP) as well.

Emerging economies like Brazil and Chile also have successfully emulated the concept of TTOs and commercialization through incubation aiming at creation of knowledge societies in the long run. In view of the changing global order, falling in line with other emerging economies, India also has adopted similar policy of initiating innovation centres and TTOs. This function of technological innovation in the development paradigm is of particular relevance to the agricultural research of countries like India, given their mandate of crop productivity enhancement linked to societal welfare.

The Indian Council of Agricultural Research (ICAR) the apex body responsible for all agricultural research and development for India adopted its Intellectual Property Rights (IPR) policy in 2006 and initiated steps towards setting up TTOs across 25 out of 90 and odd of the research institutions under its purview. Taking the concept of technology transfer and commercialization a step forward, ICAR also established 10 Business Planning & Development (BPD) facilities working on the concept of incubation with the financial support from the World Bank funded National Agricultural Innovation Project (NAIP) in 2009. Similar facility was further extended to another 12 institutes including the horticulture sector in 2013. Indian Institute of Horticultural Research (IIHR), as a premier research institution working on horticulture also started its Horticultural Technology Management & BPD, since Mid 2013.

PUBLIC PRIVATE PARTNERSHIP MODELS ADOPTED BY IIHR WHILE LICENSING

Technology commercialization through licensing is one way of strengthening Public Private Partnership. Since public sector is not equipped with mandate or infrastructure to undertake mass multiplication of seeds and planting material, it could be licensed to private sector institutions that are equipped to carry out the same, thereby creating a win-win situation. Few public sector organizations could also undertake the mass multiplication and ensure the quality supply of planting material in requisite quantity, adding value to the efforts put forth by scientists. As a policy, the institute permits license operations within the territory of India only. The following section provides a description of various such models and the prospects associated with them.

TECHNOLOGY LICENSING DIRECTLY TO PRIVATE SEED / OTHER PRODUCT COMPANIES:

IIHR has been successful in licensing some of the advanced breeding lines.. The research efforts of the scientists have also helped in developing specific crops having 'male sterility' associated with specific disease resistance characteristics. The pure line selection and standardization of such advanced breeding lines are of immense value for the private sector seed industry for reducing the time taken for production of hybrids and high yielding varieties. The institute has licensed such technologies in okra, tomato to name a few and has earned revenue to the tune of Rs. 45 lakhs from the non- exclusive licenses it has granted to over 15 private seed companies in the last two years. ITMU ensures appropriate memorandum of understanding, upfront payments, royalty and other essential features for safe guarding the interest of the scientists. IIHR also has assigned licenses to private product manufacturers. These include products such as post harvest technology products, agricultural engineering drawings *etc.*

TECHNOLOGY LICENSING TO PRIVATE SECTOR ORGANIZATIONS THROUGH A PUBLIC SECTOR MEDIATOR

Technology innovations that solicit stringent IP protection such as Patents demand financial back up. Some of the technologies or technical interventions could be simple but may not have easy access to the interested entrepreneurs for immediate commercial production. IIHR adopted the strategy to assign such technologies to a public sector agency that acts as a 'mediator' in identifying potential partners for patenting and mass multiplication followed by commercialization. The National Research Development Corporation (NRDC), New Delhi has been an effective mediator for a number of technologies for commercialization. IIHR has entered into an MOA with NRDC, valid for a period of five years. ITMU in consultation with the innovator assigns technologies to NRDC on individual case by case basis either for filing for appropriate IP protection or for commercialization. NRDC assists the institute in tracking down the entrepreneurs' sales, turn over and royalty collections *etc.*, the total emoluments so collected are then shared in the ratio of 70:30 between the Institute and NRDC.

IIHR has been able to commercialize more than 6 technologies through NRDC in the last two years. The fact that NRDC has been able to obtain patents for specific innovations and also identify firms that have commercialized products has helped IIHR improve its visibility in the market.

TECHNOLOGY LICENSING DIRECTLY TO OTHER PUBLIC SECTOR ORGANIZATIONS

The primary mandate of the institute is to ensure adequate supply of seed and planting material and other technologies at affordable prices, IIHR adopts a policy of licensing its technologies directly to other public sector organizations like the National Seeds Corporation (NSC) and KrishiVigyan Kendra's (KVKs). ITMU has also developed appropriate memorandum of understanding to be executed between two public sector organizations while sharing technical knowledge or collaborative research that could involve intellectual property rights.

OTHER PARTNERSHIP MODELS

Technology commercialization is a continuous process evolving over a period of time, institution need to keep upbreast with innovative means of arriving at partnership models. Some of the technologies warrant generation of authentication data sets, regulatory and bio safety data sets that require technical skills as well as financial support, the partnership models would need special case specific clauses that safe guard the interest of the parties involved. Though standard formats are available, case specific changes need to be made while entering into such agreements.

IMPLICATIONS DUE TO COMMERCIALIZATION THROUGH LICENSING

ICAR's Intellectual property Rights policy of 2006 has already resulted in positive monetary and value benefits to the crop breeding programmes. Opening up of research institutes for commercialization of technologies through non-exclusive licensing has helped in increasing the visibility of the efforts put forth by the researchers on one hand and making the material available in large volumes by the end users. Based on the lessons learned from the experiences of research institutions like IIHR, it has been demonstrated that more than 200 licenses have been granted to private seed companies for enhancing the supply of useful planting material. Industry useful traits such as imbibing 'male sterility' in some of the vegetable breeding lines by IIHR scientists(Breeders) has in fact helped the industry in cutting their breeding time as well as costs due to reduced cost of emasculation& pollination.

Technology commercialization through licensing also has thrown open the doors for 'assigning value' to the technological innovations undertaken by the breeders/ innovators. Public sector R&D organizations being the repositories of huge diversity of crop germplasm have been empowered through commercialization efforts to produce value added products and also to help private industry take forward the valuable genetic material for further uses. Several useful models of public private partnership are available from literature that proves the 'value amplification of genetic material' for mutual as well as societal benefits (the case of Soybean GMOs between Monsanto & EMBRAPA, 2000 (1&3)).

Huge untapped potential also is seen for these trait specific hybrids outside the country. Being hybrids, these seed material is 'safe' and can be exported once the clearances are obtained from the biodiversity authority. While the ICAR as well as the BDA are yet to get themselves equipped for speeding up the clearance processes, potential for public sector R&D based backward linkage models are ready for take off. Such efforts will not only help ensure livelihood support to a number of farm families involved in seed production activities, but also will enhance the credibility of public sector R&D efforts as well.

HYBRIDIZATION IN CROPS, AN ECONOMIC PERSPECTIVE

Plant breeding through Hybridization has been proven to be the most essential pathway for productivity enhancement since times immemorial. With benefits out numbering the fears and concerns, hybridization in crops has become a way of life in the recent years. Hybridization in crops has wider implications, especially in economic terms, much beyond those perceived by the breeders. In economic terms, hybridization in crops offers two distinct advantages measured as direct & indirect benefits, represented thus,

Total Economic benefit (TEB) = Area effect + Substitution effect

A typical hybridization programme in crops hitherto aimed at productivity enhancement. Thereby, any crop hybrid provides two way benefits to the society, which can be classified as the area effect & a Substitution effect. The increase in yield level results in increased production from a given area or higher production from reduced area, thus releasing area for the production of other crops or additional area available for the same crop. Further, seed requirement for hybrids is nearly one third of that required for improved varieties, thereby, the area required for production of total seed requirement also will be reduced, leading to the area being released for other crops, which in economic terms is referred as 'substitution effect'. Thus hybridization per se has a wide spread economic ramifications.

The modern day hybrids with 'disease resistance' focus offer even wider economic implications in terms of reduced cost of cultivation and aggregate savings in pesticides & fungicides. The economic implications get further amplified if we consider the overall benefits to the society through the reversal of environmental hazards that have been created due to unprecedented use of hazardous pesticides and fungicides.

SCOPE FOR AGGREGATE IMPACT ASSESSMENT

Public sector research organizations often face the challenge of justifying their presence and the impact of their technological innovation on the society. Unlike in the West where most of the technologies are IP protected, enabling easy tracking, developing countries like India grapple with the tough task of tracking of spread and adoption of improved varieties or technical know-how. Use of hybrids in crops makes it easier for such tracking as the quantity of seed produced and sold could be the most authentic parameter of measuring spread. Further econometric methods like the 'Economic Surplus' help ascertain the true value of technology adoption and contribution of R&D organizations to the society at large.

OPPORTUNITIES FOR EXCLUSIVE RESEARCH COLLABORATIONS

While technology commercialization can help execute non- exclusive licensing, it does not restrict the possibility for 'exclusive forms as well, in the form of research collaborations & joint ventures. Public Private Partnership models are best executed through such joint research collaborations (the case of Soybean Aphid Resistance by Michigan State University & the Michigan Soy Crop council, 2010)

CONCERNS IN PERENNIAL HORTICULTURAL CROPS

A number of success stories are readily available with respect to field crops as well as short duration horticultural crops including vegetables and flowers; the perennial fruit trees and plantations are yet to reap these benefits due to longer duration & gestation periods. While the processes of inter specific hybridization may well be possible, the technologies that help ascertain the beneficial effects need further clarity. However, the prevalent IP policies and commercialization protocols may be of value for initiating useful partnerships without the fear of losing precious material or the intellectual property rights associated with such processes.

ICAR's option of adopting Intellectual property Policy thus will help pave way for useful partnerships and ease of highly valuable genetic material transfers through authenticated Material Transfer Agreements (MTAs) between organizations within and outside the country and open up several new and assured opportunities for enhanced collaboration, as has already been demonstrated by several universities and R&D institutions aboard (Example of Chile in Grape elite genotypes exchange 2010 (2))

REFERENCES

Sudha Mysore, 2014, Technology Transfer and Commercialization – Innovative model for Strengthening Research and Industry Linkages and Valuation through Public Private Partnership in Agriculture, *Journal of Intellectual property Rights*, 19, May 2014, pp. 167-176.

World Bank, 2009. *Chile: Fostering Technology transfer and commercialisation*, Washington, D.C., USA: Latin American Region, Poverty Reduction and Economic Management Unit.

W. Lesser, Todd M. Schmit, Lilian M. Ruiz, Wilton, CT, 2001, Elite Germplasm for GMO's in Brazil: Modeling Government-Agribusiness Negotiations, International Food and Agribusiness Management Review, 2(3/4): 391–406.

www.ingramcontent.com/pod-product-compliance
Lightning Source LLC
Chambersburg PA
CBHW071554120726
48009CB00001B/38